DISTRIBUTIONAL ASPECTS OF ENERGY AND CLIMATE POLICIES

Wherever possible, the articles in these volumes have been reproduced as originally published using facsimile reproduction, inclusive of footnotes and pagination to facilitate ease of reference.

Distributional Aspects of Energy and Climate Policies

Edited by

Mark A. Cohen
*Professor of Management and Professor of Law, Vanderbilt Owen Graduate School of Management
and University Fellow, Resources for the Future, USA*

Don Fullerton
*Gutgsell Professor of Finance, Institute of Government and Public Affairs, University of Illinois
and Research Associate, National Bureau of Economic Research, USA*

and

Robert H. Topel
Isidore Brown and Gladys J. Brown Distinguished Service Professor in Urban and Labor Economics, University of Chicago, USA

Edward Elgar
Cheltenham, UK • Northampton, MA, USA

Published by
Edward Elgar Publishing Limited
The Lypiatts
15 Lansdown Road
Cheltenham
Glos GL50 2JA
UK

Edward Elgar Publishing, Inc.
William Pratt House
9 Dewey Court
Northampton
Massachusetts 01060
USA

A catalogue record for this book is available from the British Library

Library of Congress Control Number: 2013930513

This book is available electronically in the ElgarOnline.com Economics Subject Collection,
E-ISBN 978 1 78347 027 3

ISBN 978 1 78254 008 3

Printed by MPG PRINTGROUP, UK

Contents

Acknowledgements

The editors and publishers wish to thank the authors and the following publishers who have kindly given permission for the use of copyright material.

Walter de Gruyter GmbH for articles: Sebastian Rausch, Gilbert E. Metcalf, John M. Reilly and Sergey Paltsev (2010), 'Distributional Implications of Alternative U.S. Greenhouse Gas Control Measures', *B.E. Journal of Economic Analysis and Policy*, **10** (2), Article 1, i–ii, 1–44; Shanta Devarajan (2010), 'Comment on "Distributional Implications of Alternative U.S. Greenhouse Gas Control Measures"', *B.E. Journal of Economic Analysis and Policy*, **10** (2), Article 2, i, 1–3; Charles D. Kolstad (2010), 'Equity, Heterogeneity and International Environmental Agreements', *B.E. Journal of Economic Analysis and Policy*, **10** (2), Article 3, i–ii, 1–15; Scott Barrett (2010), 'Comment on "Equity, Heterogeneity and International Environmental Agreements"', *B.E. Journal of Economic Analysis and Policy*, **10** (2), Article 4, i, 1–3; Joshua Blonz, Dallas Burtraw and Margaret A. Walls (2010), 'Climate Policy's Uncertain Outcomes for Households: The Role of Complex Allocation Schemes in Cap-and-Trade', *B.E. Journal of Economic Analysis and Policy*, **10** (2), Article 5, i–ii, 1–33; Arik M. Levinson (2010), 'Comment on "Climate Policy's Uncertain Outcomes for Households: The Role of Complex Allocation Schemes in Cap-and-Trade"', *B.E. Journal of Economic Analysis and Policy*, **10** (2), Article 6, i, 1–3; Louis Kaplow, Elisabeth Moyer and David A. Weisbach (2010), 'The Social Evaluation of Intergenerational Policies and Its Application to Integrated Assessment Models of Climate Change', *B.E. Journal of Economic Analysis and Policy*, **10** (2), Article 7, i–ii, 1–32; Martin L. Weitzman (2010), 'Comment on "The Social Evaluation of Intergenerational Policies and Its Application to Integrated Assessment Models of Climate Change"', *B.E. Journal of Economic Analysis and Policy*, **10** (2), Article 8, i, 1–2; Ian W.H. Parry and Roberton C. Williams III (2010), 'What are the Costs of Meeting Distributional Objectives for Climate Policy?', *B.E. Journal of Economic Analysis and Policy*, **10** (2), Article 9, i–ii, 1–33; William Randolph (2010), 'Comment on "What are the Costs of Meeting Distributional Objectives for Climate Policy?"', *B.E. Journal of Economic Analysis and Policy*, **10** (2), Article 10, i, 1–3; Joshua Elliott, Ian Foster, Kenneth Judd, Elisabeth Moyer and Todd Munson (2010), 'CIM-EARTH: Framework and Case Study', *B.E. Journal of Economic Analysis and Policy*, **10** (2), Article 11, i–ii, 1–32; Don Fullerton (2010), 'Comment on "CIM-EARTH: Framework and Case Study"', *B.E. Journal of Economic Analysis and Policy*, **10** (2), Article 12, i, 1–3; Christoph Boehringer, Carolyn Fischer and Knut Einar Rosendahl (2010), 'The Global Effects of Subglobal Climate Policies', *B.E. Journal of Economic Analysis and Policy*, **10** (2), Article 13, i–ii, 1–33; Rodney D. Ludema (2010), 'Comment on "The Global Effects of Subglobal Climate Policies"', *B.E. Journal of Economic Analysis and Policy*, **10** (2), Article 14, i, 1–5; Don Fullerton and Garth Heutel (2010), 'Analytical General Equilibrium Effects of Energy Policy on Output and Factor Prices', *B.E. Journal of Economic Analysis and Policy*, **10** (2), Article 15, i–ii, 1–24; Samuel Kortum

(2010), 'Comment on "Analytical General Equilibrium Effects of Energy Policy on Output and Factor Prices"', *B.E. Journal of Economic Analysis and Policy*, **10** (2), Article 16, i, 1–5; Dale W. Jorgenson, Richard Goettle, Mun S. Ho, Daniel T. Slesnick and Peter J. Wilcoxen (2010), 'The Distributional Impact of Climate Policy', *B.E. Journal of Economic Analysis and Policy*, **10** (2), Article 17, i–ii, 1–26; Thomas Hertel (2010), 'Comment on "The Distributional Impact of Climate Policy"', *B.E. Journal of Economic Analysis and Policy*, **10** (2), Article 18, i, 1–5; Gary S. Becker, Kevin M. Murphy and Robert H. Topel (2010), 'On the Economics of Climate Policy', *B.E. Journal of Economic Analysis and Policy*, **10** (2), Article 19, i–ii, 1–25; Manasi Deshpande and Michael Greenstone (2010), 'Comment on "On the Economics of Climate Policy": Is Climate Change Mitigation the Ultimate Arbitrage Opportunity?', *B.E. Journal of Economic Analysis and Policy*, **10** (2), Article 20, i, 1–12.

Introduction

Preface/Acknowledgements

The idea for this book grew out of a conference held in January 2010 in Washington, DC at Resources for the Future (RFF). The conference was a joint effort of the Energy Initiative at the University of Chicago, the Center for Energy Economics and Policy at RFF, and the University of Illinois at Urbana-Champaign. It brought together a diverse group of senior policymakers and leading researchers from the fields of economics, physical sciences, computational sciences, and law. Ten original research papers were presented, along with comments and spirited discussion focused both on technical modeling strategy and direct policy implications.[1] All papers have been revised and updated into ten chapters for this book, and commentators have added their written thoughts as well.

We are most grateful to Heidi Levin, then Executive Director of Chicago's Energy Initiative, who helped to organize and superbly managed the conference logistics. We could not have managed the conference without Heidi. Thanks also to Steve Goldberg of Argonne National Laboratory who allowed us to draw on his vast knowledge of energy policy and politics, while providing wise counsel and encouragement at every stage. We also thank Aysha Ghadiali at RFF for her steadfast logistical support. Of course, we thank all of the authors in this volume for their hard work on their papers and commentaries. Finally, we thank our sponsors – the US Environmental Protection Agency Climate Economics Branch, Resources for the Future, the Energy Initiative at the University of Chicago, and three units at the University of Illinois at Urbana-Champaign: the Environmental Change Institute (ECI), the Center for Business and Public Policy (CBPP), and the Institute of Government and Public Affairs (IGPA).

Mark A. Cohen, Vanderbilt University and Resources for the Future
Don Fullerton, University of Illinois
Robert H. Topel, University of Chicago

Introduction

The chapters gathered here represent the first installment on a regular conference event, under the banner *Energy Policy Symposium*. The theme for the 2010 conference, held in Washington, DC at Resources for the Future (RFF) was *Distributional Aspects of Energy and Climate Policies*. In particular, the chapters examine policies that would 'price' carbon emissions or otherwise seek to mitigate anthropogenic climate change. Since energy and climate policies are so closely linked and are on the public agenda, our first conference focused on both. In soliciting contributions, we interpreted 'distributional' fairly broadly, to include impacts of pending or possible legislation on the living standards of households across the US income distribution, across generations, or across geographic areas, as well as international differences in the costs and benefits of climate policies that would affect countries' willingness to participate in harmonized international agreements.

Most observers are, by now, familiar with the broad outlines of global climate change. The consensus is growing among scientific and government institutions that the earth's atmosphere and oceans are warming faster than natural causes can explain, and that these changes are caused by human activity. Climate change includes not only warming, which threatens biodiversity and agricultural productivity, but also sea level rise and increased severity of droughts, floods, and other extreme weather events like hurricanes. The most significant source of this anthropogenic climate change is the emission of carbon dioxide, which comes primarily from the combustion of fossil fuels, as well as from other sources such as deforestation and changes in land use.

Many governments, either unilaterally or in the context of international agreements, either have taken action, or are considering taking action, to limit carbon emissions and so combat climate change. Serious proposals to reduce carbon emissions generally include some kind of carbon pricing system. Under a carbon pricing system, governments intervene in the market to internalize the costs of discharging carbon into the atmosphere – that is, to shift these costs from the planet as a whole to the individual emitter (for example, a 'polluter pays principle').

This 'polluter pays principle' actually has two important but very different interpretations. First, the efficiency interpretation is that the polluters should pay the right price in order to get them to abate emissions by all of the least expensive means. Second, the equity interpretation is that polluters should have to pay for the damages they cause. While most research in environmental economics is concerned with the efficiency interpretation, this conference and collection of chapters is more concerned with the equity interpretation. We note, however, that any legal requirement for the polluter to pay for damages is only the 'statutory incidence'. It does not dictate the final 'economic incidence', in terms of who really bears the burden after adjustments in the marketplace, such as changes in output prices, factor prices, and scarcity rents. The main purpose of this symposium is to focus on economic incidence.

Carbon pricing systems generally take one of two forms: a tax on emissions, or a cap-and-trade regime. A tax or price could be applied directly on the emission of carbon dioxide (CO_2) or other greenhouse gases (GHG). Emitters pay to the government a legally determined amount based on the volume of CO_2-equivalent gases that they release into the atmosphere. Under a cap-and-trade system, the government issues permits authorizing the holder to emit a certain volume of those gases into the atmosphere in a given time frame. The sum of the volumes authorized by all permits issued equals the cap – the total amount that can be emitted for the whole nation (or region or even world). Permit holders are then free to trade the permits amongst themselves, allowing the price to be set by the market.

While economists long ago extolled the virtues of a market-based system of pricing externalities such as carbon emissions to achieve an efficient allocation of resources, the distributional consequences of any pricing policy will have important implications not only for those who stand to gain and lose but also for the political feasibility of any proposed solution. Ultimately, politicians will decide both the underlying framework (for example tax, cap-and-trade, or direct regulation) and the details of the proposed solution (for example who is regulated and by how much). Many factors impact the distribution of winners and losers, including the design of the tax or cap-and-trade system, the allocation of proceeds from the tax or permit sale, and the level at which the pricing system is adopted (state, national, or international levels). The interactions of these factors are enormously complex and subject to debate. Thus, serious analytical and empirical analysis can help identify the winners and losers as well as help craft policies that balance efficiency and political feasibility.

For example, revenue from a carbon tax or from the sale of emission permits under a cap-and-trade regime can be used in any number of ways. It can be distributed to the people using a variety of allocation formulas. It can be used as general tax revenue by the government. It can be directed to other programs designed to reduce carbon emissions, such as efficiency measures or research into clean technologies. Each of these alternatives implies different winners and losers, as well as different social costs and benefits.

The ten primary chapters (and ten commentary chapters) of this book address the distributional impact of efforts to mitigate global climate change by reducing carbon or other GHG emissions. That is, they seek to answer questions such as: to the extent that climate mitigation policies have costs, who will bear these costs? Will they fall on those of us living now, or on future generations? Will they fall on the rich, the poor, the middle class, or on everyone proportionally? Which countries will benefit, and which will suffer? If a global agreement is impossible, what will happen if countries act unilaterally, or in groups? What sort of nations would participate in such a pact? Would such a subglobal arrangement succeed in reducing global carbon emissions, or would it merely push carbon-intensive industries across borders?

Chapter 1, which is written by **Gary S. Becker, Kevin M. Murphy and Robert H. Topel**, begins by introducing the economic issues that underlie the problems of climate change and climate policy. The authors describe climate as a global public good, and they identify a number of complications that make climate change more than a simple externality problem. They stress the difficulty of valuing uncertain future damages and of setting an appropriate social discount rate. Notably, they suggest that a low discount rate may be appropriate in evaluating climate policies, because such policies may function as a kind of insurance – that is, they may pay off in times when other investments do not. This approach differs from a traditional economic approach that might suggest using a market rate of interest as a social discount rate. In addition, they note that different valuations of the marginal utility of future consumption across countries will make any global agreement that much more difficult to realize.

Chapter 2 is a comment by **Manasi Deshpande and Michael Greenstone** that applauds the paper's accomplishments, and then focuses on that last observation that differences in preferences across countries will make global agreement more difficult to accomplish. These commentators argue the opposite, namely, that differences between countries provide that much more 'gains from trade'. Developed countries might have more willingness and ability to pay for greenhouse gas emission reductions, but less developed countries might find less expensive ways of doing that abatement, so an appropriate agreement can provide gains to both.

In Chapter 3, **Louis Kaplow, Elisabeth Moyer and David A. Weisbach** delve more deeply into the issue of social discounting – which has a significant impact on the distribution of costs and benefits across generations. They draw a distinction between utility functions, in which the discount rate can be determined by empirical observation of actual human behavior, and social welfare functions, which require ethical judgments. The authors point out that the question of how people discount future consumption in practice is distinct from that of how we should discount the welfare of future generations. Small changes in a model's discount rate can result in large differences in the long-range projections and policy assessments produced using that model. The authors review two existing integrated assessment models, Nordhaus's

DICE model, and the PAGE2002 model used in the British government's Stern Review, which they believe fail to distinguish adequately between the empirical and ethical judgments involved. Ultimately, the authors challenge the concept that a single discount rate is truly applicable to social welfare problems, and they recommend drawing sharp distinctions between discounting in the context of utility functions versus discounting in the context of social welfare functions.

In a brief comment to Chapter 3, **Martin L. Weitzman** (in Chapter 4) is sympathetic to the call for clearly specifying whether the underlying discount rate is behavioral or ethically driven. However, he cautions that allowing for two distinct discount rates in integrated assessment models may turn out to be unwieldy and may raise more issues than it solves.

While the first four chapters focus primarily on the global public good nature of climate change and concern themselves with big picture issues such as the tradeoff between current and future generations, the next few chapters examine the impact of climate policy on the distribution of income for households and regions in the US.

In Chapter 5, **Don Fullerton and Garth Heutel** examine the distributional impacts of a carbon tax, using analytical general equilibrium results. They present a simple model dividing the economy into a 'clean' sector that employs both labor and capital and a 'dirty' sector that also creates pollution. The analytical model is used to show effects of a pollution tax on the returns to labor and capital (the 'sources side') as well as on the prices of clean and dirty outputs (the 'uses side'). Then they use household survey data to split the population into income deciles and calculate the burden on each income group. They find a regressive burden on the uses side and a burden on the sources side that can be progressive, regressive, or U-shaped. Notably, they find that once life-cycle effects are considered, by using expenditure deciles, the net distributional impact of a carbon tax is less clear.

In his comment, Chapter 6, **Samuel Kortum** provides additional intuition for their model by use of a diagram and analogy to a simple model of international trade. He also notes a 'fundamental inconsistency' in the Fullerton–Heutel paper, which employs the simplification of a representative agent model to solve for general equilibrium outcomes analytically, but then applies those output and factor price results to a set of ten different groups' expenditure patterns.

While many analysts prefer use of a carbon tax to reduce emissions, recent policy discussion in the United States has centered on three different cap-and-trade proposals that were proposed in the 111th Congress. In 2009, the House of Representatives passed the American Clean Energy and Security Act, more commonly known as the Waxman–Markey Bill after its authors, Congressmen Henry Waxman and Edward Markey. Two competing proposals were considered by the Senate: a bill introduced by Senators John Kerry and Barbara Boxer, which was very similar to Waxman and Markey's House Bill; and an alternative plan, put forward by Congresswomen Maria Cantwell and Susan Collins, which proposed a very different formula for the initial allocation of emission permits, and for the distribution of revenues from the auction of permits. In particular, the Cantwell–Collins Bill would have rebated a significant portion of auction revenues in a lump sum form to households to offset rising energy prices. While the 111th Congress ultimately did not pass cap-and-trade legislation, the outline of any future bill – whether it involves some form of emissions fee or trading scheme – will undoubtedly consider similar provisions. The next chapters model the impacts of these alternative cap-and-trade proposals, with special attention to distributional outcomes.

Significant points of disagreement among the different cap-and-trade proposals included the methods for allocating the initial carbon emissions permits (or equivalently, for distributing the revenue from those permits that were to be auctioned by government). In Chapter 7, **Joshua Blonz, Dallas Burtraw and Margaret A. Walls** examine the permit allocation framework included in the Waxman–Markey Bill. Specifically, they highlight uncertainty in the impacts of three provisions: (1) the free allocation of some emissions permits to local electricity distributors; (2) the dedication of a portion of carbon permit auction revenues to efficiency programs; and (3) the dedication of another portion of those revenues to clean technology development. Alternatively modeling optimistic and pessimistic assumptions regarding the effects of these provisions, they identify what they consider 'substantial' uncertainty regarding the cost of the Waxman–Markey proposal on an average household. For example, they estimate the annual cost to the average household to range between $133 and $418 – with this range being caused by uncertainty in the ultimate allowance price. In contrast, they model an alternative proposal in which 75 percent of allowances are auctioned, with funds returned to households in the form of lump-sum rebates (oftentimes called a 'cap-and-dividend' program). The authors find such a policy could both reduce uncertainty and produce a progressive distribution of burdens.

In a brief comment to the Blonz, Burtraw and Walls chapter, **Arik M. Levinson**, in Chapter 8, attempts to place these uncertainties into context. In particular, he notes that households face higher expected expenditure fluctuations due to uncertainties in the treatment of offsets and banking provisions of the law, as well as in the ultimate cost of energy. For example, annual fluctuations in the price of gasoline alone could affect the average household by plus-or-minus $2,000 – far in excess of any fluctuations due to permit prices.

In Chapter 9, **Ian W.H. Parry and Roberton C. Williams III** also examine the allocation of proceeds from a carbon permit auction. Specifically, they analyze four alternatives for distributing auction revenue: (1) granting all permits to existing emitters for free; (2) a full cap-and-dividend plan in which proceeds are distributed to households in equal lump sums; (3) using auction revenues for proportional offset of income taxes; and (4) a distribution-neutral income tax offset in which benefits are steered towards lower-income households to offset the regressive nature of the carbon price itself. They find that the overall welfare cost, as well as the distributional effects of cap-and-trade, depends greatly on how auction revenues are allocated, and they identify a clear tradeoff between efficiency and progressivity. The authors also emphasize the significance of the revenue-recycling effect, realized from using auction revenues to offset income or other distortionary taxes.

In a brief comment, Chapter 10, **William Randolph** notes that an important result of the Parry–Williams Model is that the fourth method – a distribution-neutral income tax offset – can be used to make the total efficiency cost of a cap-and-trade policy almost identical to the direct cost of emission reductions. That is, virtually all distortions can be eliminated. He also reminds us that like all of the chapters in this book, their model only considers the costs of climate policies – not the offsetting benefits.

In Chapter 11, **Sebastian Rausch, Gilbert E. Metcalf, John M. Reilly and Sergey Paltsev** employ the US Regional Energy Policy Model (USREP), a computable general equilibrium (CGE) model that has been used to examine energy and greenhouse gas policies on regions, sectors, industries and household income classes. The authors extended USREP using a recursive dynamic formulation to compare the distributional impacts of the three competing proposals in the 111th Congress. The authors find that carbon pricing itself is either distribution-neutral or progressive. The seeming conflict between this conclusion and that reached by Parry and Williams, who found pricing to be regressive when considered in isolation, stems from the inclusion of government transfers in the analysis. Rausch et al. assume that government transfers, that are received mostly by poorer households, are indexed to the price level in a way that protects those households from the costs of carbon pricing, which makes a carbon price more progressive than when transfers are excluded from the model. The authors find that the Cantwell–Collins 'cap-and-dividend' plan (which called for most auction revenues to be distributed to households via lump sum rebates), is less progressive than the more complex Waxman–Markey or Kerry–Boxer proposals. Like Parry and Williams, the authors identify a tradeoff between equity and efficiency, with the more progressive plans imposing a greater welfare cost than the less progressive Cantwell–Collins proposal. This paper also devotes some attention to the regional distribution of costs and benefits; they find that the Waxman–Markey and Kerry–Boxer proposals overcompensate some affected regions in the years immediately following adoption.

In a brief comment to Chapter 11, **Shanta Devarajan** points out in Chapter 12 some important limitations of this and other CGE models used to simulate the impact of climate policy. First, he notes that all general equilibrium models assume full-employment – something that hardly seemed realistic in 2010 with US unemployment hovering at 10 percent. Similarly, he notes that their recursive model is not truly dynamic and thus ignores many important consumer and producer choices. Thus, Devarajan points out the difficulty of any such modeling exercise – the tradeoff between realism and complexity. Finally, he points out another subtle difficulty in using the recursive model to predict the political feasibility of differing policy alternatives: since any progressive aspects of these policies are short-lived, it is not clear that the winners will lobby for them; nor will losers necessarily fight them.

In Chapter 13, **Dale W. Jorgenson, Richard Goettle, Mun S. Ho, Daniel T. Slesnick and Peter J. Wilcoxen** present a new method for evaluating the effects of carbon pricing regimes. The authors have revised the Intertemporal General Equilibrium Model (IGEM), originally introduced by Jorgenson and Wilcoxen in 1990 and now employed by the Environmental Protection Agency, to include a model of household behavior developed by Jorgenson and Slesnick. Using this updated model to incorporate labor–leisure choices into their analysis, they evaluate the impact of the Waxman–Markey Bill on 244 different types of households. Their conclusion reinforces the findings of some previous studies that such a regime, if adopted, would impact different regions and demographics in very different ways, but would, on balance, be regressive.

In his comment on Jorgenson et al. in Chapter 14, **Thomas Hertel** first articulates four important contributions: intertemporal decision-making, endogenous technical progress, substantial disaggregation among households, and econometrically estimated consumer and producer behaviors. He then notes that while the model has regional households, it does not have regional production of electricity with changes in relative costs that depend on differential coal intensity of particular regions. Also, many estimated parameters are used in this complicated CGE model, so Hertel recommends that the authors provide summary tables showing overall consumer responsiveness to energy prices – and then vary those parameters in sensitivity analysis.

The remaining chapters move beyond the confines of the US and consider global aspects of climate policy. In Chapter 15, **Joshua Elliott, Ian Foster, Kenneth Judd, Elisabeth Moyer and Todd Munson** present a new computable general equilibrium model called CIM-EARTH. The authors claim that many of the limitations of previous models are due to computational and methodological constraints that can be overcome using advances in computer architecture and numerical methods. CIM-EARTH allows for a detailed representation of individual industries, income cohorts and generations. The model also has the important feature that it is open and accessible to other researchers who may modify assumptions, test different data, develop more sophisticated uncertainty scenarios and see how they affect results. They demonstrate CIM-EARTH by analyzing the hypothetical adoption of emissions reduction by the nations of the Kyoto Protocol's Annex B. Specifically they predict the 'leakage' of emissions that would result from such an agreement, that is, the extent to which carbon emissions simply shift from participating countries to nonparticipating countries.

In a short comment on Elliott et al., **Don Fullerton** notes in Chapter 16 that it is really a progress report on a newer, larger, and even more complicated model. It will therefore run smack up against the main problem facing CGE modelers: the workings of the model are opaque to readers. They cannot see the effects of alternative assumptions, only effects of assumptions the authors choose to vary. In this case, Elliot et al. vary two assumptions about the baseline and thereby generate 25 different baselines. Each is used to calculate results of the same policy change. Readers do not see the effects of thousands of other assumptions, such as alternative aggregations, elasticity parameters, functional forms, abatement technologies, imperfect competition, factor mobility and a host of possible market imperfections not discussed in the paper.

In Chapter 17, **Christoph Boehringer, Carolyn Fischer and Knut Einar Rosendahl** consider subglobal climate agreements – that is, climate reduction policies adopted by countries or groups of countries without a fully global agreement. They examine the welfare effects of such unilateral policies in the US and the EU, both on the participating countries and on their trading partners. They also evaluate the likely effects of output-based rebating and border carbon adjustments, policies aimed at reducing 'leakage'. The authors find that the situation where countries participate in subglobal climate agreements would likely result in welfare losses both for the participating countries and for petroleum exporting nations, but could result in welfare gains for other trading partners. They predict that efforts to reduce leakage would likely be ineffective – reducing such leakage by at most 22 percent – and that such policies would have little effect, either positive or negative, on overall welfare.

In his comment (Chapter 18), **Rodney D. Ludema** points out that international trade is usually thought to impose extra costs for participating countries, whereas Boehringer et al. find the opposite. Unilateral climate policy helps them, as it reduces the worldwide price of oil and improves the terms of trade for participating countries. Because the border carbon adjustments effectively turn a tax on production into a tax on consumption, Ludema points out that they also change the burdens of the policy from producers to consumers.

In Chapter 19, **Charles D. Kolstad** examines the decision countries face in choosing to join or not join an international environmental agreement (IEA). Kolstad, expanding on previous modeling efforts that assumed identical countries, models the behavior of countries that differ in size, in the damage they suffer from warming, and in the costs they assume by adopting abatement measures. He finds that this heterogeneity has implications for both the composition and welfare impacts of IEAs. That is, a more heterogeneous pool of countries could result in an IEA that includes more countries but that imposes greater welfare costs on its members. Kolstad identifies two possible equilibria, maximizing either efficiency or equity. Then, under each such goal, he explores the likely form of the predicted international agreements.

In a brief comment in Chapter 20, **Scott Barrett** refers to an alternative modeling strategy that would yield an equilibrium in which *ex ante* both fairness and efficiency are achieved. He suggests that this model could be extended to the case of country heterogeneity outlined by Kolstad. However, he also notes that while this outcome appears consistent with the Kyoto Protocol, the fact that this treaty was unable to sustain a consensus reminds us of a major shortcoming of research that relies upon such models to predict behavior – in particular, the global community has thus far lacked an effective, credible punishment mechanism to enforce any such equilibrium.

Concluding Remarks

Large-scale economic policy evaluation and calibration is always a daunting exercise, because we economists know so little about the underlying social and economic processes on which key outcomes depend. In describing the similar uncertainties that arise in military policy, former Defense Secretary Donald Rumsfeld usefully divided the space of relevant information into three parts – (1) the things we know; (2), the things we know we do not know; and (3) the things we do not know we do not know. Applied to evaluation of the economic impacts of energy and climate policies, the sad facts are that (1) is small, (2) is very large and – one suspects – (3) is biggest of all. Burdened with such ignorance, it is tempting to emulate the modesty of Mark Twain's famous response to a similarly difficult question: 'I was gratified to be able to answer promptly, and I did. I said I didn't know.'[2]

That will not occur, and not merely because economists are immodest. The development and use of abundant energy is essential to the maintenance and spread of economic well-being – rich countries use more energy per capita, and poor countries will use more if they are ever to grow rich. The fact is that new and far-reaching climate/energy policies will almost certainly be enacted – either unilaterally by certain developed countries or cooperatively through international agreements. Given the importance of energy to economic welfare, those policies have the potential to do great harm or good – with anticipated rent-seeking on a global scale, the cure could be worse than the disease. Thus, informed policy options are essential, especially with regard to economic impacts. But as Rumsfeld's partition demonstrates, part of being economically 'informed' is to acknowledge and incorporate the huge uncertainties underlying all policy choices in this area. Recognizing that we do not know much may be the most important aspect of wise energy policy.

Here is an example in the context of the distributional impacts discussed in several chapters of this book. Among the things we think we 'know' as economists are the directional 'signs' of most of the distributional effects of embedding a carbon price into the private cost of using traditional fuel sources. Economists are pretty good at signing those types of things, so we are confident that living standards of low-income households would be affected proportionally more because they spend a larger share of income on energy. And were policy designed to offset those effects in the name of equity, we will surely sacrifice some efficiency. With some nuances related to the details of various policies, that is what these papers found. The additional insights from our authors that would usefully inform policymakers come from putting numbers on the effects via tightly specified models – how large are the effects for various groups or countries, and how sensitive are they to various policy adjustments? What are the efficiency costs of redistribution? The danger is that such seeming precision gives a false sense of confidence, putting numbers on magnitudes we actually know very little about simply because our models are capable of generating them.

The point is stronger when we advance to Rumsfeld's second category – the large set of 'known unknowns'. Among these unknowns, to name just a few, are: the nature and likelihoods of catastrophic climate outcomes in the near or distant future; the pace of market-driven innovation of alternative energy sources; which countries will choose to participate in more-or-less harmonized climate policies and when; the pace of future economic growth; and the long- and short-run substitution opportunities that arise in myriad industries world-wide. As demonstrated in chapters 11, 12 and 13, the (dynamic) computable general equilibrium (CGE) models that very likely will be the workhorses of policy evaluation must assume particular structures and magnitudes for all of these possibilities, over very long horizons. And while advances in computing power allow us to build and analyze ever more sophisticated CGE models, the fact is that these models require vast amounts of empirical knowledge that is, by comparison, in the Stone Age – as if we had invented air conditioning but not electricity. After decades of empirical research, for example, economists have reached little consensus on such basics as the elasticity of labor supply or the willingness of consumers to substitute consumption over time, to say nothing of substitution over generations. Do not even ask what we know about affecting the pace of innovation or the design of research incentives. Now add in the 'unknown unknowns' and we rest our case.

We are not suggesting that the work not be done – what is the alternative? But rather that we acknowledge and even emphasize the fragility and uncertainty of what we can tell policy-makers. Twain's prompt 'I don't know' is not so far from where we are, and it can be said with great confidence because it is true.

We close with two things that we do know. First, energy and climate policies are being implemented as we speak. Even in the absence of proactive legislative or regulatory action, existing taxes, subsidies, and public and private investment decisions are being made in anticipation of future policies and climate realities. Thus, any rigorous analysis based on the best available current data and modeling capability is better than none, if appropriately caveated and undertaken with serious sensitivity analysis.

Second, people live in both Sarasota and Saskatoon, and living standards in those locations are not much different. In other words, people are resourceful and, within some bounds, are able to accommodate their lives to significant differences in climate. With what seem to be fairly dim prospects for internationally harmonized policies that would do much good, adaptation is likely to be one method, or perhaps even the most important method, for dealing with climate change.

Notes
1. All ten of these papers have been published in a special refereed issue of *The B.E. Journal of Economic Analysis & Policy*, along with a 'comment' on each of them.
2. Mark Twain, *Life on the Mississippi*.

[1]

The B.E. Journal of Economic Analysis & Policy

Symposium

| Volume 10, *Issue* 2 | 2010 | *Article* 19 |

DISTRIBUTIONAL ASPECTS OF ENERGY AND CLIMATE POLICY

On the Economics of Climate Policy

Gary S. Becker[*]   Kevin M. Murphy[†]

Robert H. Topel[‡]

[*]University of Chicago, gbecker@uchicago.edu

[†]University of Chicago, kevin.murphy@chicagobooth.edu

[‡]University of Chicago, robert.topel@chicagobooth.edu

Recommended Citation

Gary S. Becker, Kevin M. Murphy, and Robert H. Topel (2010) "On the Economics of Climate Policy," *The B.E. Journal of Economic Analysis & Policy*: Vol. 10: Iss. 2 (Symposium), Article 19.

Available at: http://www.bepress.com/bejeap/vol10/iss2/art19

On the Economics of Climate Policy*

Gary S. Becker, Kevin M. Murphy, and Robert H. Topel

Abstract

We analyze the central features of economic policies to mitigate climate change. The basic structure of Pigouvian "carbon pricing" is shown to follow from a standard Hotelling problem for the intertemporal pricing of an exhaustible resource. We extend this analysis to consider the strength and timing of research incentives, the costs of implementation delay and the impact of anticipated future technologies on current carbon prices. We study a variety of issues related to the valuation of climate investments, including uncertainty as to the future timing and distribution of climate impacts and the appropriate social rate of discount for valuing policies. Under reasonable circumstances the insurance properties of climate investments may warrant unusually low discount rates. We use the same framework to argue that policy makers in developing countries will discount the expected returns from climate investments more heavily, because such investments have weaker insurance value in the developing world.

*Prepared for a conference on "Distributional Aspects of Energy and Climate Policy" jointly sponsored by the University of Chicago, Resources for the Future, and the University of Illinois. We are grateful for comments from our discussant, Michael Greenstone, conference participants, Don Fullerton, and an anonymous referee. We also acknowledge support from George J. Stigler Center for the Study of the Economy and the State and from the University of Chicago Energy Initiative. Gary S. Becker, University Professor, Booth School of Business and Department of Economics, University of Chicago. Kevin M. Murphy, George J. Stigler Distinguished Service Professor, Booth School of Business and Department of Economics, University of Chicago. Robert H. Topel, Isidore and Gladys J. Brown Distinguished Service Professor, Booth School of Business, University of Chicago.

1. Introduction

Energy is essential to the maintenance and spread of economic welfare. At a point in time, individuals in richer countries such as the US and Canada use much more energy than individuals in poorer countries. Over time, long run growth in living standards is strongly associated with rising energy use, especially in developing countries.[1] There is little to indicate that these patterns might change, so that future growth and the escape of developing countries from current levels of poverty hinge on the existence and use of abundant energy supplies.

Yet rising worldwide demands for energy run up against new evidence of the social costs of energy use. The broad consensus of scientific research is that the continued dependence of economic activity on carbon-based fuels and their associated emission of greenhouse gases (GHG) create risks of substantial future changes in earth's climate, along with associated harm to the welfare of future generations. This "new" knowledge of anthropogenic climate change has motivated national and international efforts to regulate the use of carbon-based energy sources, and to promote the development and use of "clean" energy alternatives. For example, the Obama administration recently "committed" the US to achieve an 80 percent reduction in carbon emissions by 2050, even while enabling legislation to begin the regulation of such emissions languishes in Congress. In Europe, an incipient "cap-and-trade" market for emissions permits is in place, while the state of California is developing unilateral action along the same lines. Broadly-based international efforts have met with little success, as evidenced by the failure of the Kyoto (2000) and Copenhagen (2009) negotiations to achieve implementable frameworks for reducing GHG emissions.

These initiatives must confront several daunting challenges to the successful design and implementation of a useful energy-climate policy. First, current generations—who are the ones that get to decide—must be convinced that the future costs of climate change are worthy of current concern. This hasn't been achieved. Second, current generations must agree to forego use of abundant

---

[1] Our point is that economic development (almost) universally expands energy use. Energy use per unit of income—sometimes called the "energy intensity" of income—is generally declining, both worldwide and within countries. This reflects both technical advances in energy use and changes in the composition of GDP within countries. But GDP growth rates over long periods almost always exceed the rate of decline in energy intensity of GDP, so that overall energy demand rises with income, especially in developing countries. For example, between 1980 and 2007 GDP growth in China has averaged 10 percent per year, while energy intensity has declined at "only" 5.3 percent per year. The corresponding figures for India are 6.1 percent GDP growth and 2 percent decline in energy intensity. In the U.S. GDP growth has averaged 2.94 percent since 1980, and energy intensity of GDP has declined at 2.03 percent. For a complete tabulation of energy intensity see World Bank data at:
http://data.worldbank.org.indicator/EG.GDP.PUSE.KO.PP.KD

The B.E. Journal of Economic Analysis & Policy, Vol. 10 [2010], Iss. 2 (Symposium), Art. 19

carbon-based energy sources in order to mitigate uncertain harm to generations of the distant future. We know of no good examples of such sacrifice. Third, even if current generations are convinced that mutual sacrifice would be a good thing, effective policies require the largely voluntary yet global cooperation of nations in a setting where non-cooperation offers substantial rewards. Again, we are unaware of the success, or even the formation, of similar policies.

These problems are created by the fact that climate is a "global public good." In terms of climate impact earth's atmosphere doesn't much care where GHG emissions come from—a ton of carbon emitted in India has the same impact as one from Canada, so the atmosphere is over-used by all in a classic example of the tragedy of the commons. Meaningful efforts to successfully correct this externality must then hinge on collective and harmonized action by nations world-wide. Yet efforts at cooperation are hampered by the same free-rider incentives that created the problem in the first place—the benefits of carbon-based energy use are current and highly focused, while the social costs are greatly delayed, difficult to (currently) discern or measure, and highly dispersed. In addition, as we argue below, the social costs of climate change and the benefits of mitigation policies are not uniform, which leads to divergent valuations of social investments in "climate capital." This is especially true when, as here, the distributions of returns on such social investments are country-specific and highly uncertain. Then policies such as widely-discussed carbon taxes or cap-and-trade schemes offer much different risk-reward tradeoffs to developing countries, such as China or India, than to developed countries like the US. These divergent valuations help explain the current lack of progress in international negotiations over climate policy, such as Copenhagen (2009), and what we believe are the limited prospects for cooperation going forward.

From (very) high altitude, the economics of anthropogenic climate change is a standard problem of externality—current users of carbon-based fuels do not bear the environmental costs of energy consumption, so they use too much of the stuff, and too little of "clean" alternatives. The problem occurs because some resource—here the atmosphere—is unpriced and so overused. The idealized textbook market intervention is to price the overused resource, equating the private and social costs of its use. This can be accomplished via a Pigouvian tax (or its equivalent) equal to the marginal external cost of using a unit of carbon-based fuels. The resulting ideal "carbon-price" would exactly balance the benefits of additional carbon emissions, which occur now, against their costs, which are spread over the near and distant future.

This basic solution to the externality problem is familiar and straightforward. It is central to virtually all serious national and international policy proposals to deal with climate change, including the Waxman-Markey[2] and Lieberman-Warner[3] bills in the US House and Senate, the design of cap-and-trade policies in the EU, and the tentative framework discussed in the recent Copenhagen negotiations. But its conceptual simplicity is superficial—the actual design and implementation of such policies faces daunting challenges and unresolved questions. Our analysis seeks to contribute to a number of unresolved issues in the design and effects of policies to mitigate climate change. These include:

1. Valuing future costs: The social costs of current GHG emissions are uncertain and spread over the distant future. How should current policy value the costs of climate damage, which will fall mainly on future generations? Should the future benefits of investments in "climate capital" be discounted at market rates or, as some have argued, at much lower rates? How will these valuations differ across major countries, the large majority of which must cooperate to achieve efficient policy outcomes?

2. Uncertainty: How does the great uncertainty regarding the extent and costs of future environmental harm affect current strategies, social investments, and valuations?

3. Catastrophic climate change: Among the uncertainties is the possibility of catastrophic outcomes that could greatly reduce future living standards or endanger future populations. How should current policy value and mitigate these possibilities?

4. Market responses to climate policies: At its barest level, "carbon pricing" is a market-based solution that relies on market responses to efficiently designed incentives. How will markets respond to policy-generated incentives? Will market responses enhance or constrain the effects of policies?

5. Innovation incentives, policy design and the costs of delay: How does an efficient policy affect research incentives and the pace technical progress in alternative energy sources and in mitigation? If technical breakthroughs are likely to be the ultimate solution to the energy "problem," are incentives to innovate harmed by delays in implementing an optimal policy? How should the prospect of future innovation affect current policy?

The paper is organized as follows. Section 2 develops the basic features of optimal carbon pricing, which we relate to a standard Hotelling problem for the

---

[2] http://www.govtrack.us/congress/bill.xpd?bill=h111-2454
[3] http://www.govtrack.us/congress/bill.xpd?bill=s110-2191

intertemporal pricing of a depleting resource. We extend the analysis to consider the strength and timing of research and development incentives, the costs of delay in implementing an optimal policy, and the impact of anticipated future technologies, and their form, on current carbon prices. In Section 3 we analyze a variety of issues related to the valuation of climate policies, including uncertainty as to the future timing and distribution of climate impacts. We pay particular attention to the appropriate social rate of discount for valuing policies, showing that under certain reasonable circumstances the insurance properties of climate investments may warrant unusually low discount rates. We use the same framework to argue that policy makers in developing countries will discount the expected returns from climate investments more heavily, because such investments have weaker insurance value in the developing world. Section 4 concludes.

2. Features of Efficient Climate Change Policies

The scientific foundations for anthropogenic climate change indicate that current emissions of GHGs create environmental and other costs that are (1) greatly delayed, (2) very long-lasting, and (3) highly uncertain.[4] This is because the flow of CO_2 to the atmosphere has a long lasting impact on the stock of atmospheric CO_2, as reabsorption is very slow. In turn, the growth in global temperature lags the atmospheric stock of CO_2 because, for example, melting of ice caps reduces earth's albedo (reflectivity) and oceans warm slowly. Many costs are likely to lag a rise in temperature—for example, rising sea levels would be driven by melting of ice caps, which would follow a prolonged warming period. Finally, uncertainty as to environmental feedbacks and other impacts includes the prospects for "catastrophes" of various forms. Because of these features, policies that would mitigate these effects must balance costs and benefits over hundreds of years, and subject to large and costly contingencies, which make the problem of policy design a good deal more daunting than the usual project evaluation.

2.1 Carbon Pricing as an Exhaustible Resource Problem

To illustrate central elements of dynamic carbon pricing and its connection to key assumptions about preferences, growth and technology, consider a simple certainty framework in which the target cap on atmospheric concentration of GHGs at some endogenous future date T (say in $T=200$ years) is the goal of environmental policy. Denoting the concentration of GHGs at date t by Q_t, this

---

[4] See Archer (2007) and (2009) for useful summaries of the state of climate science research on global warming.

Becker et al.: On the Economics of Climate Policy

terminal condition is $Q_T = \bar{Q}$. Stated in this way, the optimal policy solves a Hotelling problem for allocating the use of an exhaustible resource over time—where here the exhaustible resource is the capacity of the atmosphere to "safely" hold a given concentration of GHGs.[5]

Let the private social (consumer plus producer) surplus from current emissions, q, be $V_t(q_t, b_t, y_t)$ where b_t is the unit cost of carbon-free energy sources at date t and y_t is income. Finally, let the technology for mitigating emissions be represented by the cost $C_t(s_t \mu_t^{-1})$, where s_t is the amount of period-t emissions avoided through mitigation activities and μ_t indexes the evolving efficiency of mitigation—higher values of μ_t reduce the costs of emissions mitigation. For example, s_t might be the amount of period t emissions that are eliminated by sequestration or other technologies, and μ makes the process more efficient. With these definitions, the policy problem is to maximize the present discounted value of social surplus.

(1)
$$\underset{q,s}{Max} W = \int_{t=0} (V_t(q_t, b_t, y_t) - C_t(s_t \mu_t^{-1})) e^{-rt} dt$$

$$s.t. \ \dot{Q}_T = q_t - s_t - aQ_t \ \text{and} \ Q_T = \bar{Q}$$

where r is the rate of interest used to discount future environmental costs (the rate of return on investments in environmental capital) and a is the rate at which atmospheric GHGs are reabsorbed. We shall have much more to say about r later, but for now we simply take it as given without pondering how large or small it could or should be.[6]

Letting $V_t'(\bullet)$ denote the derivative of $V_t(\bullet)$ with respect to q_t, the basic solution to (1) has a familiar structure:

---

[5] There need not be a fixed target level of GHG for this analysis to apply. As long as the effects of climate change are simply a function of the stock of GHG at some future date (200 years in our example), the optimal program will need to solve this same Hotelling problem given the optimal level of GHG at the terminal date.

[6] The appropriate rate of return on social investments in climate capital will depend on insurance properties of the investment's return. Projects that pay off by mitigating climate-related catastrophes may have discount rates that are well below the market rates of return on other risky assets, and even below the risk free rate. We take up these issues in Section 3, below.

The B.E. Journal of Economic Analysis & Policy, Vol. 10 [2010], Iss. 2 (Symposium), Art. 19

$$V_t'(q_t, b_t, y_t) = P_0 e^{(r+a)t}$$

(2a)

$$\mu_t^{-1} C_t'(s_t \mu_t^{-1}) = P_0 e^{(r+a)t}$$

In (2a) both the marginal value of using and the marginal cost of eliminating a unit of emissions are equated to the period-t "carbon price" $P_0 e^{(r+a)t} = P_t$, which represents the scarcity value of a "unit" of the otherwise unpriced absorptive capacity of the atmosphere. This price is the outcome of an ideal Pigouvian tax or cap-and-trade system, so that P_t equates the marginal benefit of q to current users and the (present value of) incremental costs imposed on future generations.[7]

2.2 Carbon Pricing, Timing and the Returns to Innovation

The fact that the socially optimal carbon price rises at the rate of interest (plus absorption) is a well-known property of this and other exhaustible resource problems—Nordhaus (2007) refers to the rate of growth $r+a$ as the "net carbon interest rate."[8]   It is a condition for intertemporal efficiency in the use and mitigation of emissions, equating the value of benefits from creating incremental emissions (due to energy use) to the present value of costs.   Less appreciated is how this property of an optimal policy impacts the social value of innovations, incentives to innovate and the cost of waiting to implement the policy.

   To fix ideas with a not-entirely-fanciful example, think of a current investment in technology that could eliminate one unit of carbon emissions at some arbitrary future date, t—a one-period "carbon eating tree."[9]   The present value of this unit reduction in future emissions is $P_t e^{-rt} = P_0 e^{(r+a)t} e^{-rt} = P_0 e^{at}$, which implies that the time profile of values is independent of the interest rate, r. If $a=0$, the present discounted value of the innovation is independent of how far in the future it pays off, t, because the value of the gain rises at the interest rate. And if $a>0$ the present value actually rises with t because the time-t value of the innovation rises faster than the interest rate. In effect, the value of the innovation is undiscounted.

   The result is even stronger if innovation is scalable.   Think of an innovation that would reduce the incremental cost of mitigation at some future t,

---

[7] Nordhaus (2007a,b) and (2008) are good summaries of the state of economic modeling applied to global warming and climate policy. Many of the same analytical tools appear in Stern (2006).

[8] The original statement is in Hotelling (1931). Treating $\bar{Q}$ as an exhaustible resource, the condition is that "owners" of the resource must be indifferent between selling a unit today and holding it for future use. Here P_t is the per-period shadow value of relaxing the constraint on GHG concentrations, $\bar{Q}$.

[9] E.g. Dyson (2008).

Becker et al.: On the Economics of Climate Policy

raising μ_t by $d \ln \mu_t > 0$. The present value of this cost reduction (per unit change in $\ln \mu$) is

$$
\begin{aligned}
& e^{-rt} \mu_t^{-1} C_t'(s_t \mu_t^{-1}) s_t \\
= & e^{-rt} P_t s_t \\
= & P_0 e^{at} s_t
\end{aligned}
$$

Here the unit value of the innovation is independent of r, and also of t if $a=0$, but the reduction in incremental cost is scalable because it applies to all units, s_t. But s_t rises over time because P_t is increasing, so the present value of the innovation also rises with t—the gain has larger present value the farther in the future it occurs. Applied to all periods, a scalable technical advance in mitigation that applies from the present day forward is worth

(3) $$ W_{\ln \mu} = P_0 \int_{t=0} s_t e^{at} dt $$

Even with $a=0$, the value of the gain applies to the quantity of mitigation in all future periods with equal weights. If $a>0$ then future periods get more weight than the present. And of course the result applies to other types of innovation, such as an advance that would improve the consumption efficiency (surplus) per unit of emissions, like a change in fuel efficiency of cars. All such gains are valued at the rising Pigouvian price P_t, which "undoes" the effect of discounting in present value calculations.

These conclusions may appear anomalous, and it is easy to extend the policy problem (1) to include factors that would cause the optimal atmospheric price P_t to rise more slowly, so that delay is more costly. For example, if we amend the optimal policy to include possible environmental damages $D(Q_t)$ from rising atmospheric concentrations along the trajectory to T, then the current optimal carbon price is continuously updated, incorporating the impact of current emissions on future damages:

$$ \frac{d \ln P_t}{dt} = r + a - P_t^{-1} D'(Q_t) $$

where $D'(Q_t) > 0$ is the current period marginal damage from emissions, equivalent to the effective current period "rental price" of atmosphere. This

The B.E. Journal of Economic Analysis & Policy, Vol. 10 [2010], Iss. 2 (Symposium), Art. 19

reduces the growth rate of P_t relative to the standard Hotelling solution, but without negating the broader point, which is that innovations yield greatest value when the carbon price is high, and in the optimal policy that price is rising over time.

The central lesson about valuing progress in (3) is simple, but it has important implications for interpreting both the form of policy responses to climate change as well as the urgency with which those policies are implemented and the costs of delay. The slow progress of international negotiations in gatherings such as Kyoto and Copenhagen is widely lamented, as is similarly slow progress in crafting and adopting enabling legislation in the US and other countries. Do these delays adversely impact incentives to find "solutions" in the form of technologies that would reduce the carbon impact of energy consumption? Is a sense of urgency warranted?

If we assume that slow progress toward implementing a policy is just that—slow progress that will eventually result in widely applied carbon pricing that reflects social costs—then equation (3) indicates that the costs of delay are small. To illustrate, use (3) for the present discounted value of a cost-saving innovation that requires substantial up-front R&D effort. Realization of these incentives requires two things: an initial incentive, P_0, that signals the current scarcity value of emissions, and a commitment to a time path for that value in the future. Given these, delaying the start of this payoff for d years would not much affect its present value, even if the initial d years of a payoff stream were foregone. And if the whole program is simply pushed back the payoff would likely rise because the initial price P_d would increase by more than simple interest (because interim unpriced emissions tighten the ultimate constraint), which also raises s. The result is that the current value of innovation incentives and the social gain from innovations are not much harmed and may even be increased by delay.

This analysis assumes that an optimal carbon pricing program is eventually implemented—that negotiations and legislation result in something useful in terms of price signals and commitment to policy. The message is not that delay is costless—the tighter constraint and necessarily higher carbon price caused by delay demonstrate the costs—but rather that the returns to climate-related innovations are not much reduced by delay in implementing well-designed incentives. And if the ultimate efficiency gain is likely to derive from currently unforeseen innovations driven by carbon pricing, rather than simply by business as usual along a rising price path, it is likely that delays of a few years don't much impact that outcome. Put differently, it is far more important to get the form of policy right—including believable commitments to the level and time path of future carbon prices—than to get a policy done quickly.

2.3 Factors Affecting the Impact of Policy

Conditions (2a) for the rate of change price of the carbon price embed the properties of evolving demand and supply for carbon-based fuels, via the social surplus $V_t(.)$, as well as changes in the availability of substitutes, z, and the evolution of technology. It's worth being explicit about these, because expectations of how demand and technology will evolve in the future are essential ingredients of current policy and the optimal level and timing of mitigation activities. Using the usual notation for time rates of change (e.g. $\dot{q} \equiv \dfrac{dq}{dt}$) displacement of (2a) gives the rates of change of emissions (q) and mitigation (s):

$$(2b) \qquad \begin{aligned} \dot{q}_t &= -\xi_t[r+a] + \sigma_t\,\dot{b}_t + \eta_t\,\dot{y}_t \\ \dot{s}_t &= \phi_t[r+a] + [1+\phi_t]\dot{\mu}_t \end{aligned}$$

Here, σ is elasticity of emissions with respect to the cost of its substitute, b, η is the income elasticity of demand for emissions generating activities, ϕ is the elasticity of mitigation supply (the inverse of the elasticity of marginal cost) and ξ is the price elasticity of current emissions, q, which embeds both supply and demand responses to changes in P.[10]

Equations (2b) have several important implications. First, absent technical progress in reducing emissions ($\dot{\mu}_t = 0$) and with negligible reabsorption ($a{=}0$), the growth rate of mitigation is proportional to the rate of interest. The factor of proportionality is the elasticity of mitigation supply, ϕ_t, so optimal mitigation grows more rapidly when supply is more elastic or when the rate of interest is high. But with $\dot{\mu}_t > 0$ the growth rate of mitigation is augmented by anticipated technical progress (the rate of decline in costs) in emissions reduction. Given the dependence of emissions mitigation on technology and research, and expectations that costs of emissions reductions actually will fall over time, this means that "waiting" to achieve emissions reductions is a central element of dynamically efficient policy. In a broader context, however, the magnitude of μ is endogenous to current policy, because it is an outcome of current and future R&D

---

[10] In a competitive market ξ will be given by the harmonic mean of supply and demand elasticities, $\xi = (\xi_S^{-1} + \xi_D^{-1})^{-1}$.

efforts, and our previous discussion indicates that incentives to innovate are powerful, provided that innovators can collect on the value of their innovations.

Similarly, the benefits of deferral are larger when the elasticity of mitigation supply is large, and especially when large values of ϕ_t are likely to evolve from future technical advances. Large values of ϕ_t mean that marginal costs of mitigation at a point in time do not rise sharply with s—mitigation activities are easily "scalable"—so there is not much cost to sharply ramping up mitigation in later periods when emissions reductions will be most valuable. But when ϕ_t is small the marginal cost of mitigation increases rapidly with s—there is a large cost penalty if mitigation efforts are concentrated in fewer periods. Then it is worthwhile to do things in smaller pieces by spreading mitigation activities over time. Then the optimal policy is to ramp up mitigation efforts sooner rather than later.

This interpretation of the elasticity ϕ_t is the "certainty" equivalent of a broader point about scalable technologies—they can be deployed as needed on large scale without much cost penalty. As we show below, development of highly scalable (high ϕ) technologies is especially valuable if we extend the analysis to incorporate uncertainty and the possibility that future environmental effects of GHGs may turn out to be much more costly than currently anticipated, or that low-probability but high-damage outcomes may occur. Then scalable technologies to reduce emissions have high option value, precisely because they can be deployed on a large scale when mitigation is most critical. Notice also that technical advances that enhance scalability (ϕ) or enhance the efficiency of mitigation ($\dot{\mu}_t$) are complementary—the social benefits from higher $\dot{\mu}_t$ are proportional to ϕ, and conversely.

Similar implications apply to the "value" side of (2b). Since the price of emitting carbon rises at the "net" interest rate $r+a$, this price rise induces conservation in proportion to the price elasticity ξ. Note that ξ embeds both production and consumption responses to carbon pricing—for example, the fact that carbon-based fuels are abundant and in fairly inelastic supply on the world market suggests that ξ is likely to be small.[11] Together with demand growth ($\dot{y}_t > 0$) on the world market, the implication is that substantial conservation relative to business-as-usual is unlikely. Then policy success is critically dependent on technical advances that would promote mitigation by enhancing ϕ

---

[11] That is, the burden of P is likely to fall on suppliers of carbon emitting energy sources, who would supply roughly the same quantities at substantially lower after-tax prices. Then imposition of emissions pricing may not much impact fuel use or emissions through conservation.

and μ, or substitution toward non-carbon-based energy alternatives with declining costs ($\sigma_t \dot{b}_t < 0$). In other words, the point of optimal emissions pricing is not so much to induce conservation on the demand side—which is likely to have small effects—as it is to guide the research incentives that will result in greater supply of clean energy alternatives in the long run.

2.4 Evolving Expectations and Changing Incentives

For a given rate of interest the rate of growth of the optimal carbon price is determined. The other key to incentives is P_0—the initial or current carbon price—which determines the level of the entire future price path. This is affected by the entire array of technology and substitution effects, and the way they are anticipated to evolve, as in (2b). Factors that reduce the anticipated growth of q or raise the growth of s will reduce P_0 and delay the ultimate date T when net additions to the stock Q optimally cease. For example, the expected emergence of technologies (ϕ) that make s more scalable allow for lower net emissions (q-s) along a flatter path, making T longer, and so on.

The prices P_t that support optimal net emissions in problem (1) could be generated by an ideal set of emissions taxes or by cap-and-trade determination of an emissions price. Though we don't wish to join a full debate over the relative merits of carbon taxes versus cap-and-trade schemes—see Nordhaus (2007c) for a good discussion—our framework does highlight some key issues that have not been emphasized in previous literature. While we have framed the optimal policy in a certainty-equivalent framework, the fact that the optimal initial price, P_0, depends on expectations of future market responses and technologies means that an efficiently updated policy should adjust the current price level, P_0, as information evolves. For example, an innovation that reduces expected future mitigation costs will reduce P_0, exactly as a new "find" that increases the future availability of an exhaustible resource (such as oil) will reduce its current price, even if the newly discovered units are not currently recoverable. In an ideal cap and trade framework in which total acceptable emissions $\bar{Q}$ (as opposed to year-by-year emissions) are fixed and unchanging, and tradable over time, the collection of information and the formation of expectations about such future innovations and technologies is decentralized to market participants—a clear advantage in terms of incentives. But the possibility of governments manipulating the variable they control, $\bar{Q}$, invites rent-seeking, which is the foundation of many economists' critique of cap-and-trade schemes and their consequent preference for tax-based incentives.

Yet tax-based schemes also have powerful disadvantages. Tax-based carbon pricing sacrifices the substantial advantage offered by market-based expectations—when carbon prices are set by governments the formation of "expectations" that determine both the level and growth of the optimal emissions tax is necessarily centralized in government, which is responsible for setting and updating the entire price (tax) path P_t. There is little reason to believe that governments would do well in this regard, and the opportunities and incentives for inefficient choices and rent-seeking appear to us just as powerful with taxes as with cap-and-trade.

2.5 Discounting Future Climate Costs and Returns

Ignoring reabsorption, a, for any given future marginal damage from incremental emissions, say P_T = \$500 per ton of CO_2 emissions in T=100 years, the strength of initial incentives, P_0, is determined by the rate of interest; $P_0 = e^{-rT} P_T$. Higher r means low P_0 and a gradual ramping up of incentives and responses. Low r means that conservation and mitigation efforts are more front-loaded. If we base r on historical market rates of return on physical and human capital, then a value in the neighborhood of r=.06 is reasonable. This yields P_0=\$1.24 if $P_{100} = \$500$. In contrast, the UK government's 2006 *Stern Review of the Economics of Climate Change* argued that policies should reflect much lower interest, r = .015, based on the *Review's* notion that that it is ethically improper to heavily discount the costs that current emissions impose on future generations. Then the current tax or price is $P_0 = \$500 \times e^{-.015 \times 100}$ = \$112, which is almost 100 times larger than with r=.06. As pointed out by Nordhaus (2007b) and Weitzman (2007), among others, this philosophical choice of a (very) low discount rate for investments in climate capital accounts for virtually all of the differences between the *Stern Review's* draconian recommendations for current action and the more gradualist policies advocated by other economists. Much then hinges on the choice of a social rate of discount for climate capital, r, which we take up below.[12]

3. Valuing Future Climate Damages

The fact that current economic activity and policy affect uncertain climate outcomes and costs over vast time periods may be the most daunting challenge of climate policy. Possible future outcomes—including the possibility of

---

[12] Moreover, since the analysis in the Stern Review assumes that the cost of GHG concentrations are proportional to GDP and that GDP grows substantially over time, the implied net discount rate is even smaller (and is essentially zero). See Nordhaus (2007b) for a clear discussion of this issue.

environmental catastrophes that could harm large populations or greatly reduce productivity—must be both envisioned and valued, and then balanced against the current cost of mitigating such harms.

We address three issues related to the current valuation of uncertain future damage. First, given the possibility of various types of catastrophe, what does economic analysis say about the costs that should be incurred today in order to avoid them? At standard discount rates, events that have even substantial impacts on productivity and population-wide living standards in the distant future have only small present value. We show that these values substantially increase, however, when climate-related damages are unequally distributed, when future lives are at risk, and when we allow for uncertainty as to <u>when</u> the damaging events might occur (holding constant the expected time to occurrence). Even at "market" rates of discount, not-implausible values for the magnitudes of future catastrophes imply substantial current willingness to pay to avoid them.

We then extend the analysis to the valuation of climate investments with uncertain future returns. We show that appropriate social discount rates for investments in climate capital may be well below market returns on other forms of capital, reflecting the insurance value of climate investments. We also find that distribution matters. The global public good nature of harmonized climate policies is challenged by heterogeneity of valuations—projects that have high insurance value to developed countries because they reduce future risks are likely to be much less valuable to developing countries, for whom the possibility of rapid economic growth is likely more important.

3.1 The Costs of Future Catastrophic Outcomes

Future catastrophic outcomes may include substantial damages to productive capacity, sustained reductions in economic growth, threats to living standards or lives of particular populations, or permanent environmental harm that reduces welfare for any given level of economic activity. To frame these possibilities, we begin with the standard infinite-horizon model of intergenerational utility that underlies most work in economic growth and climate policy.[13] Write the current value of generational welfare over the indefinite future as:

$$(4) \qquad U_0 = \int_{t=0}^{\infty} u(c_t)e^{-\rho t}dt$$

---

[13] Pindyck and Wang (2009) provide a dynamic general equilibrium approach to valuing catastrophic outcomes, including parameterized distributions for both the arrival rate and distribution of harm from catastrophes. Weitzman (2009) allows the distribution of future harm from climate change to have "fat tails", which can greatly impact the current value of avoidance.

The B.E. Journal of Economic Analysis & Policy, Vol. 10 [2010], Iss. 2 (Symposium), Art. 19

In (4) ρ is the rate of time preference, or in an intergenerational context the rate at which earlier generations discount the well being of later ones, and c_t represents the per-capita flow of goods and services (consumption) available to generations alive at future date t, which may include valuations of environmental factors. We continue to abstract for the moment from issues of uncertainty.

One form of calamity that can be represented in (4) is a permanent reduction in future living standards that is known to commence at some future date T, say in 100 years. So assume that future productivity is reduced by a constant percentage, resulting in a permanent change in future consumption of $d \ln c$ from T onward. For example $d \ln c = -.01$ represents a permanent 1-percent reduction in per-capita income and consumption. Assume a constant elasticity of the marginal utility of consumption, ω, and steady state economic growth of g. Then we can apply the Ramsey Equation linking the equilibrium interest rate to time preference and economic growth, $r = \rho + \omega g$. The current value of this harm as a fraction of current (time zero) national income is:

$$(5) \qquad \frac{1}{c_0} \frac{dU_0}{u'(c_0)} = \frac{e^{-(r-g)T}}{r-g} d \ln c$$

where $(r-g)^{-1} d \ln c$ is the damage valued at date T and $e^{-(r-g)T}$ discounts the date-T value to the present, allowing for economic growth. How large is (5)? Assume $r=.06$ and $g=.02$—fairly standard values in a growth framework—and let $d \ln c = -.01$ (a one percent permanent reduction in future incomes). Then with $T = 100$ years the right side of (5) is equal to -.0046, or about half of one percent of current income. For the US with a national income of about \$13 trillion, this implies a present discounted value of future harm of about \$59 billion. Viewed as a long term project to avoid such damage, the expenditure flow at 6 percent interest is about \$3.6 billion. Cutting the horizon to $T=50$ years substantially impacts the estimates. Then a permanent 1 percent reduction in future income is worth about 3.4 percent of current income (\$440 billion), or a flow expenditure of \$26.4 billion per year. By comparison, with these same parameters a current permanent reduction in consumption of one percent would be worth roughly \$3.25 trillion or a flow of expenditure equal to roughly \$195 billion per year.

Adding uncertainty about when such climate-related damages might occur substantially raises the present value of avoiding them.[14] To demonstrate this in a simple way, hold constant the expected time until damage occurs at $T=100$ years, but assume that the damage is equally likely to commence at any future date.

---

[14] Karp (2009) makes a related point.

This implies an arrival rate (hazard) for the damaging event of $h = T^{-1}$. The present value of future expected damages as a fraction of current income is then:

$$(6) \qquad \frac{1}{c_0} \frac{dU_0}{u'(c_0)} = \frac{1}{1 + (r - g)T} \frac{1}{r - g} d \ln c$$

Using the same values as above, a permanent 1 percent reduction in living standards with expected time to occurrence of $T=100$ years has present value equal to 5 percent of current income, which is roughly 11 times greater than when the damage was known to commence in 100 years. This is worth about $650 billion to the US in 2010, equivalent to a flow expenditure of $39 billion per year at 6 percent interest. At $T=50$ $(h=.02)$ the cost is 8.3 percent of current national income, or a flow of $65 billion. Our point is that uncertainty over the time at which climate change will have an adverse effect can greatly increase its current valuation.

Formulas (5) and (6) express the current value of marginal losses in future per capita consumption—everyone consumes one percent less than otherwise. This is consistent with most of the existing analysis of valuing climate costs, where those costs are framed in terms of reductions in future GDP, or costs as a fraction of GDP, as if the burden of climate impacts is equally spread among the future population. But much of the concern about climate-related damages has to do with the distribution of harm, where some groups are harmed much more than others. Concave $u(c)$ means that reductions in c have rising marginal cost to those who experience them, so a given reduction in aggregate income is more costly when it is highly concentrated. For example, with $\omega = 2$ a catastrophe that reduces incomes by half among 2 percent of the population is twice as costly as an across-the-board reduction in living standards of one percent, even though both events reduce overall per-capita income by the same amount (one percent). Taken a step further, a climate-related catastrophe that reduces future national incomes by one percent by killing off one percent of the population, while leaving others unharmed, may be very costly. Such catastrophes are not "marginal," reducing everyone's income proportionally. Instead they wipe out consumer surplus—or in the extreme case the value of life—for a swath of the population.

A framework for valuing such catastrophes is provided by the economic literature on the value of a statistical life (VSL), which measures people's willingness to pay for a reduction in the probability of death that would save one "statistical life." For example, if in a population of 10,000 persons each would be willing to pay $600 per year to reduce the per-year probability of accidental death by 1 in 10,000, then $VSL = \$6$ million, which is about the value used by the US Environmental Protection Agency for cost-benefit analyses of regulations or projects that would reduce mortality risks. Murphy and Topel (2006) use this value to calibrate the value of a life-year $v(c) = u(c)/u'(c)$, which is the "consumer surplus" achieved by being alive and consuming amount c, where c includes leisure and other factors that people value. They find that the value of a life year is about six times current income, so if we think of $v(c) = \psi(c)c$ the data suggest that $\psi(c) \approx 6$ at current income levels. Then the above calculations would increase by at least a factor of 6 for life-threatening events that cause an equally-calibrated reduction in future "income."

The analyses in Murphy and Topel (2006) and Hall and Jones (2007) also indicate that $\psi(c)$ rises with income, so the value of life is income elastic. Then if future generations are richer than us, the value of lives saved from mitigating future catastrophes will be proportionally greater than today. For example, an income elasticity of $\zeta = 1.2$ and long run economic growth at 2 percent yields $\psi(c) \approx 9.0$ in 100 years. The result is that a randomly occurring event that causes a "concentrated" change in future costs because of climate-related mortality has much higher current value:

$$(7) \qquad \frac{1}{c_0}\frac{dU_0}{u'(c_0)} = \frac{\psi(c_0)}{1+(r-\zeta g)T}\frac{1}{r-\zeta g}d\ln c$$

With a constant hazard rate and an expected arrival time of $T = 100$ years, with $\zeta = 1.2$ and $\psi(c_0) = 6$, a "catastrophic" event that reduces per-capita output by killing off $d\ln c = -.01$ of the population has present value equal to 36 percent of current income. Letting $T = 1000$—a catastrophe that could occur every thousand years, on average—the current value is about 4.5 percent of income. And of course the value is highly sensitive to the choice of r: a reduction in the discount rate from .06 to .04 raises the current value of avoiding such a catastrophe from 4.5 percent to 22 percent of current income.

These results indicate that uncertainty over the future timing, magnitude and distribution of losses from climate change can greatly impact our assessments of current cost, even if future costs are discounted at conventional rates of return of, say, 6 percent.

3.2 Discounting the Returns on Climate Capital

One of the most controversial aspects of debates over climate change policy is the appropriate social discount rate to be applied to future damages. At the extreme among economists, the *Stern Review's* advocacy for a very low interest rate of $r=.015$ accounts for almost all of its severe recommendations. Behind the *Stern Review* recommendations is the notion that the welfare of future generations

Becker et al.: On the Economics of Climate Policy

should be weighted equally with current ones, it being ethically repugnant (in *Stern's* view) to discount their welfare. Then the only source of a positive discount rate on real cash flows is the growth of consumption over time, because future generations are richer than we and utility is concave—$\omega > 0$ in our earlier notation. And in the *Stern* recommendations even that is given little weight in reaching the desired result. Similarly, non-economists such as Archer (2009) have argued that economic analysis is itself ill-equipped to deal with intertemporal valuations spanning a generation or more. Like *Stern,* Archer argues for effectively zero discounting because current action is a moral imperative.

The slowly ramping policy profiles offered by Nordhaus (2008) and others are based on higher discount rates that reflect historical long run returns on other types of capital. In contrast, while critiquing the analytical foundations of the *Stern* rates, Weitzman (2009) offers a "Dismal Theorem" based on the possibility of extreme catastrophes that drive consumption near zero and the marginal utility of consumption beyond the moon. Policies that can avoid such outcomes can have unbounded value under particular assumptions about the distribution of climate effects on c—they should have "fat tails"—and the rate at which marginal utility rises as consumption falls. The more general and useful point is that uncertainties about the distributions of climate damage and the payoffs from mitigation investments may greatly affect valuations. Climate policies that effectively insure against large downside risks (when the marginal utility of consumption would be large) needn't have large expected returns, so the typical market benchmarks for r might be inappropriate for valuing investments in climate capital. We return to this point shortly.

From an economic and empirical perspective the choice of a discount rate is not about the philosophical choice of the correct ethical weight to be applied to the welfare of our and other peoples' great-grandchildren, nor is it about the way we "should" discount marginal dollars of their income because they will be richer. As in all analyses that must balance costs and benefits, the issue is opportunity cost. The fact that costs and returns are so uncertain and widely spaced in time adds practical difficulties but not conceptual ones.

Consider a current project costing $1 million that would reduce the impact of climate change 100 years from now. Assume that, absent the project, the resulting climate change would impose a real cost of $20 million on future generations. The logic of *Stern* (2006) and Archer (2009) suggests that we should implement the project if, and only if, we value giving $1 to the current generation less than we value giving $20 to the future generation. That is, the question of whether the mitigation project is worthwhile allegedly depends on the relative values we place on the consumption of current and future generations.

The B.E. Journal of Economic Analysis & Policy, Vol. 10 [2010], Iss. 2 (Symposium), Art. 19

This is not correct—our choice does not depend on our relative preference for current versus future generations.

Assume we wish to give the future generation the $20 million benefit they would derive were we to implement the project today.   If undertaken, the rate of return on the current project is 3 percent, which is the solution for r in the equation $\$1 \times e^{100r} = \$20$. But if the market rate of interest is 6 percent, $1 million invested at the market rate of return would yield $403 million in 100 years, compared to the $20 million benefit generated by the climate mitigation project. This means that future generations would gain (a lot) if the current generation were to forego the climate project and invest in other assets that yield higher returns. Alternatively, it would take only $49,000 invested at 6 percent to provide the future generation with the $20 million needed to compensate them for the harm from climate change. Our point is that it is the market rate of return—not our attitudes toward future generations or our moral view of discounting—that determines the appropriate discount rate. To evaluate climate mitigation policy with a lower rate of return unnecessarily harms either current or future generations, or both. Future generations would not thank us for investing in a low-return project.

It is appropriate to discount the costs and benefits of climate change policies at a "market" rate of return because the market rate measures the opportunity cost of such investments—returns available from investing the same amount in physical or human capital—so long as such opportunities exist. But what "market rate" should we use? At the low end one might benchmark by the risk free rate as represented by the returns on government bonds.   An alternative would be the much higher historical returns on risky investments such as physical capital or equities. Offered the opportunity to invest for the benefit of our great-grandchildren in 2110—who by any reasonable expectation will be much richer than us[15]—would we opt for Treasury bills and an annual return of perhaps 3 percent when the historical equity premium consistently provides long run returns in the neighborhood of 6 to 8 percent? Most would choose equities.

Yet the fact that most of us would choose equities reflects an implicit but appropriate (in this context) belief that those assets correctly gauge the opportunity cost for long-term financial investments, including allowance for risk. The weakness in this argument is that it is not obvious that the risks and returns on climate investments align with those on other physical assets or equities. If the returns on climate investments are uncorrelated with returns on the market portfolio, or if by eliminating calamitous harm to overall productivity and living standards climate investments pay off exactly when other productive assets do

---

[15] At 1.5 percent annual growth, per capita income in 2110 will be about 4.5 times the current level. At 2 percent the multiple is 7.4. Growth rates in developing countries such as China or India are expected to be much higher.

not, then the appropriate rate of return and discount rate for climate projects should be lower than for other assets, perhaps even lower than the risk-free rate. Further, though the climate impacts of investments in climate capital are global, the risk properties of those effects, and hence their value, may differ greatly across the countries whose participation in global agreements is essential.

3.3 Expected Social Returns on Climate Capital

The risk properties of the returns on climate investments derive from at least four stochastic drivers: (1) global economic growth, because greater growth likely means greater emissions; (2) the impact of emissions on climate; (3) the impact of climate on environment, productivity and welfare; and (4) the effectiveness of current investments in mitigating future harm.

To illustrate the determinants of an appropriate discount rate for climate investments, consider a standard asset pricing framework for valuing a current (time 0) project that offers uncertain returns at some future date, F.[16] Assume that the current generation can invest in λ units of a climate project, with current cost $K(\lambda)$ and marginal cost $k(\lambda) = K'(\lambda)$. The investment offers uncertain future returns of x per unit, where x may be interpreted as the project's future impact on GHG concentrations, or other measures that would mitigate climate impacts. With this setup, the social planner's intertemporal problem is

$$(8) \qquad \underset{\lambda}{Max}\ U = u(y_0 - K(\lambda), \psi_0) + \rho E\left[u\left(y_F(\lambda x), \psi_F(\lambda x)\right)\right]$$

The representative individual in (8) derives utility from income (consumption) y and the state of the environment ψ, both of which can be affected by current climate investments. The factor $\rho < 1$ reflects pure time preference between the present $(t=0)$ and future $(t=F)$, and E is the expectations operator reflecting uncertainty over the joint distributions of y, ψ and x. We interpret this social valuation problem as country-specific, so that the distributions of outcomes may be quite different for, say, China than for the US.

The choice of investment in the climate project solves

$$(9) \qquad \begin{aligned} k(\lambda) &= E\left[m_F X_F\right] \\ &= e^{-r_f} E(X_F) + \text{cov}\left(m_F, X_F\right) \end{aligned}$$

---

[16] See Cochrane (2005) for a clear presentation of asset pricing and discounting issues.

The B.E. Journal of Economic Analysis & Policy, Vol. 10 [2010], Iss. 2 (Symposium), Art. 19

where r_f is the risk-free rate of return, m_F is the marginal rate of substitution between future and current consumption, and X_F is the future generation's value of the income and environmental payoffs on the investment:

(10) $$X_F = (y'_F + v_F \psi'_F)x$$

where $v_F = u'_\psi / u'_y$ is future willingness to pay for environmental improvements. Divide (10) by $k(\lambda)$ to obtain marginal returns per dollar invested ($R_F = X_F / k(\lambda)$) and solve for the required return on the environmental asset, which yields the familiar CAPM form for required expected returns on the investment, r_E:

$$r_E = r_f - \text{cov}(m_F, R_F)$$
(11) $$= r_f - \beta_{m,R}(r_M - r_f)$$
$$= r_M - (1 + \beta_{m,R})(r_M - r_f)$$

In (11), r_M is the market rate of return on equities and $r_M - r_f = \text{var}(m)$ is the equity premium. The term $\beta_{m,R} = \text{cov}(m, R) / \text{var}(m)$ is the environmental project's "beta." We have expressed β in terms of the covariance of R with m instead of the more traditional covariance with growth in income because of the presence of environment in welfare, $u(.)$.

According to the third line in (11), the required expected return on the environmental asset will be smaller than the market rate so long as its market "beta" $(-\beta_{m,R})$ is smaller than 1.0. This has the usual risk-return interpretation—if the environmental asset offers greater payoff than the market when m is high, then it reduces risk and should have a lower than market expected return. While this may seem likely, so $r_E < r_M$ is plausible, the first line of (11) offers a more aggressive point about risk and return for investments in climate projects—the expected return on an environmental project may fall below the risk-free rate if $\text{cov}(m, R) > 0$. Because m falls with income, positive covariance of m and R is not relevant for most financial assets. But climate projects are alleged to have the potential of averting disasters, so they may pay off precisely in states of the world where willingness to pay, m, is greatest. For example, if climate change may greatly reduce future productivity and living standards, or cause widespread harm and death in some states of nature, then projects that avert such outcomes (see equation (10)) may be highly valued even if the payoff is rare—they have low expected return but high market value because they pay off when mitigation of damage is most valuable.

3.4 Scalable Technologies and the Effectiveness of Current Investments

Our earlier discussion emphasized the importance of "scalable" technologies in mitigating extreme climate outcomes.   Scalable technologies are particularly likely to warrant low rates of discount because they can greatly reduce the risk of extreme climate outcomes. This point can be demonstrated in the current context by putting a bit more structure on the form of future environmental harm and the technology of mitigation.

Let future income in the absence of mitigation be $y_F = \bar{y}_F - \theta_F Q_F$, where Q_F is the future stock of global atmospheric GHGs above some base and $\theta_F > 0$ is the extent of future economic damage per unit of Q_F.[17]   Both Q_F and θ_F are currently unknown—Q_F is determined by global economic growth and carbon-based energy use, while the distribution of θ_F represents current uncertainty about the future cost of GHG concentrations.   Their interaction means that extreme values of θ_F can cause future environmental "catastrophes" when Q_F is large.   In anticipation of such damage, we assume that the current generation can invest in units of climate capital, λ, which can be combined with future variable inputs Z_F to mitigate $G_F(\lambda, Z_F)$ units of environmental harm once Q_F and θ_F are known.   For example, G_F may represent units of Q_F removed from the atmosphere, or emissions that are avoided by deploying clean energy technologies.   Assume   that   mitigation   has   constant   returns,   so $G_F(\lambda, Z_F) = \lambda g_F(z_F)$, where $z = Z/\lambda$.   Let Δ_F be the fixed future cost of deploying the technology. If deployed, future income net of mitigation is:

$$(12) \qquad y_F = \bar{y}_F - \theta_F\left[Q_F - \lambda g(z_F)\right] - \lambda z_F - \Delta_F$$

Let $g_F(z)$ be iso-elastic; $g_F(z_F) = A_F \alpha^{-1} z_F^{\alpha}$, where A_F is the unknown future productivity of currently chosen environmental capital. Then the optimal date-F choice of z yields

$$(13) \qquad y_F = \bar{y}_F - \theta_F\left[Q_F - \lambda \phi^{-1} A_F^{1+\phi} \theta_F^{\phi}\right] - \Delta_F$$

---

[17] If future harm is convex in Q_F then average harm per unit, $\theta_F(Q_F)$, will be increasing in Q_F. We ignore this so not to complicate the analysis.

The B.E. Journal of Economic Analysis & Policy, Vol. 10 [2010], Iss. 2 (Symposium), Art. 19

where $\phi = \alpha(1-\alpha)^{-1}$ is the elasticity of supply of mitigation, which indexes the scalability of the investment project—a perfectly scalable (constant marginal cost) project has $\alpha = 1$. The marginal return on the environmental asset is then

$R_F = \dfrac{\left(A_F \theta_F\right)^{1+\phi}}{\phi k(\lambda)}$.[18] According to (13), the extent of mitigation increases with the

scale of the investment in climate capital, λ, the state of future productivity, A_F, and with the damage from future GHG concentrations, θ_F. For given values of A_F, and θ_F, mitigation is greater when the project is more "scalable"—that is, when ϕ is large.

   Some investments such as reducing current GHG emissions are not scalable in that they cannot be cheaply adjusted once the values of A_F and θ_F are realized; ϕ is small and marginal cost rises sharply with mitigation efforts. Others—such as development of clean energy technologies or investments in the capacity to remove carbon from the atmosphere or sequester emissions—can be deployed in large scale based on the future demand for climate mitigation. A perfectly scalable technology, $\alpha = 1$, would provide a great deal of insurance by effectively truncating the distribution of harm in what would otherwise be the most damaging states of nature, when θ_F is large, yielding $y_F = \overline{y}_F - \Delta_F$ because excess concentrations of GHG are eliminated. Even technologies that are not perfectly scalable can have substantial value. Larger values of $\phi = \alpha(1-\alpha)^{-1}$ mean that the payoff from implementing the technology is more sensitive to the realized marginal value of environmental improvements, θ_F, which enhances the positive covariance between m and R. This further reduces the implied rate of discount for such projects because highly scalable technologies provide additional insurance—they are deployable as needed by varying Z_F. The magnitude of this advantage depends on the uncertainty about θ_F (and A_F). Greater uncertainty raises the (current) value of ex-post scalability.

   This is our earlier point about the value of scalable technologies—research and development investments in mitigation technologies that can be deployed in large scale in the event that damages are large can offer important insurance against looming catastrophe. Such projects should not be heavily discounted.

---

[18] With fixed cost, Δ_F, $R_F = 0$ for low values of A and θ because the technology will not be deployed. At the other extreme, *marginal* returns are zero when A and θ are very large, or when ϕ is large, because all excess emissions are mitigated. It is also plausible that scalable (high ϕ) technologies are more costly to develop, so current costs are $K(\lambda, \phi)$.

More broadly, much of the discussion surrounding climate investments stresses that they are meant to avert future catastrophes, which we interpret to mean that they yield dividends when willingness to pay, m, is highest. Our analysis provides a positive economic case for discounting them lightly. This is, we think, the less extreme and more relevant implication of Weitzman's "Dismal Theorem" analysis.

3.5 Heterogeneous Valuations

Much has been made of the fact that effective climate policies require harmonized international efforts, because earth's atmosphere is a global public good. Free-riding and the tragedy of the commons aside, our analysis suggests an additional impediment to cooperation and harmonization, based on heterogeneous valuations of climate investments between developed and developing countries.

Consider China (CH) and the United States (US) as extremes of the relevant development scale. In terms of our notation above, any reasonable growth scenario implies $m^{CH} < m^{US}$; that is, China will continue to grow faster than the US, so the ratio of future to current marginal utility of consumption is lower in China. And states of nature where China grows fastest correspond to the smallest values for m^{CH} —these are the good states of nature from China's perspective, because they are rich. The danger of harmful climate outcomes increases with global GHG emissions, which increase with economic growth. So it is reasonable to assume that future GHG concentrations will be greatest if China (and India, and others) grows rapidly, possibly approaching the living standards and energy consumption now observed in the US. This means that "good" states of nature from China's growth perspective are, climate-wise, most damaging to the US, especially if US living standards are harmed. Climate projects valued highly by the US because of strong insurance properties ($\text{cov}(m^{US}, R)$ is high) may have little current value to China because $\text{cov}(m^{CH}, R)$ is weak or even negative. In effect, a world with greater climate damage is one in which China gets rich, and they are more willing to bear the future cost that the US would like to avoid.

This discussion can be framed in terms of current efforts to establish a harmonized price for carbon emissions. With rapid economic growth, China's real future willingness to pay for a unit reduction in GHG concentrations may be equal to that of the US. But China discounts this return more heavily, because it only occurs when China prospers. Expressed as a preference for a current tax on GHG emissions, policy makers in China and the rest of the developing world will rationally prefer a lower (or no) tax that is less of a hindrance to attaining prosperity, while the US and other developed countries prefer a higher tax that

The B.E. Journal of Economic Analysis & Policy, Vol. 10 [2010], Iss. 2 (Symposium), Art. 19

insures the prosperity they already have. Even ignoring other challenges to international accord, the likelihood of an effective global policy in such a world appears to us slim.

4.   Conclusions

The basic designs of economic policies to mitigate anthropogenic climate change and its effects are not novel. They are rooted in well-understood methods for dealing with externalities, which remedy market failure by pricing an otherwise over-used resource—here the capacity of the atmosphere to safely absorb GHG emissions. Implementing such policies is more challenging, for two basic reasons. First, the external costs of GHG emissions are global rather than local, so useful policies that would price or regulate current emissions require harmonized action worldwide. Second, the harms that policies seek to value and internalize are both highly uncertain and spread over future generations—all of the benefits of using carbon-based energy occur today, while the possible social costs are far removed.

Our analysis has sought to extend previous work on both the form and substance of climate policies, particularly in the area of valuing uncertain future costs of GHG emissions and the benefits of policies that would mitigate those costs. An important finding is that "gradualist" policies advocated by most economists—setting a low initial emissions price that would rise at a "market" rate of interest—are based on the implicit assumption that returns on climate investments have a similar payoff structure to other forms of investment in physical or human capital. This assumption ignores the possible insurance value of social investments in climate capital, which may pay off precisely when other forms of capital do not.

References

Archer, David, *Global Warming: Understanding the Forecast*, Malden, MA: Blackwell Publishing, Ltd., 2007

Archer, David, *The Long Thaw*, Princeton, N.J.: Princeton University Press, 2009

Cochrane, John, *Asset Pricing*, Princeton, N.J.: Princeton University Press, 2005

Dyson, Freeman, "The Question of Global Warming", *The New York Review of Books,* June 12, 2008

Hall, Robert E. and Charles I. Jones, "The Value of Life and the Rise in Health Spending, *Quarterly Journal of Economics,* 122, 2007, pp 39-72

Hotelling, Harold, "The Economics of Exhaustible Resources", *Journal of Political Economy,* April, 1931, pp 137-175

Karp, Larry, "Sacrifice, Discounting and Climate Policy: Five Questions", Working Paper, U.C. Berkeley, July 2009

Murphy, Kevin M. and Robert H. Topel, "The Value of Health and Longevity", *Journal of Political Economy,* October, 2006, pp 871-904

Nordhaus, William, "To Tax or Not to Tax: Alternative Approaches to Slowing Global Warming", *Review of Environmental Economics and Policy,* 1 (1), Winter, 2007, pp 26-44. (2007a)

Nordhaus, William, "A Review of the Stern Review on the Economics of Climate Change", *Journal of Economic Literature* 45 (3), 2007, pp 686-702. (2007b)

Nordhaus, William, *The Challenge of Global Warming: Economic Models and Environmental Policy*, Working Paper, Yale University, 2007. (2007c)

Nordhaus, William, *A Question of Balance: Weighing the Options on Global Warming Policies,* New Haven, CT: Yale University Press, 2008

Pindyck, Robert S. and Neng Wang, "The Economic and Policy Consequences of Catastrophes", Working Paper, Massachusetts Institute of Technology, September, 2009

Stern, Nicolas, *The Stern Review on the Economics of Climate Change,* H.M. Treasury, London, 2006.

Weitzman, Martin, "The Stern Review of the Economics of Climate Change", *Journal of Economic Literature* 45 (3), 2007

Weitzman, Martin, "On Modeling and Interpreting the Economics of Catastrophic Climate Change," *Review of Economics and Statistics*, February, 2009, pp 1-19

[2]

The B.E. Journal of Economic Analysis & Policy

Symposium

Volume 10, Issue 2	2010	Article 20

DISTRIBUTIONAL ASPECTS OF ENERGY AND CLIMATE POLICY

Comment on "On the Economics of Climate Policy": Is Climate Change Mitigation the Ultimate Arbitrage Opportunity?

Manasi Deshpande*   Michael Greenstone[†]

*MIT, Department of Economics, manasi@mit.edu
[†]MIT, Department of Economics, and NBER, mgreenst@mit.edu

Gary Becker, Kevin Murphy, and Robert Topel's (BMT), "On the Economics of Climate Policy," is an important paper that deepens our understanding of what some have referred to as the challenge of our generation. It is an excellent example of how basic price theory can be used to gain a better understanding of the world. The paper is filled with insights, including how to determine the appropriate discount rate to value investments in climate mitigation, the importance of the distribution of damages across the population, the smaller than expected costs of delay in undertaking research and development programs for clean energy sources, and a cogent defense of analyzing investments in carbon mitigation with standard cost-benefit tools.

One of BMT's observations is that heterogeneity across countries in the marginal rate of substitution between current and future consumption poses a major, indeed possibly insurmountable, challenge to a solution to the problem of climate change. In this brief note, we draw a different conclusion. Our argument is that this heterogeneity in the marginal rate of substitution provides the basis for a Pareto-improving trade that could in principle facilitate an international agreement to mitigate climate change.

As BMT observe, in developed economies it is likely that the value of marginal changes in consumption today is not dramatically different from the value of marginal changes in the future. Thus, the great risk from climate change for developed countries is the possibility that it will cause a dramatic decline in their consumption; they want to mitigate climate change for its insurance properties to protect their existing high standards of living.

In contrast, marginal consumption today has a very high value in China, India, and other developing countries, relative to marginal consumption in the future. This reflects the relatively low levels of consumption in these countries today and their high expected growth rates, both of which make them reluctant to reduce consumption today in order to protect future consumption. As just one measure of the urgency of greater consumption today, Burgess et al. (2011) find that the mortality impact of a day with the temperature exceeding 90 degrees F is 20 times greater in India than in the United States. To put it plainly, India and other developing countries are *literally dying* to increase consumption today. By comparison, reducing the risk of future consumption losses is not as pressing of an issue in these countries.

The result is that, at least for the foreseeable future, developing countries' primary focus is likely to be economic growth, a strategy that is at odds with emissions reductions. BMT present the case for why developing countries are unlikely to agree to meaningful emissions reductions. With economic growth in the developing world as the primary driver of greenhouse gas emissions, the state of the world in which global emissions are high and the earth is warming is the state of the world in which developing countries become wealthy. The state of

The B.E. Journal of Economic Analysis & Policy, Vol. 10 [2010], Iss. 2 (Symposium), Art. 20

the world in which global emissions are stabilized and climate change is curbed is one in which these countries remain below their growth potential.

This drive for economic growth is troubling for climate change mitigation strategies because reducing the degree of climate change to levels recommended by scientists and economists will require significant abatement from the developing world. The United States' commitment at Copenhagen—to reduce its emissions by 17 percent below 2005 levels by 2020—would account for around 20 percent of the global reduction in emissions needed to stabilize the atmospheric greenhouse gas concentration at 550 ppm, a recommended and possibly feasible target. Together, the combined Copenhagen commitments of the United States and the European Union would account for less than 40 percent of necessary reductions, based on current emissions modeling projections.[1]

One potential solution is compensation from the developed world to the developing world. This could happen directly through a lump-sum transfer. Or it could happen indirectly, through investments by the developed world in the research and development of technologies—for example, carbon capture and storage—that would allow the developing world to grow more cleanly. Though the latter approach may be more politically palatable, both involve substantial risk: the former in verifying reductions and the latter in managing the uncertainties of technological innovation.

How much would the developed world need to pay to compensate developing countries for their foregone consumption? In this note, we develop calculations of the lower and upper bounds of a Pareto-improving transfer from the developed world to the developing world, using China as an example and using a series of assumptions about growth rates and the impact of carbon reductions on Chinese growth.

The heterogeneity in the intertemporal elasticity of the marginal utility of consumption features prominently in these calculations. In particular, this heterogeneity leads to different social discount rates in the developed and developing worlds, with two important implications for a Pareto-improving transfer. First, the value of foregone Chinese consumption using Chinese discount rates is far less than one would expect using conventional developed world discount rates.[2] Depending on the assumptions, the present value of

---

[1] Authors' calculations based on estimates and projections in Clarke et al. (2007) and U.S. EPA (2011).

[2] Another useful, though distinct, exercise would be to compare the cost of emissions reductions in China (using Chinese discount rates) to the cost of emissions reductions in the United States (using U.S. discount rates). The cost of emissions reductions in the United States is likely to be much larger than the cost of emissions reductions in China, and therefore the United States would be willing to pay—and China would be willing to accept—some amount between these two costs for China to reduce emissions. The wedge between different countries' marginal abatement costs forms the basis of international offset agreements.

Chinese consumption loss using the U.S. discount rate generally exceeds the present value using the Chinese discount rate by an order of magnitude or more. For example, if China were to participate in the type of global agreement outlined by the Energy Modeling Forum (Clarke et al. 2009) to stabilize atmospheric concentrations of CO_2 at 550 ppm, the value of Chinese consumption loss through the end of the century would be $1.6 trillion, or $1.60 per ton of CO_2 abated over the course of the 21st century, using China's discount rate. This compares to $50.1 trillion using a U.S. discount rate. The Chinese valuation of Chinese consumption loss is the lower bound of a Pareto-improving transfer from the U.S. to China.

Second, the social cost of carbon, which is the monetized damages associated with an incremental increase in carbon emissions, takes on different values depending on whether U.S. or Chinese discount rates are used. Using a Chinese discount rate would lead to a social cost of carbon of effectively zero. In contrast, the U.S. government has determined the global social cost of carbon using U.S. discount rates to be $21 per ton of CO_2.[3] The social cost of carbon determines the upper bound of the Pareto-improving transfer—in other words, the maximum amount the U.S. would be willing to pay China to reduce emissions.

The large disparity between the U.S. and Chinese marginal rates of substitution appears to provide an opportunity for developed and developing countries to make a mutually beneficial deal that provides both sides with substantial surplus.   Hence, in contrast to BMT, we conclude that the heterogeneity in the marginal rate of substitution may provide a way forward for a mutually beneficial climate agreement, rather than being the death of one.

I. Lower Bound: Estimating the Value of China's Loss in Consumption Due to CO_2 Abatement

This section outlines a method for determining the value of the loss in China's consumption associated with adopting a system of emissions cuts. This value is the minimum amount the U.S. would have to pay for China to reduce emissions—

---

[3] Two issues with the number $21 per ton are relevant for this exercise. First, it is an estimate of the global damages of climate change, yet Americans' utility functions may not include losses outside of the United States. In this case, it would be useful to have a separate estimate of the domestic damages. Later, we show that the paper's qualitative conclusions generally remain when U.S. domestic damages are used, rather than global damages. Second, $21 per ton is the value of the averted damages associated with a marginal reduction in carbon emissions, but this paper's exercise is about non-marginal reductions. Damages are likely increasing in the atmospheric concentration of greenhouse gases, so the $21 figure is likely an overestimate of the averted damages associated with the substantial reductions in Chinese emissions contemplated here.

The B.E. Journal of Economic Analysis & Policy, Vol. 10 [2010], Iss. 2 (Symposium), Art. 20

in other words, the lower bound of a Pareto-improving transfer. The exercise is conducted from the present through 2100, which is approximately when the Chinese economy is expected to equal the size of the U.S. economy in a baseline scenario (Goldman Sachs 2007). We assume that China begins reducing emissions immediately, in 2012, so the relevant equation is:

NPV of China's Consumption Loss from CO_2 Abatement

$$= \sum_{t=2012}^{2100} \frac{c_{0t} - c_{1t}}{(1 + r_t)^{t'}} \tag{1}$$

where c_{0t} is China's baseline consumption in year t and c_{1t} is China's consumption in year t under the mitigation scenario. The parameter r_t is the social discount rate in year t.

The first step is to estimate the percentage reduction in emissions that China would have to make under a global climate agreement. Figure 1 presents China's baseline emissions trajectory and its hypothetical emissions trajectory under a standard 550 ppm stabilization path without overshooting from 2012 through 2100, based on emissions models from the Energy Modeling Forum (Clarke et al. 2009). Under the mitigation target, China's emissions are more than 70 percent below baseline levels in 2050 and more than 90 percent below baseline in 2100. Of course, the amount of abatement needed from China depends on the amount of abatement undertaken by the rest of the world. As the world's largest emitter, however, China will have to reduce emissions substantially under any cost-effective scenario.

The next step is to translate the reduction in China's emissions to a reduction in GDP, which allows for estimates of c_{0t} and c_{1t} in equation (1). Estimates of the effect of emissions reductions on China's growth vary considerably, reflecting different assumptions about the technological evolution of the Chinese economy and its emissions intensity. We consider three scenarios for a reduction in Chinese economic activity relative to baseline, modeled by the Energy Modeling Forum (Clarke et al. 2009). In the first scenario, CO_2 reductions in China have a small impact on Chinese GDP; GDP losses are 0.7 percent below baseline in 2020 and peak at 1.9 percent below baseline in 2050. In the second scenario, CO_2 reductions in China have a moderate impact on GDP; GDP losses are 1 percent below baseline in 2020 and peak at 7 percent below baseline in 2080. In the third scenario, CO_2 reductions have a large impact on GDP; GDP losses are 2 percent below baseline in 2020 and peak at 18 percent below baseline in 2100. The "low impact" scenario corresponds to a world in which technological breakthroughs allow China to reduce the carbon intensity of its economy significantly. The "high impact" scenario corresponds to a world in which China experiences a gradual decrease in carbon intensity that lags behind its baseline growth.[4]

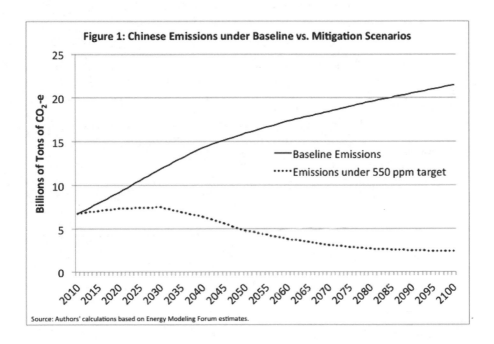

Figure 1: Chinese Emissions under Baseline vs. Mitigation Scenarios

Source: Authors' calculations based on Energy Modeling Forum estimates.

The final step in estimating the value of China's foregone consumption is to discount the difference in expected consumption under the two scenarios by the social discount rate. Figure 2 shows the annual value of foregone Chinese consumption using a U.S. discount rate; this value increases over time because abatement costs rise faster than the discount rate. Assuming a "moderate" impact of carbon reduction on Chinese GDP, the U.S. would have to pay a minimum total amount of $50.1 trillion to compensate China for its consumption loss between 2012 and 2100. Even if this amount were less than the social cost of carbon, such an enormous transfer would almost undoubtedly be a political non-starter in the U.S.

---

[4] The 550 ppm target considered here is one in which atmospheric concentrations never exceed 550 ppm (i.e., stabilization). For "low impact," we use the Energy Modeling Forum (EMF) model from Clarke et al. (2009) predicting the smallest nonzero impact on GDP (the MESSAGE model). For "high impact," we use the EMF model predicting the largest impact on GDP (the SGM model). For "moderate impact," we use the median over all nonzero-impact EMF models in each year.

The B.E. Journal of Economic Analysis & Policy, Vol. 10 [2010], Iss. 2 (Symposium), Art. 20

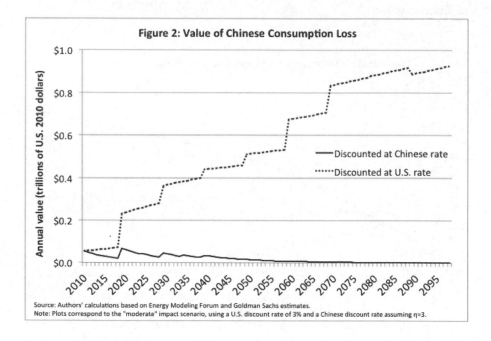

Figure 2: Value of Chinese Consumption Loss

Source: Authors' calculations based on Energy Modeling Forum and Goldman Sachs estimates.
Note: Plots correspond to the "moderate" impact scenario, using a U.S. discount rate of 3% and a Chinese discount rate assuming η=3.

However, the U.S. discount rate is not the appropriate one for valuing Chinese consumption loss. Instead, we calculate the Chinese discount rate using the Ramsey formula for the social discount rate:

$$r = \rho + \eta g, \qquad (2)$$

where ρ is the pure rate of time preference, η is the intertemporal elasticity of the marginal utility of consumption, and g is the economy's growth rate (Ramsey 1928).

Of particular importance in this context is the parameter η, since economic theory suggests that the marginal utility of consumption is higher when consumption is lower—as is the case under the 550 ppm target. Given that the developing world is clamoring for consumption now, we assume relatively large intertemporal substitution elasticities—between 2 and 4—meaning that China would be willing to give up substantial future consumption to consume more today. We assume a pure rate of time preference of 2 percent.

These large values of η, as well as the high growth rate of the Chinese economy, lead to social discount rates for China that are well above conventional

U.S. discount rates. Figure 3 shows Chinese discount rates under three values of η (2, 3, and 4). The Chinese discount rates decrease over time because of declines in the projected Chinese economic growth rate. In all cases, the Chinese discount rate is significantly higher than conventional U.S. discount rates, which generally range from 2.5 to 7 percent.

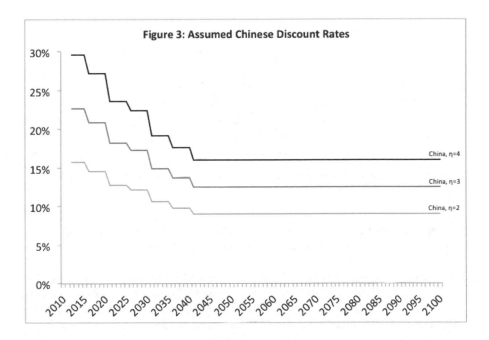

As shown in Figure 2, the annual value of foregone Chinese consumption using Chinese discount rates declines over time and approaches zero by the end of the century. This valuation contrasts sharply with the valuation obtained using conventional U.S. discount rates (the figure uses a discount rate of 3 percent for the U.S.). Note that the large gap between Chinese and U.S. valuations does *not* mean that China would not consider its consumption loss economically meaningful. Instead, it simply means that the value of foregone consumption to the Chinese is far less than what one would expect using conventional U.S. discount rates.

Table 1 provides estimates of the net present value of foregone Chinese consumption between 2012 and 2100 under the 550 ppm target, given different assumptions about the impact of CO_2 reductions on Chinese GDP. Estimates of the total value of foregone Chinese consumption over this period range from $400

The B.E. Journal of Economic Analysis & Policy, Vol. 10 [2010], Iss. 2 (Symposium), Art. 20

billion to $5.3 trillion using Chinese discount rates.[5] These amounts are the minimum transfer that China would be willing to accept to reduce its CO_2 emissions and undergo the resulting loss in consumption.   The amount corresponding to "moderate impact" and $\eta=3$ is the $1.6 trillion mentioned in the introduction as a reasonable estimate for the lower bound acceptable to China.

Table 1: Value of Chinese Consumption Loss, 2012-2100 (trillions)

		Intertemporal Elasticity of MU of Consumption		
		$\eta=2$	$\eta=3$	$\eta=4$
Impact of CO_2	Low	$1.37	$0.68	$0.40
reduction on GDP	Moderate	$3.88	$1.56	$0.81
in China	High	$5.31	$2.16	$1.19

Source: Authors' calculations based on estimates from the Energy Modeling Forum and Goldman Sachs.

Notes: "Low" uses the EMF model predicting the smallest nonzero impact on GDP; "high" uses the EMF model predicting the largest impact on GDP; "moderate" uses the median over all nonzero impact models in each year. η is the intertemporal elasticity of the marginal utility of consumption for China; it is used to calculate a social discount rate for China assuming a 2% rate of time preference.

II. Upper Bound: Estimating the U.S. Willingness to Pay for Chinese Carbon Reductions with the Social Cost of Carbon

What is the maximum amount that the U.S. would be willing to pay China for emissions reductions? In this section, we calculate the upper bound on a Pareto-improving transfer from the U.S. to China. This upper bound is determined by the social cost of carbon, since the U.S. would be willing to pay only up to the present value of expected future climate damages.

Here too, the disparity in U.S. and Chinese discount rates plays a key role. Using Chinese discount rates, the social cost of carbon is effectively zero. The Chinese place little value on climate damages that are expected to occur far in the future; they are instead focused on increasing today's consumption.   In determining how much the U.S. would be willing to pay, however, it is the U.S. discount rate, rather than the Chinese discount rate, that matters. Using a U.S. discount rate, the U.S. government values the global social cost of carbon at $21 per ton of CO_2 (Greenstone et al. 2011).

How does this amount compare to the Chinese value of Chinese consumption loss?   Under the 550 ppm stabilization target and division of emissions reductions laid out by the Energy Modeling Forum, China would

---

[5] All dollar amounts are in 2010 dollars.

Deshpande and Greenstone: Is Climate Change Mitigation the Ultimate Arbitrage Opportunity?

reduce emissions by nearly 1000 Gt CO_2 between 2012 and 2100. Table 2 presents the results from dividing the present value of China's foregone consumption in Table 1 by the 1000 Gt CO_2 avoided, which is the cost in lost Chinese consumption per ton of abatement. For example, the $1.6 trillion corresponding to the "moderate" impact and $\eta=3$ translates to $1.60 per ton CO_2. The maximum cost per ton of CO_2 abated is $5.33, which comes from the "high" impact and $\eta=2$ combination of assumptions.

In all cases, the Chinese valuation of its consumption loss is far less than $21 per ton of CO_2. This huge difference suggests an opportunity for a Pareto-improving transfer from the developed world to the developing world for emissions reductions. By paying China some amount between the Chinese valuation of its consumption loss and the U.S. value of the social cost of carbon, the U.S. would get the emissions reductions necessary to insure their current standard of living against climate damage—at less than its willingness to pay for such reductions. And by fulfilling China's strong desire for more consumption today, this transfer would fully compensate China for the consumption it must forgo tomorrow in reducing emissions.

Table 2: Cost per Ton of CO_2 Abated

		Intertemporal Elasticity of MU of Consumption		
		$\eta=2$	$\eta=3$	$\eta=4$
Impact of CO_2 reduction on GDP in China	Low	$1.38	$0.68	$0.40
	Moderate	$3.89	$1.57	$0.81
	High	$5.33	$2.17	$1.19

Source: Authors' calculations based on estimates from the Energy Modeling Forum and Goldman Sachs.

Notes: "Low" uses the EMF model predicting the smallest nonzero impact on GDP, "high" uses the EMF model predicting the largest impact on GDP; "moderate" uses the median over all nonzero impact models in each year. η is the intertemporal elasticity of the marginal utility of consumption for China; it is used to calculate a social discount rate for China assuming a 2% rate of time preference. Cost per ton calculated using 1000 Gt CO_2 avoided.

It is important to emphasize that while the $21 per ton figure is an estimate of the global damages, the U.S. is likely more interested in paying to avoid the U.S. domestic damages associated with Chinese emissions. Although the research on regional impacts is not as advanced as on global damages, the available research suggests that U.S. domestic damages are about 10 percent of the global benefit with a 3 percent discount rate (Greenstone et al. 2011). Alternatively, if the fraction of GDP lost due to climate change is assumed to be similar across countries, the domestic benefit would be proportional to the U.S.

share of global GDP, which is currently about 23 percent.[6] In many, although not all, of the scenarios, there are still opportunities for a Pareto-improving transfer because the U.S. valuation of the reductions in emissions exceeds the Chinese valuation of the lost consumption.

Of course, the U.S. is just one of many developed countries. The resources available to compensate China would be greater if all developed countries banded together with the U.S. in an international agreement with China. Each of them could contribute an amount less than the avoided damages in their country per ton of CO_2 abated. The sum of these payments is likely to easily exceed the present value of China's lost consumption, even with the assumptions that yield high estimated costs for China. In principle, similar deals could be struck with other developing countries.

III. Conclusions

The analysis presented here has numerous limitations and is intended to provide only a back-of-the-envelope method for estimating the value of foregone Chinese consumption. In particular, these calculations reflect a highly simplified view of the nature of climate damages and national interests.

Perhaps the most important limitation of this calculation is the failure to account for China's avoided climate damages. The available research suggests that the benefits of climate change mitigation are heavily concentrated in the developing world. For example, hundreds of millions of people living near the coastline are threatened by rising sea levels. Although the biggest beneficiaries of mitigation will be small island nations, China may also experience significant benefits, depending on the severity of climate change in the baseline scenario and the effectiveness of emissions reductions in mitigating climate change. As we highlighted, however, a high social discount rate for China reduces the value of the benefits that emerge further in the future. Regardless, a full accounting of these benefits would change the calculations presented here.

These numbers come with at least three additional caveats. The first is the time period of compensation. We calculate the value of foregone consumption until 2100, when China's GDP is expected to catch up with U.S. GDP under baseline assumptions. In reality, China's GDP would continue to be lower than

---

[6] Although the mortality impacts of climate change are likely to be unequally distributed across the globe, the associated welfare loss may be more equally distributed because the available evidence suggests that the value of a statistical life increases faster than income. Further, there may be greater scope and devotion of resources to defensive expenditures in high income countries. For these reasons, the assumption that climate change induced losses will be proportional to GDP may not be unreasonable.

baseline into the 22nd century—though as demonstrated above, the value of those losses approach zero using Chinese discount rates. Second, we have not considered trades with other rapidly developing countries, including India and Indonesia. The amounts would be smaller than compensation to China but likely substantial. Third, the world currently lacks the technological capability to verify a country's emissions, which would be an essential ingredient in a credible trade of resources for emissions reductions. Fourth, even with a credible and internationally recognized method to measure country-level emissions, the immediate transfer of resources to China in exchange for emissions reductions over the next 90 years would need to be accompanied by a credible enforcement mechanism (e.g., large and enforceable penalties for failing to meet the agreed upon emissions targets).

We recognize the possibility of technological innovation that would allow developing countries to develop at much lower carbon intensity than is possible today, although such innovation usually requires a certain and substantial stream of resources for basic research and development that likely exceeds the amounts being invested today. Perhaps the most likely scenario, however, is that developing countries will face a tradeoff between their own economic development and emissions reductions for the global good. The differences in the developed and developing countries' marginal rates of substitution between present and future consumption outline the contours of a trade that would reduce the degree of climate change and benefit both sets of countries.

REFERENCES

Becker, G., K. Murphy, and R. Topel, 2010. "On the Economics of Climate Policy." *The B.E. Journal of Economic Analysis and Policy*, Volume 10, Issue 2.

Burgess, R., O. Deschenes, D. Donaldson, M. Greenstone, 2011. "Weather and Death in India: Mechanisms and Implications for Climate Change." Mimeograph, Cambridge, MA: MIT.

Clarke, L., J. Edmonds, H. Jacoby, H. Pitcher, J. Reilly, R. Richels, 2007. "Scenarios of Greenhouse Gas Emissions and Atmospheric Concentrations." Sub-report 2.1A of Synthesis and Assessment Product 2.1 by the U.S. Climate Change Science Program and the Subcommittee on Global Change Research. Department of Energy, Office of Biological & Environmental Research, Washington, DC., USA, p. 154.

The B.E. Journal of Economic Analysis & Policy, Vol. 10 [2010], Iss. 2 (Symposium), Art. 20

Clarke, L., C. Bohringer, and T. Rutherford, 2009. "International, U.S., and E.U. Climate Change Control Scenarios: Results from EMF 22." *Energy Economics*, Volume 31, Supplement 2, pp. S63-S306.

Goldman Sachs, 2007. "BRICs and Beyond."
http://www2.goldmansachs.com/ideas/brics/book/BRIC-Full.pdf.

Greenstone, M., E. Kopits, A. Wolverton, 2011. "Estimating the Social Cost of Carbon for Use in U.S. Federal Rulemakings: A Summary and Interpretation." NBER Working Paper No. 16913.

Ramsey, F., 1928. "A Mathematical Theory of Saving." *Economic Journal*, Volume 38, No. 152, pp. 543-559.

U.S. EPA, 2011. Global Greenhouse Gas Data.
http://www.epa.gov/climatechange/emissions/globalghg.html.

[3]

The B.E. Journal of Economic Analysis & Policy

Symposium

Volume 10, *Issue* 2	2010	*Article* 7

DISTRIBUTIONAL ASPECTS OF ENERGY AND CLIMATE POLICY

The Social Evaluation of Intergenerational Policies and Its Application to Integrated Assessment Models of Climate Change

Louis Kaplow*   Elisabeth Moyer[†]

David A. Weisbach[‡]

*Harvard Law School, meskridge@law.harvard.edu
[†]University of Chicago, moyer@uchicago.edu
[‡]University of Chicago, d-weisbach@uchicago.edu

Recommended Citation
Louis Kaplow, Elisabeth Moyer, and David A. Weisbach (2010) "The Social Evaluation of Inter-generational Policies and Its Application to Integrated Assessment Models of Climate Change," *The B.E. Journal of Economic Analysis & Policy*: Vol. 10: Iss. 2 (Symposium), Article 7.
Available at: http://www.bepress.com/bejeap/vol10/iss2/art7

The Social Evaluation of Intergenerational Policies and Its Application to Integrated Assessment Models of Climate Change[*]

Louis Kaplow, Elisabeth Moyer, and David A. Weisbach

Abstract

Assessment of climate change policies requires aggregation of costs and benefits over time and across generations, a process ordinarily done through discounting. Choosing the correct discount rate has proved to be controversial and highly consequential. To clarify past analysis and guide future work, we decompose discounting along two dimensions. First, we distinguish discounting by individuals, an empirical matter that determines their behavior in models, and discounting by an outside evaluator, an ethical matter involving the choice of a social welfare function. Second, for each type of discounting, we distinguish it due to pure time preference from that attributable to curvature of the pertinent function: utility functions (of consumption) for individuals and the social welfare function (of utilities) for the evaluator. We apply our analysis to leading integrated assessment models used to evaluate climate policies. We find that past work often confounds different sources of discounting, and we offer suggestions for avoiding these difficulties. Finally, we relate the standard intergenerational framework that combines considerations of efficiency and distribution to more familiar modes of analysis that assess most policies in terms of efficiency, leaving distributive concerns to the tax and transfer system.

KEYWORDS: individual discount rate, social discount rate, intergenerational distribution, climate policy, integrated assessment models

[*]Louis Kaplow, Harvard University and National Bureau of Economic Research. Elisabeth Moyer, The University of Chicago. David A. Weisbach, The University of Chicago. Please send correspondence to: David Weisbach, the University of Chicago Law School, 1111 E. 60th St., Chicago, IL 60637; email: d-weisbach@uchicago.edu; phone: (773) 702-3342. We thank Martin Weitzman, the editor, and the referees for comments as well as Harvard's John M. Olin Center for Law, Economics, and Business, the Daniel and Gloria Kearney Fund, the Walter J. Blum Faculty Research Fund, and the Walter J. Blum Professorship in Law for financial support.

Evaluating climate change policies requires aggregating costs and benefits that accrue over long time periods and to different generations. The standard procedure in evaluating these costs and benefits is to discount future outcomes to a present value. Because of the large timescales involved, however, seemingly small changes in the discount rate can have dramatic impacts on policy choices. In a ranking of central uncertainties in evaluating climate policies, the Intergovernmental Panel on Climate Change (2007) (p. 823) listed two components of the discount rate as second and fourth in importance; only climate sensitivity ranked higher, and both discount rate factors outranked the estimated uncertainty surrounding the economic impact of a 2.5° temperature increase. Accordingly, it should not be surprising that the widely contrasting policy recommendations of different models, which range from prompt, aggressive action to little immediate response, are largely due to their different choices of discount rates.

The controversy over discounting in the climate context makes it useful to step back and consider intergenerational policy evaluation from its fundamentals before applying it to the specific case of climate change. The standard welfare economic framework for policy evaluation, which we analyze in section 1, holds that the optimal policy is that which maximizes social welfare. Social welfare is determined by a function W of the utilities, U, of all individuals in society. Utilities in turn are (under a simplification frequently employed in assessing climate change) a function of individuals' levels of consumption, c. How consumption affects utility is an empirical question. Observations of decisions that individuals make, especially decisions made under uncertainty, can be used to estimate the rate at which individuals' marginal utility falls with consumption. By contrast, how utility affects social welfare is a normative question, that is, a social judgment made by an outside evaluator.

These two distinct concepts, however, are frequently conflated both in actual policy evaluation and in discussions of the choice of the discount rate. Instead of separately specifying the social welfare function $W(U)$ and individuals' utility functions $U(c)$, analysts often use a single, reduced-form representation, $Z(c)$, under which social welfare is taken to be a direct function of consumption. There are a number of problems with using a reduced form that combines ethical parameters (those relating to W) and empirical parameters (those relating to U). This conflation of the ethical and the empirical makes discussions of the appropriate shape of $Z(c)$ confusing and sometimes misleading. For example, what many would take to be sensible parameter choices for the reduced-form, when combined with standard estimates of the underlying empirical evidence on individuals' utility functions, can imply ethical judgments regarding the social welfare function that most observers would, on reflection, reject. A related problem is that empirical arguments are sometimes mistakenly taken to be ethical

The B.E. Journal of Economic Analysis & Policy, Vol. 10 [2010], Iss. 2 (Symposium), Art. 7

and thus a matter that the outside evaluator is free to choose without regard to existing evidence.

In section 2, we extend this general welfare economic framework to the intergenerational setting, such as that involved in evaluating climate change policies. Corresponding to the two elements in social policy evaluation, there are two domains of discounting that should be considered separately. Individuals maximizing their utilities will discount their own future consumption. Their consumption and savings choices will be determined by the specification of the utility function U, by their expectations, and by government policies. For example, if a model is being used to evaluate a carbon tax or a cap on emissions, it must estimate how individuals will respond to such a policy. The pertinent features of the utility function are entirely an empirical matter. We will call this *predictive* discounting because the role of an assumption about discount rates is to predict individuals' behavior.

These utilities (themselves derived from individual maximizing behavior) are aggregated by social evaluators outside the model, reflected in the choice of W. W may (but need not) involve further implicit discounting because it weights increments to the utilities of individuals who are better off relatively less. A social welfare function also may (but need not) discount utilities simply because individuals live in the future (which is referred to as pure time discounting). We call any discounting based directly on the social welfare function *evaluative* discounting. Evaluative discounting is an ethical exercise, and one of a particular sort that indicates what types of argument are appropriate and what sorts of considerations are not.

After developing these arguments in sections 1 and 2, we apply this analysis in section 3 to discounting practices in the context of integrated assessment models (IAMs). IAMs are the central tool used in climate policy to estimate future costs and benefits of policies. Moreover, it is common in these assessments to present results in terms of an overall present value welfare equivalent. Accordingly, it is important to understand how discounting has been used in IAMs. To illustrate our main arguments, we discuss the widely known models used by William Nordhaus (Nordhaus (2008)) and Nicholas Stern (Stern (2007)) and show how each of their analyses conflates empirical and ethical parameters, making interpretation of their results difficult and potentially misleading. Clear separation between the two domains in which discounting is performed, between the empirical, predictive roles of IAMs and the ethical judgments about the predictions, would make model interpretation and discussion of appropriate discount rates more transparent.

The existing literature has two principal approaches to making the requisite separation. The first is to do sensitivity analysis with respect to the parameters of the social welfare function. The second is to directly present the time series of outcomes in terms of what each generation receives (without offering any social assessment). Although sensitivity analysis and the direct presentation of outcomes are standard in many areas of policy assessment, they are less common with respect to the discount rates in the integrated climate assessment literature.

In section 4, we return to a question that, although rarely asked in this context, is central to discounting and climate policy assessment: namely, why should analysts or evaluators focus on social discounting, which is concerned with the intergenerational distribution of income, rather than on efficiency? After all, in the intragenerational policy context, it is conventional to assess policy using cost-benefit analysis and largely leave distributive matters to the tax and transfer system. An analogous approach in the climate context has much to commend it, and following this different course would significantly affect the role of evaluative discounting in the analysis.

1   Framework for Social Assessment of Policies – The Single Generation Case

We consider in this section the welfare economic framework for assessment of policies that has been developed (implicitly) in the single-generation context, drawing on Kaplow (2010). To motivate the analysis, it is useful to remember why it is that social assessments of consumption increments to different individuals are needed. If all individuals had constant marginal utility of income and society was indifferent to the distribution of utility, simple cost-benefit analysis would suffice: projects should be adopted if and only if benefits exceed costs. Moreover, even when one or both of these conditions fail, so that distribution does matter to total social welfare, it would still be correct to employ cost-benefit analysis (and without any distributive weights) if policies were accompanied by tax adjustments that keep the overall distribution of income constant.[1] It is necessary to compare the social welfare contribution of consumption increments to different individuals only when distribution will not be held constant, including the important case of questions of pure income distribution, whether within or between generations.

Suppose that we are considering a project that increases consumption to some individual *i* by an increment, but reduces consumption to some other individual *j*. We should engage in the project if and only if the social welfare gain attributable to the former exceeds the social welfare loss due to the latter. To make this determination, we have to specify both how much individuals' utilities change because of an increment to consumption – determined by individuals'

---

[1] We will discuss this issue in more detail in section 4.

The B.E. Journal of Economic Analysis & Policy, Vol. 10 [2010], Iss. 2 (Symposium), Art. 7

utility functions – and how society evaluates that increase – indicated by the social welfare function.

Let each individual's utility be a concave function only of consumption, $U(c)$. Social welfare is taken to be the integral of a concave function of utility, so we can write social welfare as:

$$SWF = \int W\left(U(c_i)\right) f(i) di, \tag{1}$$

where $f(i)$ is the density of levels of consumption (that is, the relevant fraction of individuals at each consumption level). The marginal change in social welfare from a change in individual i's consumption is:

$$\xi_i = \frac{\partial W}{\partial U}\frac{dU}{dc_i}, \tag{2}$$

which depends on both the U and W functions and, in general, on the consumption level c_i since the U and W functions, and accordingly their derivatives, are evaluated at the current levels of their respective arguments. Integrating expression (2) over all individuals gives us the change in social welfare from a given change in policy.[2] Therefore, our hypothetical policy is desirable if and only if the increment to i's consumption, weighted by ξ_i, exceeds the fall in j's consumption, weighted by ξ_j.

The marginal change in social welfare for an increment of consumption, ξ_i, ordinarily declines with consumption for two possible reasons: because marginal utility declines as consumption increases and because marginal social welfare (in some formulations) declines as utility increases. The overall effect depends on the curvature of both the utility function and the social welfare function. Because U and W depend on their arguments, these factors affect ξ_i differently: the faster marginal utility declines with consumption, the more equal utility levels will be for given differences in levels of consumption and, therefore, the less the curvature of the social welfare function matters. To illustrate, suppose that $U(c)=ln(c)$ and c=\$10,000 for a poor person and c=\$100,000 for a rich person. The ratio of their marginal utilities (equal to $1/c$) is then 10 to 1. If social welfare is linear (utilitarian), this is also the ratio of the ξ_i; a dollar is 10 times more socially valuable to the poor person than to the rich person. Now suppose instead that social welfare depends on the log of utility: $W(U)=ln(U)$. The ratio of the ξ_i increases to 12.5 to 1 (not to 100 to 1). Curvature of the social welfare

---

[2] Typically, projects are assumed to be small enough that the ξ_i's are assumed to be constant; in some cases, possibly including climate change policies, this may not be the case.

Kaplow et al.: Discounting in Integrated Assessment

function does matter, but for many typically contemplated U and W functions, it matters less than that of individuals' utility functions.

Following Atkinson (1973), it is common to combine $W(U)$ and $U(c)$ into a single, composite function. To illustrate, consider the commonly used constant relative risk aversion (CRRA) utility function. These take the form

$$U(c) = c^{1-\alpha} / (1-\alpha), \qquad (3)$$

where α is the coefficient of relative risk aversion.[3] Social welfare may be taken to have a similar form:

$$SWF = \int \frac{U(c_i)^{1-\beta}}{1-\beta} f(i) \, di, \qquad (4)$$

where β can be interpreted as the coefficient of relative utility inequality aversion. A reduced form of the social welfare function that is a composite of (3) and (4) is:[4]

$$SWF = (1-\alpha)^\beta \int \frac{c_i^{1-\eta}}{1-\eta} f(i) \, di, \qquad (5)$$

where $\eta = 1 - (1-\alpha)(1-\beta)$. We might then interpret η as a measure of the social aversion to inequality in consumption.[5]

There are several problems with this approach.[6] One difficulty is that expression (5) is not well defined if α exceeds 1 because the first term before the

---

[3] Throughout, for all such functions considered, we will set aside without further remark the special case in which the curvature parameter equals 1, in which event one substitutes the natural log functional form, as in our preceding example.

[4] Most analysts who use a constant-elasticity reduced form do not include the expression before the integral.

[5] We find this manner of expression most transparent in suggesting the origin of η. One could instead write $\eta = \alpha + \beta - \alpha\beta$. The presence of the latter term is strongly suggestive of some of the points that follow.

[6] An additional serious difficulty is that, strictly speaking, equation (5) is a violation of welfarism because what matters for social welfare assessments are not just individuals' utility levels but also how they come about. For example, changing someone's coefficient of relative risk aversion and their consumption simultaneously so as to leave utility constant will nevertheless change the social valuation of that utility; hence, social decisions do not depend solely on individuals' utility profiles. Kaplow and Shavell (2001) show that this sort of violation of welfarism implies Pareto violations.

The B.E. Journal of Economic Analysis & Policy, Vol. 10 [2010], Iss. 2 (Symposium), Art. 7

integral then involves raising a negative number to a non-integer exponent.[7] More centrally, η is not the sum of, or similar combination of, α and β. For example, suppose that η is taken to be 2, and α is determined, through observation, to be, say, 0.9. The implied value of β is 11, a very high level of aversion to utility inequality. If others interpreting the data favored a slightly higher estimate of α of 0.99, then the implied value of β would rise to 101. But if new observational data show that α is even higher, actually 2, taking η to be 2 implies that we are now utilitarian, implying that β must be 0. Furthermore, if, as many economists believe, we have α greater than 1 but less than 2, say 1.5, β is !1, which seems to mean that we would actually have a preference for inequality. Figure 1 shows the implied value of β for $\eta=2$ and a given observation of α.

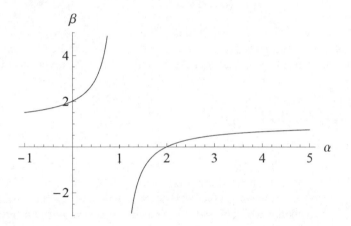

Figure 1 – Implied β for a given observation of α, when $\eta=2$

As is evident, we cannot simply chose η through ethical reflection on aversion to consumption differences and expect the resulting curvature of the social welfare function to make sense.

As mentioned, utilitarians take social welfare to be the unweighted sum of utilities, so $\beta=0$. In that case, $\eta = \alpha$, and the composite form in this case is unproblematic. However, the value of η in this case is a purely empirical matter (because, after all, α is an empirical parameter), its value having nothing to do with ethical reflection.

---

[7] The same problem arises in expression (4) if U is negative, which occurs, again, when α exceeds 1 and the utility function takes the form in expression (3). For further discussion, see Kaplow (2010) (p. 34).

This point about the utilitarian case has larger implications. Specifically, it indicates that there is something fundamentally mistaken about any notion that social welfare can be taken to depend directly on consumption inequality rather than on utility inequality. Social welfare functions are functions of utility, after all, and there is good reason for this: individuals' utilities are taken to be of normative importance, whereas consumption is purely a means to an end, that end being utility. For example, if at some point further consumption reduced utility rather than increased it, consumption beyond that level would be socially undesirable, and a formulation that deemed it socially beneficial regardless of its adverse consequences for individuals' utility would make no ethical sense. The same reasoning indicates that the *extent* to which consumption should be seen to augment utility is an empirical, pragmatic question, one that depends on the fact of the matter (however difficult it may be to discern) rather than on a priori ethical reasoning.[8] The reduced-form parameter η in the composite functional form (5), which is a direct measure of our aversion to consumption inequality, can only be given ethical meaning if it is *derived*, taking as the two key inputs the manner in which consumption actually affects well-being, an empirical matter indicated by the utility function, and the manner in which utility is deemed to affect social welfare, a legitimate subject for ethical reflection by an outside evaluator.

2 Framework for Social Assessment of Policies – Future Generations

To assess problems of climate change, we need to extend the framework of section 1 to encompass future generations. For concreteness, suppose that a project produces an increment of consumption for future generations and the question is what reduction in consumption the present generation should appropriately be expected to incur to generate that benefit. The analysis would in many ways be similar to that in section 1. We still need to specify the individual utility function and the social welfare function so that we can determine the marginal social value of an increment of consumption to individuals in different generations. To keep matters simple, we will adopt the common practice of using a representative agent for each generation, a simplification that is not innocuous.[9] In that case, the social maximand is:

---

[8] Put another way, it seems difficult to defend the view on ethical grounds that the empirical relationship between consumption and utility can be ignored so long as marginal utility is positive, but that it would become central if marginal utility ever fell slightly below zero rather than staying infinitesimally above zero.

[9] Following the analysis in section 1, it will matter which individuals within a generation are affected by the consumption increments since, in general, each will have a different effect on social welfare. One might make generalizations about intragenerational distribution, but in that

The B.E. Journal of Economic Analysis & Policy, Vol. 10 [2010], Iss. 2 (Symposium), Art. 7

$$SWF = \int_{t=0}^{\infty} W\left(U(c_t)\right)dt, \tag{6}$$

where c_t is the consumption (of the representative individual) at each time period (generation) t. Equation (6) is the same as equation (1) except that we are now integrating over time. As in the single-generation case, we still must separately specify U and W. The former is empirical and the latter is ethical, and their curvatures affect the social evaluation of outcomes differently.

Expression (6) does not allow pure time discounting of utilities. (Discounting, if any, arises in (6) because of the curvature of W, as we elaborate below). We can introduce pure time discounting of utilities by altering expression (6) as follows:

$$SWF = \int_{t=0}^{\infty} W\left(U(c_t)\right)e^{-\delta t}dt. \tag{7}$$

The discount rate in this case, δ, is an element of the social welfare function and is determined through ethical reflection.

What does the social welfare function (7) imply about how *consumption* is discounted? Beginning with the first term in the integrand, we can see that, as before, the weight on increments of consumption in any generation t will depend on the curvature of both U and W. Then, when that weight is determined (using consumption levels, c_t, for generation t), the result is further discounted, to an extent determined by δ, to produce the overall impact of the increment in generation t on social welfare. (This combination of factors will be further elaborated below, when we discuss the Ramsey equation.)

Expression (6) also did not explicitly include discounting by individuals as they choose their savings patterns to maximize their utilities. Suppose that instantaneous utility (the felicity function) at time τ is $u(c_\tau)$, and the individual discounts future utility at a constant rate θ. (Individuals might discount future utility because, for example, they are uncertain about when they will die or they are impatient.) Suppose that each individual lives for n periods and that utility is time-separable. Then an individual maximizes:

---

case any summary would itself reflect some combination of a social welfare function and individuals' utility functions, whereas the standard representative individual framework takes the summary to be a utility function. Such complications are not considered further here, but in light of the large intragenerational disparities in consumption, especially across countries (if the maximization is in terms of global welfare), this complication warrants further attention. (The analysis in section 4 does, however, bear on how this problem might best be addressed.)

Kaplow et al.: Discounting in Integrated Assessment

$$U = \int_{\tau=0}^{n} u(c_{\tau})e^{-\theta\tau}d\tau, \qquad (8)$$

subject to that individual's lifetime budget constraint. (It might be helpful to think of the representative individual in each generation t living for n subperiods, denoted by τ, which would be appropriate in a non-overlapping-generations model.) Analogous to our discussion of expression (7) for the social welfare function, we can now ask what the representative individual's utility function (8) implies about how *the individual* discounts consumption over a lifetime. The answer is similar. The first term in the integrand indicates that the weight on an increment of consumption in time period τ will affect utility by an amount that reflects the magnitude of c_{τ} in a way that depends on the curvature of u. This weight is then further discounted, to an extent determined by θ (not δ) to produce the overall impact of the increment in year τ on the representative individual's utility, U. (This combination of factors will likewise be examined below in connection with our discussion of the Ramsey equation.)

The social optimization problem is to maximize (7) given that individuals maximize (8) under the policy being evaluated. An important point is, as just mentioned, that the pure time utility discount rate used by individuals when maximizing utility, θ, is not the same as the pure time discount rate used by the outside evaluator when aggregating utilities, δ. For example, one may plausibly believe that θ is a positive number based on empirical studies while setting $\delta=0$ based on the ethical view that all individuals should be weighted equally. One discount rate, θ, is empirical, while the other, δ, is ethical. (Note, however, that, depending on the ethical basis for δ, it may depend in whole or in part on empirical evidence.[10])

Unfortunately, analysis usually proceeds without attention to the difference between utility functions and the social welfare function and, relatedly, between empirical and ethical discounting. Specifically, it is common to use a reduced form similar to equation (5):

$$SWF = \int_{t=0}^{\infty} Z(c_{t})e^{-\delta t}dt. \qquad (9)$$

---

[10] For example, following Parfit (1984), one might hold that the only legitimate basis for pure discounting of future utility is to reflect the less-than-certain probability that future individuals will exist. In that case, δ would be determined purely empirically (in this instance, rather speculatively).

The B.E. Journal of Economic Analysis & Policy, Vol. 10 [2010], Iss. 2 (Symposium), Art. 7

$Z(c)$ is a measure of social aversion to consumption inequality. As in the single-period case, analysts often take Z to be an isoelastic function of consumption, which yields:

$$SWF = \int_{t=0}^{\infty} \frac{c_t^{1-\eta}}{1-\eta} e^{-\delta t} dt. \tag{10}$$

For example, Stern's discussion of discounting, which we review in section 3B, uses this functional form (which can be seen by combining equations (3) and (6) in Chapter 2A of the Stern Review, Stern (2007)). Similarly, the IPCC's only substantial discussion of discounting, found in Arrow et al. (1996), uses this form (represented by combing their equations 4A.1 and 4A.3). Partha Dasgupta (2008) also uses this form, although in discrete time (his equations 2 and 3). Nordhaus (2008) (equations A.1, A.2, and A.3) likewise appears to use this form.

Equation (10), however, has the same problems as does equation (5), the reduced-form maximand in the single-generation case. As before, we cannot readily interpret η because it is a complex composite of the coefficient of relative risk aversion from individuals' utility functions and of society's aversion to utility inequalities from its social welfare function. Just as in the single-generation case discussed in section 1, it combines empirical facts and ethical views in an opaque and unintuitive manner. For a given reduced form parameter η, small changes to empirical facts imply dramatic and in some cases seemingly bizarre changes in ethical views, as we illustrated for the case in which η was taken to equal 2.

An additional complication is that the individual discount rate, θ, does not appear in equation (10). It is not clear whether θ is arbitrarily set to zero or whether it is now included in the value of δ. If θ is included as part of the value of δ, then δ, like η, embodies both empirical and ethical components and cannot be set based solely on either empirical measurements or ethical reflection.

One explanation for the prevalence of equation (10) is that it was adopted from growth models: it is the same as the maximand in the canonical neoclassical growth model with δ representing the net-of-population-growth discount rate of an infinitely-lived, representative household.[11]   Barro and Sala-I-Martin (1995)

---

[11] The justification for the use of an infinitely-lived individual requires two steps. First, we must believe that the conditions are met for the use of a representative individual at any given time. So-called Gorman preferences are sufficient to ensure that maximizing (10) produces a Pareto outcome. Acemoglu (2009) (chapter 5). Second, we must believe that it is appropriate to treat the representative individual as infinitely lived. The framework is justifiable only if either of two conditions hold: (i) the likelihood of death is a constant probability, or (ii) there are multiple generations but each is purely altruistic with a constant weight on future utility. Both of these conditions are in principle empirically testable. If they hold, θ cannot be chosen freely but should have the value that best captures the justifying conditions. For example, if we believe altruism is

Kaplow et al.: Discounting in Integrated Assessment

(Equation 2.1), Acemoglu (2009) (Equation 8.49). If the individual is fully informed and there are no externalities, the individual will make the appropriate trade-offs over time to maximize equation (10). In this framework, no valid social welfare function would choose anything other than what the infinitely-lived individual would choose. That is, η and δ in expression (10) become α and θ respectively as they now are parameters of the utility function, not the social welfare function. Adjusting the notation, if there is a representative infinitely-lived individual, expression (10) becomes:

$$SWF = \int_{t=0}^{\infty} \frac{c_t^{1-\alpha}}{1-\alpha} e^{-\theta t} dt. \qquad (11)$$

Suppose that the representative individual for some reason fails to maximize his utility – say he fails to fully consider the future and therefore sets his personal discount rate, θ, too high. We cannot simply adjust θ to a preferred value because θ is empirically determined. In this case, we must instead consider the full model represented by expressions (7) and (8). For purposes of predicting individuals' behavior, we cannot change the empirically determined utility function in (8) to reflect some normative view of correct behavior because the model's results will no longer conform to reality. Treatment of the social welfare function in this case is more controversial with regard to the representation of individuals' utility to be employed in the W function in (7), although not regarding the functional form of W itself.[12]

The interpretation suggested by some analysts is that expression (10) is a utilitarian social welfare function with the addition of pure time discounting (i.e., as part of the social welfare function, not as part of the utility function).[13] That is, in the notation used above, $\beta=0$. This view implies, as noted previously, that $\eta = \alpha$. In this case, η is then properly understood to be (entirely) a parameter of the utility function. Therefore, it must be determined empirically, not posited by stipulation according to the ethical views of a social evaluator. Nevertheless, authors suggesting that expression (10) is utilitarian sometimes also suggest that η is an ethical parameter rather than an empirical parameter, which is internally inconsistent.[14] For example, Dasgupta (2008) explicitly states that he is adopting

the appropriate justification for using an infinitely-lived representative individual, θ should represent that altruism.

[12] However, if behavioral utility differs from normative utility (say, due to self-control problems), many would deem it appropriate to substitute the normative utility function's assessment of the individual's situation when aggregating utilities in the social welfare function.

[13] Stern (2007), Heal (2005), Dasgupta (2008), Arrow (1982), Stiglitz (1982).

[14] A partial reconciliation is that the term "utilitarian" is sometimes used loosely to refer to the general class of social welfare functions of the sort in expression (4), thus encompassing as well

The B.E. Journal of Economic Analysis & Policy, Vol. 10 [2010], Iss. 2 (Symposium), Art. 7

a utilitarian interpretation of η but then devotes much of his article to ethical arguments about the correct value of η.[15] (For further examples, see section 3.)

Moreover, unlike in the single-generation case, where viewing expression (5) as utilitarian resolved the problems with using the composite form, a discounted utilitarian reconciliation raises some problems in the multiple-generation case. The reason is that, if δ in expression (10) is part of the social welfare function, we have implicitly made the empirical assertion that $\theta=0$. Unless individual time preferences can be explained solely through an empirically plausible choice of α, however, this is not a good interpretation of (10). Put another way, viewing (10) as a discounted utilitarian social welfare function is not simply a matter of interpretation; it also requires making an empirical claim that may or may not be true.[16]

An additional problem with interpreting (10) as embodying discounted utilitarianism is that the use of a utilitarian social welfare function is a substantial restriction on the social welfare function being imposed by the modeler. Some readers will not agree with this form of the social welfare function or, perhaps more importantly, even realize that it is being prescribed.[17]

It is useful to apply the foregoing analysis to the so-called Ramsey equation for the discount rate since this equation is so commonly used in analyses of discounting in the climate context. The Ramsey equation, which derives from the neoclassical growth model, is an expression for the equilibrium return ρ demanded by consumers as they optimize their consumption patterns over time. It is derived (in the case of CRRA utility) by taking expression (11) as the social maximand, subject to the budget constraint and the laws of motion for capital. If we make use of pertinent first-order conditions, we can produce:

$$\rho=\alpha\dot{c}+\theta, \tag{12}$$

---

all those in which $\beta > 0$. However, in that case we are back to the problems with which we began (and which are only avoided by a social welfare function that is utilitarian in the formal, literal sense).

[15] Heal (2005) asserts that, in a full CGE model, we only need to specify δ because the other parameters of the social optimization are determined internally to the model; the model produces utility flows that we then simply aggregate according to the social welfare function which is specified through the choice of δ. We take Heal to be saying that the only *ethically* chosen parameter is δ because in a CGE model, the parameters of the utility function (in the case of CRRA utility, α and θ) must still be specified, albeit, empirically. Heal is explicitly operating in with a discounted utilitarian social welfare function so he is, in fact, choosing β (by setting $\beta=0$) as well as δ. By contrast, Heal (2009, p. 281) asserts that the choice of form for the utility function (in our notation, α) is an ethical choice, creating problems similar to those discussed in the text.

[16] If expression (10) is taken to be undiscounted utilitarian, so that both η and δ are elements of the utility function, this problem does not arise.

[17] A utilitarian social welfare had both strong adherents and detractors. See, for example, the competing views in Sen and Williams (1982) and Kaplow (2008) (chapter 14).

where $\dot{c}$ is the growth rate of consumption. This equation says that households allocate consumption over time so that the rate of return is equal to the rate of decrease in the marginal utility of consumption due to consumption growth, plus the representative individual's pure rate of time preference (θ). Note that this equation and explanation maps directly to our discussion of how expression (8) for the representative individual's utility relates to discounting. In the climate policy assessment literature, it is common to discuss the proper discount rate by considering the appropriate values for the parameters on the right side of expression (12). That is, numerous authors skip equation (7) for social welfare (or analogues thereto) and instead go directly to the Ramsey equation. Prominent examples include Nordhaus (2007) and Weitzman (2007).

The Ramsey equation originates with growth models that consider a representative, infinitely-lived individual. The parameters α and θ are parameters of the utility function of that representative individual. To consider distributional issues, however, we need to have more than a representative individual and incorporate a separate social welfare function that, if appropriate, includes the pure discounting of utilities. To illustrate, if we treat equation (10) as the social maximand rather than equation (11), the Ramsey equation becomes:

$$\rho = \eta\dot{c} + \delta, \text{ or}$$
$$\rho = \left(1 - (1-\alpha)(1-\beta)\right)\dot{c} + \delta. \tag{13}$$

Not surprisingly, the problems with using the reduced form expression (10) arise when one works with the Ramsey equation, since, after all, the latter is derived from expression (10). (This encompasses the point that including δ while omitting θ implies that $\theta{=}0$, which may be inconsistent with empirical evidence.)

Whether operating through the Ramsey equation, an infinitely-lived representative individual framework, or otherwise, we have emphasized the difference between $U(c)$ and θ, on one hand, and $W(U)$ and δ, on the other hand, in policy evaluation. The overall discount rate on the representative individual's consumption implied by $U(c)$ and θ (and the set of policies) is the discount rate used by such individuals to compare their own consumption in different periods of their lives when they are maximizing their own utilities. The discount rate implied by the choice of $W(U)$ and δ, by contrast, is the discount rate an outside evaluator applies to the utility of representative individuals of different generations when aggregating their utilities to assess the outcome of a policy. Aggregating $U(c)$ and $W(U)$ into a single composite creates confusion between the empirical selection of the parameters of $U(c)$ and the ethical choice of the parameters of $W(U)$. Likewise, the pure time discount rates, θ for individuals'

The B.E. Journal of Economic Analysis & Policy, Vol. 10 [2010], Iss. 2 (Symposium), Art. 7

discounting of their own utilities during their lifetimes and δ for society's discounting of each generations' utilities, need to be clearly distinguished.

3   Discounting in Integrated Assessment Models

The evaluation of policies aimed at mitigating climate change is a particular case of an intergenerational policy evaluation as described in section 2. Those IAMs that solve for optimal policies are, in theory, maximizing equation (7) for social welfare subject to the individual's utility-maximizing behavior (8).[18] Therefore, conflation of parameters of the social welfare function and of the individuals' utility functions in these models, and in climate policy evaluation generally, introduces precisely the problems discussed thus far. In particular, policy evaluation in the IAM context requires both the specification of the utility function and an evaluation of outcomes by aggregating utilities. Individuals in the model will discount as they allocate consumption over time. Evaluation of model output may separately involve discounting of different generations' utilities as part of the social aggregation process. Discussions of "the discount rate" in an IAM are often ambiguous – referring to either or both individuals' (predictive) discounting and social (evaluative) discounting, and for each of them referring to the discounting implied by the curvature of the pertinent function (utility function or welfare function), to pure time discounting (θ and δ, respectively), or to some sort of aggregate of these four distinct components. As a consequence, it is often difficult to interpret the models and the debate about the correct discount rate to use in the models.

    We illustrate this problem by considering two models that have been prominent in the recent debates over discounting: Nordhaus's DICE model (Nordhaus (2008)), and the model used in the Stern Review (Stern (2007)), which is a modification of the PAGE2002 model found in Hope (2006). Neither cleanly separates predictive from evaluative discounting.

    Once predictive and evaluative discount rates are appropriately separated in policy analysis, another question arises: should modelers be choosing evaluative discount rates and embedding them in bottom-line results or policy prescriptions? While specifying predictive discount rates is a necessary part of the modeler's task if the model is to predict outcomes based on individuals' maximizing behavior, choosing an evaluative discount rate is an ethical exercise that could well be left to the reader. We discuss two procedures that allow modelers to avoid choosing evaluative discount rates: sensitivity analysis on the parameters of the social welfare function and direct presentation, by which we

---

[18] Not all IAMs attempt to solve for optimal policies. Some merely simulate policies chosen by the modeler and do not attempt to provide any assessment of the estimated results.

mean the display of the time series of model outcomes in terms of welfare in different periods without any attempt to apply intergenerational weighting to those values. Neither of these procedures is novel and both are sometimes used in the climate IAM context as well as being regularly employed elsewhere. Our suggestion is that more consistent and prominent adoption of these procedures in IAMs would help clarify the issues and reduce some of the disagreements over discounting.

A. DICE

The DICE model, a standard in climate policy analysis since the early 1990s, is essentially a neoclassical growth model coupled with a climate module.[19] Production creates emissions that (via the climate equations) cause a change in global mean temperature, which in turn causes damages, reducing production. The model finds the optimal policy by maximizing:

$$\omega = \sum_{t=1}^{T \max} \frac{c_t^{1-\eta}}{1-\eta}(1+\delta)^{-t}. \tag{14}$$

This expression is a combination of equations A.1, A.2, and A.3 in Nordhaus (2008), with the notation modified to match ours. It differs from our formulations by using discrete time and having a finite endpoint, neither of which are consequential for present purposes. We do not account for population growth, which is included in DICE.

In expression (14), we use ω instead of U or SWF on the left side because it is unclear how to interpret expression (14) in light of the analysis in sections 1 and 2. Is ω an individual utility function, a social welfare function, or a composite and are the parameters accordingly empirical, ethical, or some combination? The answer appears to be that it varies and in ways that are possibly inconsistent.

On one hand, a representative individual in DICE optimizes the savings rate to maximize (14).[20] Under this view, expression (14) is a utility function, which implies that the parameters are entirely empirical. If they were ethical and the empirical parameters were different, the behavior in the model would be different.

---

[19] A model description can be found in Nordhaus (2008). The model itself can be downloaded at http://nordhaus.econ.yale.edu/DICE2007.htm.
[20] More precisely, individuals find the constant savings rate that maximizes expression (14). In a more complete optimization, of course, they might vary their savings rate over time.

The B.E. Journal of Economic Analysis & Policy, Vol. 10 [2010], Iss. 2 (Symposium), Art. 7

On the other hand, Nordhaus (2008) explicitly posits that the parameters in expression (14) are normative. For example, he states (p. 33) (emphasis added):

> In the DICE model, the world is assumed to have a well-defined set of preferences, represented by a '*social welfare function*,' which ranks different paths of consumption. The social welfare function is increasing in the per capita consumption of each generation, with diminishing marginal utility of consumption. . . . The relative importance of different generations is affected by two central *normative* parameters: the pure rate of time preference and the elasticity of the marginal utility of consumption (the 'consumption elasticity' for short).

Further, he cautions (p. 172):

> It must be emphasized that the variables analyzed here apply to comparisons of the welfare of different generations and not to individual preferences. The individual rates of time preference, risk preference, and utility functions do not, in principle at least, enter into the discussion or arguments at all. An individual may have high time preference, or perhaps double hyperbolic discounting or negative discounting, but this has no necessary connection with how social decisions weight different generations. Similar cautions apply to the consumption elasticity.

Given these sharp, unambiguous statements and the fact that he uses expression (14) to undertake social evaluation, the parameters have to be normative. It is not clear, if this is the case, what the individual utility function is or why the parameters of (14) would affect savings rates.

One might partially reconcile these conflicting interpretations by supposing that expression (14) represents discounted utilitarianism (i.e., $\eta = \alpha$, δ is part of the social welfare function, and $\theta=0$). As we noted, this interpretation necessarily includes an empirical assertion ($\theta=0$) that may not be true. An alternative reconciliation is that the integrand in expression (14) is the composite function $Z(c)$. For the reasons we discuss at length in sections 1 and 2, this interpretation is problematic for different reasons.

A reconciliation of these two conflicting views is to take expression (14) be an *undiscounted* utilitarian social welfare function with a single, infinitely-lived representative individual (i.e., with individual discounting but *without* discounting as part of the social welfare function). Nordhaus himself at one point explicitly adopts this interpretation, stating that "the social welfare function is

taken to be an additive separable utilitarian form, $W = \int_0^\infty U[c(t)]e^{-\rho t}dt$ " (p. 172).
(Under this interpretation, Nordhaus's ρ would, in our notation, be θ.) In that
event, his expression (14) should be interpreted to be the same as our expression
(11). Its parameters are then components of the representative individual's utility
function and, as such, are entirely empirical; $\eta = \alpha$ and $\delta = \theta$. This view, however,
flatly contradicts the above quotations. Moreover, as we noted, this view is
restrictive. Some readers may not agree with this choice of a social welfare
function and, given the language used to describe (14), many will not realize that
it is being imposed. Also, some may accept Nordhaus's posited pure time
preference as empirically representing individuals' actual discount rates but reject
it as a social discount rate (or conversely) but not appreciate what is being
assumed. Finally, as we noted, the infinitely-lived representative individual
framework is applicable only under certain conditions and the discounting
parameters used must be tied to the justifying conditions (see footnote 11).

B. Stern/PAGE2002

The integrated assessment model used in the widely known Stern Review (Stern
(2008)) is a modification of PAGE2002.[21] PAGE2002 is a partial equilibrium
model. It assumes an exogenous GDP growth rate; computes the resulting
damages, adaptation costs, and, for a chosen policy, mitigation costs; and then
calculates the present value of GDP for the chosen policy. Baseline emissions
appear to be (otherwise) exogenous, but spending on mitigation reduces
emissions, which in turn reduces damages.

　　　Note that because PAGE2002 is a partial equilibrium model that uses
exogenous GDP as an input, there are no individuals, no individual utility
functions, and no individual maximization, discounting, or savings choices.
Instead, these are implicit in the assumed GDP growth rate and the overall
structure of the economy. We might imagine that, if the implicit utility function
that is consistent with the assumptions about GDP growth were isoelastic, there
would be an implied α and θ, but these parameters are not (and probably cannot
easily be) specified.

---

[21] Stern refers to Hope (2003) for a description of the model. This paper does not have the model
equations, however. We use Hope (2006), which describes the PAGE2002 model as of 2003 and
which does have the model equations. Because the 2003 paper might describe an earlier version
of the model, there could be some differences between our description and the model Stern
actually used.

The B.E. Journal of Economic Analysis & Policy, Vol. 10 [2010], Iss. 2 (Symposium), Art. 7

Stern uses PAGE2002 to compute optimal climate policy.[22] To do this, he divides annual GDP produced by the model by exogenous annual population to get GDP per capita. He then imposes a CRRA utility function (with exponent η) and an assumed, fixed savings rate of 20% to compute utility per capita. He sets $\eta=1$, based on his view of the empirical data (pp. 183-184).

This methodology seems problematic because he is imposing a utility function on the model ex post when there is no reason to suppose that it is consistent with the (unspecified) utility function that implicitly generates the model's results. That is, he is choosing empirical parameters for the utility function (both a value of α and of θ), which involves predictive discounting, but is not attending to whether the resulting predictions are in line with his results (which would seem to be true only by coincidence given how each was generated).

The conflict between empirical and ethical parameters arises in connection with Stern's statement of his social welfare function. He specifies a social welfare function that is explicitly utilitarian, with discounting of utilities only for the possibility of extinction. Adjusting his notation to conform to ours, he maximizes:[23]

$$SWF = \int_{t=1}^{\infty} \frac{c_t^{1-\eta}}{1-\eta} e^{-\delta t} dt. \tag{15}$$

This representation is the same as our expression (10) where, because the social welfare function is utilitarian, $\beta=0$, which implies that $\eta = \alpha$. Furthermore, based on ethical reasoning, Stern takes the view that δ should be set equal to zero with the exception of the possibility of extinction. He estimates this possibility to be such that $\delta=0.1$.

Stern's interpretations of both η and δ are problematic. Stern repeatedly states that the value of η is an ethical variable, contradicting the claim that (15) is a utilitarian social welfare function. For example, he states: "η which is the elasticity of the marginal utility of consumption [in] this context . . . is essentially a value judgment." Stern (2007) (p. 52). He confirms this view in Stern (2008), stating that "η is the elasticity of the *social* marginal utility of consumption. . . . Thinking about η is, of course, thinking about value judgments – it is a prescriptive and not a descriptive exercise" (pp. 14-15) (emphasis added). The subsequent discussion in the article (pp. 15-17) is then devoted to ethical reflections on the proper value of η.

---

[22] Our sources for the following discussion are, in Stern (2007): box 6.3, where he describes his modifications to PAGE2002, and chapter 2A, where he describes his approach to discounting.

[23] This is a combination of equations (3) and (6) in chapter 2A.

The difficulties with this view should, at this point, be relatively clear. If his η is truly ethical, then $\eta = \beta$ and $\alpha = 0$, which is empirically implausible and contrary to his claim that the social welfare function is utilitarian. He could be using the composite form $Z(c)$, but this is contrary to his statements and introduces the problems of interpretation described in section 1. Moreover, if one accepts his apparent view in favor of a social welfare function that is strictly concave – and notably more egalitarian than a utilitarian SWF – and combines this judgment with an empirically plausible estimate for α, individuals' coefficient of relative risk aversion, it seems that one would then be endorsing a significantly higher overall social discount rate on future generations' consumption. If so, one who agreed with both the Stern Review's predictions and Stern's value judgments would reach significantly different policy conclusions, notably, ones favoring much less immediate action to address global warming.

Problems also arise in interpreting (15) because of the omission of θ (the individual pure time discount rate). Stern takes δ to be purely ethical, which implies that $\theta=0$. This implication, however, entails an empirical assertion that may not be borne out.

The tension between the empirical and the ethical is illustrated when Stern discusses savings rates. Some authors, notably Arrow (1999), have criticized setting δ equal to *0.1* because they claim it implies an implausibly high savings rate. Stern (2007) (p. 54) responds to this criticism by arguing that the model on which it is based is restrictive and that in reality many things affect savings. Likewise, Stern (2008) (p. 16) points to the possibility of technical progress as reducing the savings rate implied by a pure time discount rate of (or near) zero. It is not clear, however, how to make sense of any of this discussion because it suggests that a parameter of the social welfare function can affect savings rates. But δ can only affect savings rates if it is part of individuals' utility functions, that is, if it is really what we are calling θ. But in that case, the parameter would be purely empirical and thus properly determined in an entirely different manner.[24]

Both DICE and Stern's use of PAGE2002 make valuable contributions to our understanding of the policies that might be adopted to address the consequences of greenhouse gas emissions. Unfortunately, in each model and presentation, confusion about what is in the utility function and what is part of the social welfare function impedes our understanding of these models and their implications. It is difficult to relate competing views about discount rates to each other given the mix of the empirical and the ethical – of the predictive and the evaluative – as well as discounting of consumption that arises from curvature (of the utility or social welfare function) on one hand and pure time discounting (of

---

[24] The shape of the social welfare function might indirectly affect savings because it might affect government policy toward savings. Stern does not seem to have this in mind.

The B.E. Journal of Economic Analysis & Policy, Vol. 10 [2010], Iss. 2 (Symposium), Art. 7

consumption or of utility, respectively) on the other hand.   Regardless, the difficulties in interpretation that we identify may help explain how modelers with otherwise similar analyses can nevertheless end up choosing widely different discount rates, resulting in large differences in policy prescriptions.

C. Separating Predictive and Evaluative Discounting

An essential first step in avoiding the interpretative difficulties just discussed is for modelers to state clearly the empirical parameters used for their utility functions and the ethical parameters employed in their social welfare functions. The former are necessary for any predictions that are based on individuals' maximizing behavior, and both are essential for offering social evaluations or for deriving an optimal policy path.   Because ethical parameters are more controversial and because their choice is not particularly within the expertise of climate or economic modelers, it is helpful if these parameters are either left unspecified or are presented in a manner that allows readers to apply their own value judgments. Accordingly, in this section we suggest how this goal might be accomplished, following the familiar tradition of policy analyses in other areas. In particular, we examine the use of sensitivity analysis and the direct presentation of time series results as useful ways of separating the unavoidable specification of empirical parameters from the more discretionary selection of ethical parameters.

Sensitivity analysis

Sensitivity analysis is, of course, widely used to examine how results change in response to changes in parameters.[25] Typically, sensitivity analysis is used when empirical parameters are uncertain. It can also be used for ethical parameters. That is, modelers would show how optimal results change or how results of a given policy are assessed differently as the parameters of the social welfare function are altered. This type of presentation allows readers to see directly how the choice of ethical views affects policy decisions. Sensitivity analysis on the parameters of the social welfare function is distinct from sensitivity analysis on parameters that affect the discounting employed by individuals when maximizing utility. As one changes the parameters of the utility function, individuals in the model will change their behavior because they are now optimizing a different function.   The social welfare function remains fixed, but it will evaluate the results of a given policy differently because the predicted behavior has changed

---

[25] There are numerous references on the topic, including books such as Saltelli, Chan, and Scott (2000).   In the environmental context, the EPA recommends sensitivity analysis for its environmental models. Council for Regulatory Environmental Modeling (2009).

(and because the level of individual utility associated with a given outcome changes). Contrariwise, as one changes the parameters of the social welfare function, behavior and utility will remain fixed but the evaluation will nevertheless be different.

We are not aware of any sensitivity analysis that cleanly separates the sensitivity of predictive discounting from the sensitivity of evaluative discounting. Sensitivity analysis confined to the implied discount rate in composite function $Z(c)$ is difficult to interpret because we cannot tell which parameters, the ethical, empirical, or both, are being changed. Nevertheless, even sensitivity analysis on composite discount rates is not common. Ferenc Toth, in a 1995 survey of discounting in integrated assessment models of climate change, concluded that "the majority of integrated assessments do not include sensitivity tests for the discount rate, or do not hold it sufficiently important to report them in the literature." Toth (1995) (p. 408) Fifteen years later, this unfortunately remains the case.[26]

There are, however, some examples. Tol (1999) analyzes the marginal damages from emissions using the FUND integrated assessment model. FUND estimates a time series of damages from greenhouse gasses in a number of sectors and then aggregates them into a single, monetized amount to estimate the marginal cost of emissions. In presenting results, he displays calculations for varying discount rates, ranging from 0% to 10%. Anthoff and Tol (2009), reporting additional results from FUND, follow a similar procedure. In both cases, these appear to be aggregated discount rates, representing some combination of predictive and evaluative rates.[27]

Nordhaus (1994) performs sensitivity analysis on the results of the 1994 version of DICE for changes to both η and δ. In his more recent book (Nordhaus (2008)), however, he does not do sensitivity analysis on the parameters of the discount rate. He does do sensitivity analysis for other variables, and also presents a comparison of his results to Stern's results by running DICE with Stern's discount rates (showing that the choice of discount rates is the primary driver of the differences in policy recommendations from the two models), but he does not consider other values. As with Tol, however, we still have the difficulty of untangling predictive and evaluative discounting.

A number of papers are devoted to exploring the sensitivity of IAM results to parameters of the discount rate. Anthoff, Tol, and Yohe (2009), Hope (2008),

---

[26] Dasgupta (2007) (p. 6), commenting on the Stern Review, also calls for more sensitivity analysis: "What we should have expected from the [Stern] Review is a study of the extent to which its recommendations are sensitive to the choice of eta. (Many economists would expect a sensitivity analysis over the choice of delta too.)"

[27] Not all papers presenting results of FUND, however, follow this practice of sensitivity analysis. E.g., Tol (2007).

The B.E. Journal of Economic Analysis & Policy, Vol. 10 [2010], Iss. 2 (Symposium), Art. 7

Nordhaus (1997), Nordhaus (1999) and Gerlagh and van der Zwaan (2004). These papers are valuable because they suggest the direction and magnitude of change we might expect from changing various discounting parameters. Like the other papers we discuss, however, they mix up the parameters of the social welfare function and the parameters of the utility function, making interpretation difficult. Moreover, our suggestion here is for more sensitivity analysis with respect to ordinary use of IAMs; when reporting the results of a study, say about the effect of technology subsidies, border taxes, or any other object of analysis, authors should report how those results are sensitive to the parameters in the social welfare function used in the model. Studies focused primarily on the discount rate are not a substitute for this.

Direct presentation

Sensitivity analysis of particular components of discounting allows one to compare bottom-line assessments under different social welfare functions. An alternative to sensitivity analysis is to forgo such assessments. Rather than aggregating model results into a single number, one can instead display a time series of model results (i.e., the resulting time path of consumption or utility across years or generations). This procedure avoids any ethical choice by the modeler, leaving the ultimate assessment to the reader.[28] By presenting the model's predicted outcomes of a policy, the analyst gives his or her audience the full information content of the modeling effort – who gets what, when – rather than effectively compressing that information into a single policy recommendation in which the analyst's choice of ethical parameters plays a major

---

[28] Many may be concerned about the difficulty of choosing an evaluative discount rate because of the unintuitive power of exponential discounting to the uninitiated. For example, few (non-economist) readers of IAM reports understand that doubling the discount rate from 2% to 4% changes present values over a 100 year period by almost sevenfold and that doubling them again from 4% to 8% means a further change by a factor of 43. In addition, these multiples change in nonlinear ways with the time period. Koopmans (1965) (p. 226) noted "[T]he problem of optimal growth is too complicated, or at least too unfamiliar, for one to feel comfortable in making an entirely a priori choice of [a time discount rate] before one knows the implications of alternative choices." However, our analysis of the foundations of evaluative discounting in section 1 suggests another view on this problem, one that derives a social welfare function, much as in the single-generational context, from first principles. For example, one might use arguments that ground welfarism to adopt a social welfare function of the form in expression (7); adopt the reasoning of Parfit (1984) and others to suggest that pure time discounting of utility should reflect only the probability of extinction, which (along with difficult-to-ascertain empirical evidence) determines δ; and then engage in further ethical analysis to choose the curvature of $W(U)$, perhaps adopting the utilitarian (linear) form, following the argument of Harsanyi. The resulting overall discount rate of utility, or of consumption, then just is what it is. No choice of a composite discount rate, which implicitly incorporates numerous value (and empirical) choices is involved.

role. Predictive discounting will still be involved in model calculations since outcomes depend on individuals' maximizing behavior (unless, e.g., savings behavior is stipulated), but there is no evaluative discounting since there is no evaluation. By avoiding the choice of a social welfare function altogether, this method eliminates the possibility of confusion over what is empirical and what is ethical.

This procedure is not uncommon: it is used frequently in policy analyses outside the climate context and also appears in some climate IAMs.[29] Results from MIT's integrated assessment model, EPPA, are most often stated as a time series of changes to either utility or consumption.[30] For example, Paltsev et al. (2009) uses the EPPA model to analyze the costs of meeting specified emissions reduction targets. Rather than presenting an aggregate evaluation of their results, they present the change in utility in various time periods as a result of the policies being simulated. The IPCC presented results this way in its Third Assessment Report, which shows the percent change over time in world GDP under a business-as-usual scenario for three different IAMs. Smith, Schellnhuber, and Mirza (2001). The Stern Review also presents some results in this format, showing changes in per capita GDP under various scenarios. Stern (2007) (p. 178).

There are some limitations to this approach. One is that, without specifying a social welfare function, one cannot optimize policies. To address this problem, one might employ a compromise procedure under which one chooses the ethical parameters of the social welfare function in order to perform optimization and then displays the raw results for the optimal policy and for other regimes, such as some status quo baseline. One might also perform sensitivity analysis, deriving multiple optimal policies for different ethical parameters and displaying the raw results for each.

Another problem is that it may be difficult for a decision-maker to come to a policy conclusion from a visual inspection of different paths of consumption or utility over time unless there is a near dominance relationship among them. As a consequence, the decision-maker would ultimately have to employ some particular social welfare function in order to compare policies. Nevertheless, the separation of these processes – prediction and evaluation – will achieve great clarification, and there is substantial benefit of allowing one to choose which

---

[29] Examples in the general context include Altig et al. (2001), Heckman, Smith, and Clements (1997), and Fullerton and Rogers (1993).

[30] Paltsev et al. (2005) provides a model description. They use the term welfare, but they do not mean social welfare as we use the term here. Instead, welfare in their model description refers to utility. The use of direct presentation rather than sensitivity analysis might be connected to model structures. For example, EPPA does not attempt to estimate damages from climate change, so it could not easily be used for optimization. For broad descriptions of model types, see Weyant et al. (1996), Weyant and Hill (1999), and Hope (2005).

The B.E. Journal of Economic Analysis & Policy, Vol. 10 [2010], Iss. 2 (Symposium), Art. 7

integrated assessment seems most credible, independently of particular modelers' ethical beliefs, and then to debate which is the proper mode of social evaluation, independently of particular evaluators' assessment models.

4   Intergenerational Distribution versus Efficiency

As we note at the outset of section 1, distributional issues are assumed to be implicated in the climate policy literature and in the discounting debate in particular.   Indeed, this distributive dimension motivates our framework that explicitly and systematically derives discount rates using a social welfare function that depends on individuals' utility functions.   We also remarked that, in economic policy assessment, it is far more common to perform cost-benefit analysis (without distributive weights) and leave distributive concerns to the tax and transfer system.   This familiar separation of efficiency and distribution has much to commend it in general: notably, the pursuit of efficiency in the domain under consideration leads to a larger pie that can be distributed, for example, to make all income groups better off.[31]   In this section, we briefly sketch how the main ideas apply in the climate policy context, which involves distribution across generations.[32]

The problem in the context of cost-benefit analysis for climate change can be roughly approximated by a model in which the cost of reducing greenhouse gases is a fraction of GDP for all generations, but the benefits of that action – the damages that were averted by that action – rise over time.   Undertaking immediate, aggressive abatement of greenhouse gases then involves the present generation sacrificing its consumption in order to benefit future generations. The intergenerational distribution problem is commonly framed by considering costs borne in a single present period and benefits in a single future time period. Suppose that the marginal dollar of present sacrifice provides a benefit of B dollars at future time t.   According to expression (7), this marginal sacrifice is socially desirable if and only if:

$$W'\big(U'(c_t)\big)e^{-\delta t} \cdot B > W'\big(U'(c_0)\big). \tag{16}$$

However, it is also natural to ask whether the return of B at time t is a good payoff relative to other ways to invest the funds.   Alternatives include other public projects (e.g., infrastructure), R&D, human capital, or conventional private sector investments.   If the return at time t on any alternative is greater than B, then

---

[31] See, for example, Kaplow (2004) and Kaplow (2008) and, for formal analysis in the context of controlling externalities, see Kaplow (2006).
[32] See Kaplow (2007) and Samida and Weisbach (2007) for elaboration.

it would be better (regardless of whether the social criterion (16) is satisfied) to invest in such alternative instead of in additional greenhouse gas mitigation. Indeed, this strategy is Pareto superior, and it can be made strictly so. If the present generation invests slightly less, making itself better off, yet still invests enough that the future generation is also better off, both generations gain. Similarly, if the return on another alternative is less than B, both generations could be made better off (again, regardless of whether the social criterion (16) is satisfied) by reducing the investment in that alternative and shifting the resources to greenhouse gas reductions.

In other words, it is in principle socially best to choose the level of control of greenhouse gas emissions that equates the marginal return to those of all alternative investments. This decision rule is based entirely on the rates of return on other investments and depends not at all on the social criterion (16) that (in some form) constitutes the standard basis for assessment for climate policy.[33] One should also note that, although framed in terms of rates of return, the underlying idea is familiar from analysis confined to a single period, where it is optimal to allocate scarce resources to their highest-valued uses.

There are a number of complexities and subtleties associated with determining the rate of return on other investments that it is appropriate to employ. One important point is that the rate of return is endogenous when nonmarginal policies are under consideration. Because of the large potential impact of climate change as well as the possible magnitude of mitigation strategies, the market rate of return is endogenous to climate policy. It nevertheless remains true that we should invest in climate change mitigation up to the point where the marginal return is equal to that on other projects. For example, the initial benefit from mitigation might be very high but subject to diminishing returns. It is also possible that, as mitigation efforts increase and thus climate damages to the economy are reduced, the returns on other projects will be boosted. (And they may also rise for the simple reason that, as more resources are shifted away from alternative projects toward climate mitigation, the now-marginal projects will be more productive ones.) At the optimum, we would invest in climate mitigation up to the point where the marginal returns are equal to those available elsewhere.[34]

---

[33] Important relationships between these concepts do exist. Notably, individuals' own discount rates for consumption guide their savings behavior, producing an equilibrium relationship between personal discount rates and the market interest rate. Moreover, to the extent that society has intergenerational distributive preferences that determine policy, the resulting intergenerational distribution will itself influence market returns.

[34] There are other important complications that may be pertinent. For example, looking to historic market rates of return may not well predict future rates of return if temperatures rise significantly (because the resulting damages may nontrivially alter average returns to other investments). Rates of return should in principle be calculated within the model based on estimates of market

The B.E. Journal of Economic Analysis & Policy, Vol. 10 [2010], Iss. 2 (Symposium), Art. 7

Another controversy concerns whether to use a before- or after-tax rate of return when discounting returns on public projects. Indeed, some analysts have argued that discount rates should be chosen on the basis of ethical reflection because of divergences between the market and social discount rates. Stern (2008) (pp. 12-13). Similarly, appropriate adjustments for risk must be made because the benefits of emissions reductions are uncertain. These issues and many others are beyond the scope of the present investigation.[35] For present purposes, we emphasize that the previous logic is nevertheless sound; various complications or subtleties call for appropriate adjustments, not abandonment of the valid core idea. For example, if there is a positive externality so that some project has a greater social return than its market return, that greater social return should be employed; indeed, this case is precisely the justification for government action to reduce greenhouse gases. But, once it is determined that the full social return in generation t is B, we should still wish to know whether this return exceeds or falls short of the full social return on alternative projects.

Relatedly, it is difficult to understand how distortions or other factors affecting the market rate of return are best addressed by ethical reflection that leads to choosing different behavioral parameters (notably the values of α or θ), as these are empirically determined, or how changing the shape of W, the method of aggregating utilities, is a coherent response. In the single-generation setting, no one would suggest that the existence of monopoly, tariffs, tax wedges, or other distortions should be addressed in whole or in part by employing different utility functions from those that would otherwise be used or a different social welfare function from that otherwise deemed proper. This point reinforces the broader theme of this section: although adding an intergenerational dimension introduces interesting and important complications, it is not a reason to abandon well-established and justified techniques of policy assessment, replacing them with methods that, when simplified to the single-generation context, would not make sense. Climate change is a particularly complex setting, involving long time horizons, irreversibilities, high levels of uncertainty, and economy-wide effects. The appropriate response to those complications is increased care and transparency.

---

conditions, including climate damages. Similarly, climate change may affect sectors of the economy differently, for example, making agricultural production more expensive relative to other sectors. Proper estimation of these effects – adjusting the relative price of goods because of climate damages – is also needed.

[35] See, for example, Arrow et al. (1996) and Lind et al. (1982). These and other issues are addressed further and related to the present argument in Kaplow (2007).

5   Conclusion

The choice of the discount rate for integrated assessment models is one of their most consequential decisions, one that can readily sway the policy prescription from near inaction to immediate, aggressive, and costly intervention. Unfortunately, agreement ends there. Furthermore, much of the debate over the discount rate resembles ships passing in the night.   In order to advance understanding, improve analysis, and foster intelligent policy-making, we present a conceptual framework that clarifies a number of issues by unpacking the concept of discounting into qualitatively distinct components.

The central distinction is between features of individuals' utility functions and of the social welfare function. Utility functions must be specified both for purposes of what we call predictive discounting – that is, to analyze models in which individuals' maximizing behavior determines outcomes – and for purposes of ultimate social welfare assessments since social welfare is taken to be a function of individuals' utilities. For this latter purpose, which includes what we label evaluative discounting, the social welfare function must be specified as well.

To further complicate matters, each of these domains – specification of utility functions for predictive discounting and of the social welfare function for evaluative discounting – involves two dimensions. First, discounting depends on the curvature of the pertinent functions, how utility depends on consumption, $U(c)$, in the former case and how social welfare in a given generation depends on utility, $W(U)$, in the latter case.   As we explain and demonstrate through examination of some of the leading models and writings on climate change, the common use of a single reduced form, $Z(c)$, often leads to conflating these distinct phenomena, one (utility) empirical and the other (social welfare) ethical.

Second, each domain can also involve pure time discounting: for predictive discounting, the rate at which individuals discount their own future utility, which we denote by θ; and for social evaluation, the rate at which an outside evaluator discounts the contribution of a generation's utility to social welfare, which we denote by δ. Again, the former, involving individuals' utility functions, is empirical, and the later, regarding the social welfare function, is ethical.

We examine some of the most notable contributions in the IAM literature and find that discussions frequently conflate the two domains, individual utility (empirical) and social welfare (ethical). Sometimes the result is ambiguity, and at other times, arguments are affirmatively off base, such as in attempting to offer ethical arguments over what are, in principle, empirical parameters (matters of fact) or suggesting that certain empirical facts call into question choices of ethical parameters.

The B.E. Journal of Economic Analysis & Policy, Vol. 10 [2010], Iss. 2 (Symposium), Art. 7

We advocate that consistent and sharp distinctions be drawn between these two domains (utility and social welfare) and between the two sources of discounting (curvature and pure time preference) that can arise in each. Ceasing altogether the use of the term discounting without further modifiers might be a good place to start. Additionally, to avoid modelers' imposition of ethical views on consumers of their work, often in ways that are opaque, we suggest more widespread use of techniques familiar in other realms: sensitivity analysis (here, to the choice of ethical parameters of the social welfare function) and direct presentation of the time series of outcomes of IAMs, to permit readers to perform their own assessments.

Conceptual clarification is an important goal of economists and other analysts attempting to inform policy-making. It should also aid future work unrelated to discounting issues per se by making IAMs – and, in particular, the bases for different conclusions – more transparent to other investigators. Finally, by focusing debate about components of "the discount rate" so that empirical questions are separately identified and then resolved using appropriate evidence, with ethical argument confined to and more sharply focused on determination of an appropriate social welfare function, disagreement may or may not subside, but at least it will be concentrated on the right questions and lead to proposed methods of deriving answers that are appropriate.

References

Acemoglu, Daron. 2009. *Introduction to Modern Economic Growth*. Princeton: Princeton University Press.

Altig, David, Alan J. Auerbach, Laurence J. Kotlikoff, Kent A. Smetters, and Jan Walliser. 2001. Simulating Fundamental Tax Reform. *American Economic Review* 91 (3):574-599.

Anthoff, David, and Richard S.J. Tol. 2009. The Impact of Climate Change on the Balanced Growth Equivalent: An Application of FUND, *Environmental and Resource Economics* 43 (3):351-367.

Anthoff, David, Richard S.J. Tol, and Gary W. Yohe. 2009. Discounting for Climate Change. *Economics: The Open-Access, Open-Assessment E-Journal* 3 (2009-24).

Arrow, Kenneth J. 1982. The Rate of Discount on Public Investments with Imperfect Capital Markets. In *Discounting for Time and Risk in Energy Policy*, edited by Robert. C. Lind. Washington DC: Resources for the Future.

————. 1999. Inter-Generational Equity and the Rate of Discount in Long-Term Social Investment. In *Contemporary Economic Issues, Economic Behavior and Design*, edited by Murat Sertel. New York: MacMillan Press.

Arrow, Kenneth, J., William R. Cline, Karl-Göran Mäler, Mohan Munasinghe, Ray Squitieri, and Joseph E. Stliglitz. 1996. Intertemporal Equity, Discounting, and Economic Efficiency. In *Climate Change 1995: Economic and Social Dimensions of Climate Change, Contribution of Working Group III to the Second Assessment Report of the Intergovernmental Panel on Climate Change*, edited by James P. Bruce, Hoesung Lee, and Erik F. Haites. Cambridge: Cambridge University Press.

Atkinson, Anthony B. 1973. How Progressive Should Income Tax Be? In *Essays in Modern Economics*, edited by Michael Parkin. London: Longman.

Barro, Robert J., and Xavier Sala-I-Martin. 1995. *Economic Growth*. New York: McGraw-Hill.

Council for Regulatory Environmental Modeling, EPA. 2009. *Guidance on the Development, Evaluation, and Application of Environmental Models*, edited by EPA. Washington, DC.

Dasgupta, Partha. 2007. Commentary: The Stern Review's Economics of Climate Change. *National Institute Economic Review* 199 (1):4-7.

————. 2008. Discounting Climate Change. *Journal of Risk and Uncertainty* 37 (2): 141-169.

Fullerton, Don, and Diane Rogers. 1993. *Who Bears the Lifetime Tax Burden?* Washington DC: Brookings Institution.

Gerlagh, Reyer, and Bob van der Zwaan. 2004. A Sensitivity Analysis of Timing and Costs of Greenhouse Gas Emission Reductions. *Climatic Change* 65 (1):39-71.

Heal, Geoffrey. 2005. Intertemporal Welfare Economics and the Environment. In *The Handbook of Environmental Economics*, edited by Karl-Göran Mäler and Jeffery R. Vincent. Amsterdam: Elsevier.

————. 2009. The Economics of Climate Change: A Post-Stern Perspective. *Climatic Change* 96:275-297.

Heckman, James, Jeffrey Smith, and Nancy Clements. 1997. Making the Most Out Of Programme Evaluations and Social Experiments: Accounting for Heterogeneity in Programme Impacts. *Review of Economic Studies* 64 (4):487-535.

Hope, Chris. 2003. The Marginal Impacts of CO_2, CH_4 and SF_6 Emissions. *Judge Institute of Management Research Papers, Number 10/2003*. Cambridge: Cambridge University, Judge Institute of Management.

————. 2005. Integrated Assessment Models. In *Climate-Change Policy*, edited by Dieter Helm. Oxford: Oxford University Press.

The B.E. Journal of Economic Analysis & Policy, Vol. 10 [2010], Iss. 2 (Symposium), Art. 7

————. 2006. The Marginal Impact of CO_2 from PAGE2002: An Integrated Assessment Model Incorporating the IPCC's Five Reasons for Concern. *The Integrated Assessment Journal* 6 (1):19-56.

————. 2008. Discount Rates, Equity Weights and the Social Cost of Carbon. *Energy Economics* 30 (3):1011-1019.

Intergovernmental Panel on Climate Change. 2007. *Climate Change 2007 – Impacts, Adaptation and Vulnerability, Contribution of Working Group II to the Fourth Assessment Report of the IPCC*, edited by UNFCCC. Cambridge, Cambridge University Press.

Kaplow, Louis. 2004. On the (Ir)Relevance of Distribution and Labor Supply Distortion to Government Policy. *Journal of Economic Perspectives* 18 (4):159-175.

————. 2006. Optimal Control of Externalities in the Presence of an Income Tax. *NBER Working Paper 12339*. Cambridge, MA: NBER.

————. 2007. Discounting Dollars, Discounting Lives: Intergenerational Distributive Justice and Efficiency. *University of Chicago Law Review* 74 (1):79-118.

————. 2008. *The Theory of Taxation and Public Economics*. Princeton: Princeton University Press.

————. 2010. Concavity of Utility, Concavity of Welfare, and Redistribution of Income. *International Tax and Public Finance* 17:25-42.

Kaplow, Louis, and Steven Shavell. 2001. Any Non-welfarist Method of Policy Assessment Violates the Pareto Principle. *Journal of Political Economy* 109 (2):281-286.

Koopmans, Tjalling C. 1965. *On the Concept of Optimal Economic Growth, The Econometric Approach to Development Planning*. Amsterdam: North-Holland.

Lind, Robert C, Kenneth J. Arrow, Gordon R. Corey, Partha Dasgupta, Amartya Sen, Thomas Stauffer, Joseph E. Stiglitz, J.A. Stockfisch, and Robert Wilson, eds. 1982. *Discounting for Time and Risk in Energy Policy*. Washington DC: Resources for the Future.

Nordhaus, William D. 1994. *Managing the Global Commons: The Economics of Climate Change*. Cambridge, MA: MIT Press.

————. 1997. Discounting in Economics and Climate Change: An Editorial Comment. *Climatic Change* 37:315-328.

————. 1999. Discounting and Public Policies that Affect the Distant Future. In *Discounting and Intergenerational Equity*, edited by Paul R. Portney and John P. Weyant. Washington DC: Resources for the Future

————. 2007. A Review of the Stern Review on the Economics of Climate Change. *Journal of Economic Literature* 45 (3):686-702.

————. 2008. *A Question of Balance*. New Haven, CT: Yale University Press.

Paltsev, Sergey, John Reilly, Henry D. Jacoby, and Jennifer F. Morris. 2009. The Cost of Climate Policy in the United States. Report No. 173, in *Joint Program for the Science and Policy of Global Change*. Cambridge, MA: MIT Press.

Paltsev, Sergey, John Reilly, Henry D. Jacoby, Richard S. Eckhaus, James McFarland, Marcus Sarofim, Malcolm O. Asadoorian, and Mustafa Babiker. 2005. The MIT Emissions Prediction and Policy Analysis (EPPA) Model: Version 4. Report No. 125, in *Joint Program for the Science and Policy of Global Change*. Cambridge, MA: MIT Press.

Parfit, Derek. 1984. *Reasons and Persons*. New York: Oxford University Press.

Saltelli, Andrea, Karen Chan, and E.Marian Scott. 2000. *Sensitivity Analysis*. Chichester: John Wiley & Sons, LTD.

Samida, Dexter, and David Weisbach. 2007. Paretian Intergenerational Discounting. *University of Chicago Law Review* 74 (1):145-170.

Sen, Amartya, and Bernard Williams, eds. 1982. *Utilitarianism and Beyond*. Cambridge: Cambridge University Press.

Smith, Joel, Hans-Joachim Schellnhuber, and M. Monirul Qader Mirza. 2001. Vulnerability to Climate Change and Reasons for Concern: A Synthesis. In *Climate Change 2001: Impacts, Adaptation and Vulnerability, Intergovernmental Panel on Climate Change*. Cambridge: Cambridge University Press.

Stern, Nicholas. 2007. *The Economics of Climate Change, The Stern Review*. Cambridge: Cambridge University Press.

———. 2008. The Economics of Climate Change. *American Economic Review, Papers and Proceeding* 98 (2):1-37.

Stiglitz, Joseph E. 1982. The Rate of Discount for Benefit-Cost Analysis and the Theory of the Second Best. In *Discounting for Time and Risk in Energy Policy*, edited by Robert C. Lind. Washington DC: Resources for the Future.

Tol, Richard S.J. 1999. The Marginal Costs of Greenhouse Gas Emissions. *Energy Journal* 20 (1):61.

———. 2007. The Double Trade-off between Adaptation and Mitigation for Sea Level Rise: An Application of FUND. *Mitigation and Adaptation Strategies for Global Change* 12:741-753.

Toth, Ferenc L. 1995. Discounting in Integrated Assessments of Climate Change. *Energy Policy* 23 (4-5):403-409.

Weitzman, Martin L. 2007. A Review of the Stern Review on the Economics of Climate Change. *Journal of Economic Literature* 45 (3):703-724.

The B.E. Journal of Economic Analysis & Policy, Vol. 10 [2010], Iss. 2 (Symposium), Art. 7

Weyant, John P., Ogunlade Davidson, Hadi Dowlatabadi, James A. Edmonds, Michael Grubb, E.A. Parson, Richard G. Richels, Jan Rotmans, P.R. Shukla, Richard S.J. Tol, William R. Cline, and Samuel Fankhauser. 1996. Integrated Assessment of Climate Change: An Overview and Comparison of Approaches and Results. In *Climate Change 1995: Economic and Social Dimensions – Contribution of Working Group III to the Second Assessment Report of the Intergovernmental Panel on Climate Change*, edited by James P. Bruce, Hoesung Lee and Erik F. Haites. Cambridge: Cambridge University Press.

Weyant, John P., and J. N. Hill. 1999. Introduction and Overview, The Costs of the Kyoto Protocol: A Multi-Model Evaluation. *Energy Journal* 1999:R7-44.

[4]

The B.E. Journal of Economic Analysis & Policy

Symposium

| Volume 10, Issue 2 | 2010 | Article 8 |

DISTRIBUTIONAL ASPECTS OF ENERGY AND CLIMATE POLICY

Comment on "The Social Evaluation of Intergenerational Policies and Its Application to Integrated Assessment Models of Climate Change"

Martin L. Weitzman*

*Harvard University, mweitzman@harvard.edu

Recommended Citation
Martin L. Weitzman (2010) "Comment on "The Social Evaluation of Intergenerational Policies and Its Application to Integrated Assessment Models of Climate Change"," *The B.E. Journal of Economic Analysis & Policy*: Vol. 10: Iss. 2 (Symposium), Article 8.
Available at: http://www.bepress.com/bejeap/vol10/iss2/art8

Weitzman: Comment on "The Social Evaluation of Intergenerational Policies"

This paper by Weisbach, Moyer and Kaplow (2010) makes an interesting and oft-overlooked point about discounting in Integrated Assessment Models (hereafter IAMs) applied to climate change. There are really two kinds of discounting occurring under one umbrella. The first, which the authors call *predictive* discounting, concerns the discounting that the representative agent within the model uses to make decisions about inter-temporal tradeoffs like saving and investing. In principle at least, this kind of discounting is empirically measurable. The second kind of discounting, which the authors call *evaluative* discounting, concerns the ethical comparison of utility trajectories by an outside observer with a social welfare function. This kind of discounting is subjective and probably not measurable by observed behavior. In the IAM literature, according to the authors, these two meanings of discounting are often confounded, or even confused.

The distinction is useful and offers some insights, but I wonder if the issue is not being stretched a bit too far. Presumably the IAM creators and users cited in the paper [like Nordhaus (2008), Stern (2007), and so forth], have something in mind that they consider to be relatively coherent and consistent. In the case of Nordhaus, I think what he has in mind (notwithstanding the confusing quotes attributable to him in the paper) is that the representative agent and the ethical observer ought to be considered one and the same, or at least two different sides of the same coin. And then, Nordhaus uses the typically-high discount rate of the predictive interpretation to pin down some of the relevant utility or welfare parameter values as inferred by market behavior. In the case of Stern, I believe that he also has in mind (again, notwithstanding the confusing quotes attributable to him in the paper) that the representative agent and the ethical observer ought to be considered one and the same, but in this case the dominant personality is the evaluative observer-planner whose typically-low discount rate and parameter values are inferred from ethical considerations concerning the welfare of future generations. I could be wrong, but this is what I think these two prototype interpreters have in mind.

When one introduces an extra layer of complexity into a model (here the distinction between predictive and evaluative discounting), one also introduces an extra degree of freedom into the model. Such a layered model can then make more nuanced distinctions between related but distinct concepts. However, all models are practical compromises, and then the question becomes whether the extra complexity is worth it. The authors of this paper are making a valuable point: the current batch of IAMs is inadequately unclear about what is the interpretation of the distinction between the representative agent within the model and the ethical observer outside the model. But the proposed solution introduces a lot more complexity into the situation. Allowing readers to insert their own judgments about ethical parameters, while the IAM modeler specifies the

The B.E. Journal of Economic Analysis & Policy, Vol. 10 [2010], Iss. 2 (Symposium), Art. 8

behavioral parameters, may in practice turn out to be more unwieldy than working with the current Nordhaus-Stern distinction, as I understand it.

So, this paper is prompting us to think more clearly about an important distinction between normative evaluation (by people outside the model) and positive description (of people within the model) that tends to get blurred in most climate-change IAMs. This is all to the good. However, my own take is that the practical import of such a separation remains to be worked out in more operational detail. Although it doesn't always happen this way, hopefully time will tell.

References

Nordhaus, William D. (2008), *A Question of Balance*, New Haven, CT: Yale University Press.

Stern, Nicholas (2007), *The Economics of Climate Change, The Stern Review*, Cambridge, UK: Cambridge University Press.

Weisbach, David A., Elisabeth Moyer, and Louis Kaplow (2010), "The Social Evaluation of Intergenerational Policies and Its Application to Integrated Assessment Models of Climate Change", *The B.E. Journal of Economic Analysis & Policy*, Vol. 10, No. 2.

[5]

The B.E. Journal of Economic Analysis & Policy

Symposium

Volume 10, Issue 2	2010	Article 15

DISTRIBUTIONAL ASPECTS OF ENERGY AND CLIMATE POLICY

Analytical General Equilibrium Effects of Energy Policy on Output and Factor Prices

Don Fullerton*   Garth Heutel[†]

*University of Illinois at Urbana-Champaign, dfullert@illinois.edu
[†]The University of North Carolina at Greensboro, gaheutel@uncg.edu

Recommended Citation
Don Fullerton and Garth Heutel (2010) "Analytical General Equilibrium Effects of Energy Policy on Output and Factor Prices," *The B.E. Journal of Economic Analysis & Policy*: Vol. 10: Iss. 2 (Symposium), Article 15.
Available at: http://www.bepress.com/bejeap/vol10/iss2/art15

Analytical General Equilibrium Effects of Energy Policy on Output and Factor Prices*

Don Fullerton and Garth Heutel

Abstract

Using an analytical general equilibrium model, we find solutions for the effect of energy policy on factor prices as well as output prices. We calibrate the model to the U.S. economy, and we consider a tax on carbon dioxide. By looking at expenditure and income patterns across household groups, we quantify the uses-side and sources-side incidence of the tax. When households are categorized either by annual income or by total annual consumption as a proxy for permanent income, the uses-side incidence is regressive. This result is robust to sensitivity analysis over various parameter values. The sources-side incidence can be progressive, U-shaped, or regressive. Results on the sources side are sensitive to parameter values.

KEYWORDS: incidence, climate policy, relative burdens, uses-side, sources-side

---

*This paper was prepared for a conference on January 20-21, 2010, organized by the University of Chicago, Resources for the Future, and the University of Illinois. For comments, we thank Sam Kortum, Gilbert Metcalf, Jacob Vigdor, and anonymous referees. We also thank Dan Karney and Matt Trombley for valuable research assistance.

Fullerton and Heutel: Effects of Energy Policy

Energy is an integral input to nearly all aspects of economic life. Energy policies, especially policies aimed at curbing greenhouse gas emissions associated with energy consumption, thus have sizable effects on nearly all participants in our economy. The distribution of these effects, both costs and benefits, across participants is an important consideration of policy design.

The incidence of the costs of energy or climate policy manifests itself in at least two major ways. First, policy affects the "uses side" of income, through product prices. A carbon tax may disproportionately increase the price of gasoline and electricity, two goods that represent a higher share of expenditure for poorer households. The uses side incidence is then regressive. Second, policy affects the "sources side" of income, through factor prices. A carbon tax may be more burdensome to capital-intensive industries and disproportionately reduce the return to capital. If so, and if capital provides a higher share of income for richer households, then the sources side incidence may be progressive.

Many studies of the distributional impacts of energy policy focus on the uses side only, through a partial equilibrium approach. The purpose of this paper is to analyze both the uses side and the sources side incidence of domestic climate policy using an analytical general equilibrium model, highlighting conceptual issues by showing the general effects of each parameter on each result.[1]

Our model is based on the standard Harberger (1962) tax incidence model, with two factors of production (labor and capital) and two sectors of production (a "dirty" or polluting sector, and a "clean" sector). We add pollution, modeled as a third input to production in the dirty sector. In earlier papers, we show analytically how output prices and the returns to capital and labor are affected by changes in several types of pollution policy, including a pollution tax, tradable permits, performance standards, or technology mandates.

In this paper, we quantify those analytical results numerically. We calibrate the model to the US economy, and we distinguish households using expenditure and income data from the 2008 Consumer Expenditure Survey (CEX), supplemented by capital income data from the 2007 Survey of Consumer Finances (SCF). We then consider effects of carbon policy. We solve for the impacts on the prices of carbon-intensive goods relative to clean goods, and on the wage and the capital rental rate. We then apply these price changes to the households in our data to calculate the burdens across income groups and regions. In this paper, we find distributional effects on the uses side (commodity price changes) and sources side (factor price changes). We do not calculate effects through the use of the revenues by government, either for rebates to households or for the indexing of government transfers (as in Rausch et al. 2010).

---

[1] Besides these two effects, Fullerton (2009) lists and discusses four other distributional effects of environmental policy not considered here: (3) scarcity rents, (4) transition effects, (5) land or stock price capitalization, and (6) distribution of the benefits of environmental protection.

The B.E. Journal of Economic Analysis & Policy, Vol. 10 [2010], Iss. 2 (Symposium), Art. 15

When families are categorized either by annual income or by total annual consumption, the uses-side incidence of a carbon tax is regressive. Lower-income households spend a higher fraction of their expenditures on carbon-intensive goods than do higher-income households. This result is robust and corroborates many other papers (Burtraw, et al., 2009, or Hassett, et al., 2009). When categorized by region, the uses-side incidence is again robust; regions that spend more than average on carbon-intensive goods bear a disproportionately high burden (especially the Midwest and the South).

On the sources side, however, incidence results are sensitive to chosen parameter values. In particular, the regressivity or progressivity on the sources side depends on the elasticities of substitution in production for polluting industries. These elasticities have not been estimated, and thus we present incidence calculations for several alternative values. A partial equilibrium analysis that focuses only on output prices might understate or overstate the extent to which carbon policy is regressive, by neglecting general equilibrium effects on factor prices.

A disadvantage of our methodology lies in its aggregation to only two sectors and two or three factors of production. A more disaggregated model could be more realistic and could be used to calculate more specific effects on prices of each different good and factor. However, more disaggregation and other features would require a numerical solution. For us, the aggregation and other simplifications provide the advantage that we can derive analytical solutions for general equilibrium effects on both output and factor prices that hold for any parameters in the model, not just for particular numerical implementations. Our model can be interpreted as a complement to a more detailed computational general equilibrium (CGE) model, to examine more closely what drives certain results. As a referee put it, we provide a "model of the model."

In a special case where the two sectors have the same factor intensity and the same substitution parameters, we show that carbon pricing has no effect on the wage-rental ratio. If so, then analysts could focus on product prices alone. With other values for these unknown parameters, however, changes in the wage/rental ratio can offset or exacerbate regressivity on the uses side. We conclude that these production parameters need to be estimated, before these effects on the sources side can be dismissed.

The next section presents the model and analytic solutions. Section 2 describes the calibration, and section 3 presents the simulation results.

I. Model

This model is based on an earlier one from Fullerton and Heutel (2007), which itself is an extension of Harberger (1962). The model is solved by log-linearizing about the initial equilibrium, so it is valid for small changes in the tax rate. We briefly summarize the model here.

The economy consists of two sectors producing two different final goods. One sector, X, uses only capital K_X and labor L_X as inputs; it is labeled the clean sector. The dirty sector, Y, uses both capital and labor (K_Y and L_Y) and a third input, pollution (Z). Production functions have constant returns to scale:

$$X = X(K_X, L_X)$$
$$Y = Y(K_Y, L_Y, Z).$$

Total capital and labor resources are fixed:

$$K_X + K_Y = \overline{K},$$
$$L_X + L_Y = \overline{L}.$$

By totally differentiating these two constraints, we get:

$$\hat{K}_X \lambda_{KX} + \hat{K}_Y \lambda_{KY} = 0, \tag{1}$$
$$\hat{L}_X \lambda_{LX} + \hat{L}_Y \lambda_{LY} = 0, \tag{2}$$

where variables with a hat denote a proportional change (e.g. $\hat{K}_X = dK_X/K_X$), and where λ_{ij} denotes sector j's share of factor i (e.g. $\lambda_{KX} = K_X/\overline{K}$).

Producers of the clean good X face a rental price for capital (r) and a wage price for labor (w). Their factor demand choices are defined by their elasticity of substitution in production, σ_X:

$$\hat{K}_X - \hat{L}_X = \sigma_X(\hat{w} - \hat{r}). \tag{3}$$

Producers of the dirty good Y face prices for all three of their inputs, including a tax or other price on emissions τ_Z. Their factor demand choices can be defined in terms of Allen elasticities of substitution between their inputs, e_{ij}, and revenue shares of inputs (e.g., $\theta_{YK} = rK_Y/p_Y Y$). These relationships follow Mieszkowski (1972) and Allen (1938):

$$\hat{K}_Y - \hat{Z} = \theta_{YK}(e_{KK} - e_{ZK})\,\hat{r} + \theta_{YL}(e_{KL} - e_{ZL})\,\hat{w} + \theta_{YZ}(e_{KZ} - e_{ZZ})\,\hat{\tau}_Z, \tag{4}$$
$$\hat{L}_Y - \hat{Z} = \theta_{YK}(e_{LK} - e_{ZK})\,\hat{r} + \theta_{YL}(e_{LL} - e_{ZL})\,\hat{w} + \theta_{YZ}(e_{LZ} - e_{ZZ})\,\hat{\tau}_Z. \tag{5}$$

We assume perfect competition and constant returns to scale in production.[2] These yield zero profit conditions that can be differentiated to get:

$$\hat{p}_X + \hat{X} = \theta_{XK}(\hat{r} + \hat{K}_X) + \theta_{XL}(\hat{w} + \hat{L}_X), \tag{6}$$
$$\hat{p}_Y + \hat{Y} = \theta_{YK}(\hat{r} + \hat{K}_Y) + \theta_{YL}(\hat{w} + \hat{L}_Y) + \theta_{YZ}(\hat{Z} + \hat{\tau}_Z). \tag{7}$$

---

[2] These assumptions may be questionable, especially for a dirty industry that is composed in large part by regulated electric utilities. For example, an emissions tax may not fully be passed through to ratepayers if it reduces infra-marginal rents on base-load generating units. In the conclusion, we mention some extensions to the model that could incorporate these concerns.

The B.E. Journal of Economic Analysis & Policy, Vol. 10 [2010], Iss. 2 (Symposium), Art. 15

Totally differentiating each sector's production function and using the assumption of perfect competition yields:

$$\hat{X} = \theta_{XK}\hat{K}_X + \theta_{XL}\hat{L}_X, \tag{8}$$

$$\hat{Y} = \theta_{YK}\hat{K}_Y + \theta_{YL}\hat{L}_Y + \theta_{YZ}\hat{Z}. \tag{9}$$

Lastly, we model consumer preferences for the two goods with the elasticity of substitution in utility, σ_u:

$$\hat{X} - \hat{Y} = \sigma_u(\hat{p}_Y - \hat{p}_X). \tag{10}$$

These ten equations constitute the model. Because the model has eleven unknowns, we choose good X as numeraire, setting $\hat{p}_X = 0$. Then, the linearized system of equations can be solved to consider how a small exogenous change in the pollution tax τ_Z affects factor prices w and r and output prices, given by p_Y. The choice of normalization means that all price changes are relative to the price of X. Thus, if $\hat{p}_Y > 0$, the price of good Y increases relative to the price of good X, so consumers who spend more than average on good Y are burdened relatively more than are other consumers on the uses side. Furthermore, the normalization implies that $\hat{w}$ and $\hat{r}$ are always of opposite sign (subtract equation (8) from equation (6)). Sector X has only two inputs, so if one input price rises then the other must fall for those firms to break even, with no change in output price. Yet, this does not imply that owners of one factor will gain and owners of the other will lose. Rather, if $\hat{w} > 0$ and $\hat{r} < 0$, it means that the burden on capital is proportionally greater than capital's share in national income. As in Harberger (1962), we assume that pollution tax revenue is used to purchase the two goods in the same proportion as the consumer does, so that tax revenue reallocation has no impact on relative prices.

The model's solution for output and factor prices is presented below.[3] See our earlier paper for the steps to derive this solution (Fullerton and Heutel 2007).

$$\hat{p}_Y = \frac{(\theta_{YL}\theta_{XK} - \theta_{YK}\theta_{XL})}{D}[A(e_{ZZ} - e_{KZ}) - B(e_{ZZ} - e_{LZ}) + (\gamma_K - \gamma_L)\sigma_u]\hat{t}_Z$$
$$\quad + \theta_{YZ}\hat{t}_Z$$

$$\hat{w} = \frac{\theta_{XK}\theta_{YZ}}{D}[A(e_{ZZ} - e_{KZ}) - B(e_{ZZ} - e_{LZ}) + (\gamma_K - \gamma_L)\sigma_u]\hat{t}_Z$$

$$\hat{r} = \frac{\theta_{XL}\theta_{YZ}}{D}[A(e_{KZ} - e_{ZZ}) - B(e_{LZ} - e_{ZZ}) - (\gamma_K - \gamma_L)\sigma_u]\hat{t}_Z$$

These equations use additional definitions for convenience. Let $\gamma_K \equiv \lambda_{KY}/\lambda_{KX} = K_Y/K_X$ and $\gamma_L \equiv \lambda_{LY}/\lambda_{LX} = L_Y/L_X$, $A \equiv \gamma_K\gamma_L + \gamma_L\theta_{YK} + \gamma_K(\theta_{YL} + \theta_{YZ})$, $B \equiv \gamma_K\gamma_L + \gamma_K\theta_{YL}$

---

[3] We omit expressions for the remaining seven endogenous variables, including the change in pollution ($\hat{Z}$), since our focus here is on incidence through price changes.

4

Fullerton and Heutel: Effects of Energy Policy

$+ \gamma_L(\theta_{YK} + \theta_{YZ})$, and $D \equiv (\theta_{XK}\gamma_K + \theta_{XL}\gamma_L + 1)\sigma_X + A[\theta_{XK}\theta_{YL}(e_{KL}-e_{LZ})-\theta_{XL}\theta_{YK}(e_{KK}-e_{KZ})] - B[\theta_{XK}\theta_{YL}(e_{LL}-e_{LZ})-\theta_{XL}\theta_{YK}(e_{KL}-e_{KZ})] - (\gamma_K - \gamma_L)\sigma_u(\theta_{XK}\theta_{YL} - \theta_{XL}\theta_{YK})$.

We briefly identify and interpret the effects present in these rather complex equations. In the equations for the factor price changes, $\hat{w}$ and $\hat{r}$, the last term in the bracket, $(\gamma_K - \gamma_L)\sigma_u$, represents the "output effect." The expression $(\gamma_K - \gamma_L)$ is positive whenever the dirty sector is capital-intensive. If so, and assuming the denominator is positive ($D > 0$), then an increase in the pollution tax ($\hat{\tau}_Z > 0$) will tend through this term to decrease the return to capital r relative to the wage w. The extent to which capital is burdened depends both on $(\gamma_K - \gamma_L)$, the degree of capital intensity, and σ_u, the consumer's ability to substitute away from the taxed sector's output.

The first two terms in the bracket of these equations represent "substitution effects." The signs of these terms depend on the values of the Allen elasticities e_{ij}. In the case of equal factor intensities where the output effect disappears, it can be shown that the substitution effect burdens capital more than it burdens labor whenever $e_{LZ} > e_{KZ}$, that is, whenever labor is a better substitute for pollution than is capital. When the price of emissions rises, and a firm wants to reduce emissions, it may do so and retain the same output level by altering its labor and capital inputs. If it increases labor more than it increases capital, then we say that labor is a better substitute for pollution than is capital. For example, a firm may switch from operating capital machinery that creates pollution and toward using more relatively clean labor inputs.

In the case with equal factor intensities ($\gamma_K = \gamma_L$) *and* equal cross-price terms ($e_{LZ} = e_{KZ}$), it can further be shown that $\hat{w} = \hat{r} = 0$. In this knife's edge case, the sources side can be ignored (as in Burtraw et al. 2009, or Hassett et al. 2009). We look below at actual parameters where the sources side effects may offset or exacerbate regressive effects on the uses side.

In the equation for $\hat{p}_Y$, the final term is $\theta_{YZ}\hat{\tau}_Z$. This term represents a "direct" effect from an increased pollution tax: the increased cost of the pollution input is passed into the output price in proportion to pollution's share in production, θ_{YZ}. The rest of the expression represents all of the general equilibrium effects, or "indirect" effects, which include output and substitution effects described above.[4]

II. Calibration

We now calibrate this model to the US economy in a way that allows us to consider a tax on carbon dioxide (CO_2) emissions. Our model only has two

---

[4] The incidence of this tax in this model includes the efficiency cost of tax-induced changes in consumption bundles, but not the efficiency cost of tax distortions in factor markets, since total factor supplies are fixed.

The B.E. Journal of Economic Analysis & Policy, Vol. 10 [2010], Iss. 2 (Symposium), Art. 15

sectors, so we must decide which industries of the economy are aggregated under the dirty sector and which under the clean sector. Because CO_2 is emitted in the generation of electricity, an intermediate input used by all industries, no output is completely "clean" as is output X in this model. Instead, we choose to label as dirty industries those that emit the most CO_2 relative to their output.

For information on factor intensities by industry we use data from Jorgenson et al. (2008).[5] These data divide the US economy into 35 sectors (roughly corresponding to two-digit SIC codes). They present the value of capital and labor inputs for each sector through 2006. Most CO_2 emissions come from three industries: electricity generation (38.7%), transportation (30.6%), and manufacturing (23.3%).[6] Of manufacturing industries, the highest emitter of CO_2 is petroleum refining, both absolutely and as a fraction of the value of output.[7] We use our data to isolate petroleum refining, along with electricity and transportation. We include these in the dirty sector, and all remaining industries are aggregated to the clean sector.

These definitions give us total factor inputs of labor and capital in the clean and dirty sectors. One more datum is needed to determine the factor intensity parameters, and that is θ_{YZ}, the share of sector Y's output that derives from pollution. Since polluting industries do not pay an explicit price for CO_2 emissions, this parameter cannot be calibrated from data. Instead, we perform a back-of-the-envelope calculation based on estimates of the optimal price on CO_2 from prior papers. A price of $15 per metric ton of CO_2 is often recommended, and we use this price as a starting point (Hassett et al. 2009). The value of θ_{YZ} is 0.0723, based on this price and our definition of the dirty sector of the economy.[8]

That calibration and the data from Jorgenson et al. (2008) jointly determine the factor intensity parameters shown in Table 1. Without loss of generality we define a unit of each good as the amount that sells for $1 in the

---

[5] Available at http://www.economics.harvard.edu/faculty/jorgenson/.

[6] See http://www.epa.gov/climate/climatechange/emissions/downloads09/ExecutiveSummary.pdf, Table ES-2.

[7] See http://www.eia.doe.gov/oiaf/1605/ggrpt/pdf/industry_mecs.pdf. Table 1. Petroleum refineries emitted 277.6 million metric tons of CO_2 in 2005; the next highest manufacturing industry group was iron and steel mills with 126.0 million metric tons. As a fraction of output, the petroleum industry's emissions are 30% higher than the next highest industry (primary metals).

[8] Total U.S. GNP in 2008 is $14.3 trillion. Our definition of the dirty sector accounts for 6.7% of total factor income, or $0.954 trillion. Annual carbon dioxide emissions from the dirty sector total 4.589 billion metric tons. At $15 per metric ton, the value of these carbon emissions is $68.8 billion, or 7.23% of the value of the dirty sector. This calculation combines current emissions with the $15/ton price, and so it may overstate the initial level of spending on emissions, but we don't observe actual emissions with that tax. Instead we use this hypothetical initial equilibrium as a starting point from which to calculate the effects of a small change in tax.

initial equilibrium. The total factor income of the economy $(\overline{K} + \overline{L})$ is also normalized to one. Using these parameters, the clean sector represents about 93% of factor income of the economy. The dirty sector is relatively capital-intensive. Labor accounts for about 61% of total factor income.

Table 1: Base Case Factor Intensity Parameters	
$K_Y = 0.0375$	$L_Y = 0.0291$
$K_X = 0.3515$	$L_X = 0.5819$
$\lambda_{KY} = 0.0963$	$\lambda_{LY} = 0.0477$
$\lambda_{KX} = 0.9037$	$\lambda_{LX} = 0.9523$
$\theta_{YK} = 0.5220$	$\theta_{YL} = 0.4057$
$\theta_{XK} = 0.3765$	$\theta_{XL} = 0.6235$
$\theta_{YZ} = 0.0723$	

The elasticity of substitution in production for the clean sector, σ_X, is set to one. This value is consistent with estimates of the economy-wide elasticity, and since the clean sector is 93% of the economy, it is a decent approximation of the elasticity we seek. In the "base case", we also set the elasticity of substitution in consumption, σ_u, to one.[9]

The last set of parameters needed are the Allen elasticities of substitution in production for the dirty sector, e_{ij}. Only three of these can be set independently; the rest are determined by these three values and the factor intensities, using equations from Allen (1938). We use estimates of a translog KLEM model of a 35 sector US economy from Jin and Jorgenson (2010) to calculate these elasticities.[10] From this, we find $e_{KL} = 0.1$, $e_{KZ} = 0.2$, and $e_{LZ} = -0.1$. These suggest that capital is a slightly better substitute for pollution than is labor. We use these values in our base case, and we vary them to test the sensitivity of results.[11]

Our aggregated model gives us the change in input and output prices for any given policy change. We want to translate those aggregate price changes into effects on real people, to calculate the uses- and sources-side incidence of the policy. To do so, we gather data on the expenditure and income of households

---

[9] We can show that the price elasticity of demand for the dirty good is $-[\alpha + \sigma_u(1-\alpha)]$, where α is its expenditure share. With our base case $\sigma_u = 1$, this elasticity is -1. We vary this parameter in sensitivity analysis.

[10] To get a single set of parameters, we weight the estimated elasticities across the three dirty industries by the value of total output in each (electricity, transportation, and petroleum refining).

[11] For comparison, de Mooij and Bovenberg (1998) review data from Western European countries and find $e_{KL} = 0.5$, $e_{KZ} = 0.5$, and $e_{LZ} = 0.3$. Their elasticities are all higher than the ones calculated from Jin and Jorgenson (2010), but both sets of elasticities find that capital is a better substitute for pollution than is labor.

The B.E. Journal of Economic Analysis & Policy, Vol. 10 [2010], Iss. 2 (Symposium), Art. 15

with various demographic characteristics. For example, we divide all households into ten deciles by annual income. For each decile, we calculate the fraction of income spent on clean vs. dirty goods, and we calculate the fraction of income from capital vs. labor. We can then quantify the burden of this policy change on each group. A potential inconsistency is that we use a model with a representative consumer to get price changes and then use those price changes to explore implications of consumer heterogeneity in expenditure and income patterns. The required assumption is that the overall effect on factor and output prices with heterogeneous consumers is the same as in the aggregate model.

We use expenditure and income data from the 2008 Consumer Expenditure Survey (CEX) and the 2007 Survey of Consumer Finances (SCF).[12] The CEX data come from a representative sample of the U.S. population; the SCF oversamples rich households but includes sampling weights. These micro-level data provide much information on expenditures and income sources. We can define groups numerous ways, such as by annual income, race, and region of residence. For each group (say, the lowest income decile), we calculate from the CEX the annual average expenditure on fairly detailed categories, including foods of various types (beef, pork, etc.), housing (mortgage interest, property tax, rent, etc.), and clothing (mens, womens, footwear, etc.). The CEX provides information on the distribution of income sources, including income from wages and salaries, self-employment, and interest, dividends, rental income, and other property income. Yet the CEX's income data are limited, especially on capital income. In particular, capital gains are omitted. When aggregating total capital and wage income reported in the CEX, we find that capital income is less than 5% of total factor income, which indicates that much capital income is missing.

To supplement these capital income data we use the SCF, which provides much more complete capital income data (but no expenditure data, as in the CEX).[13] We impute capital income for each household in the CEX based on the distribution of capital income in both data sets. For instance, for the household in the 75th percentile of the capital income distribution in the CEX (with a reported capital income of about $230), we assign the value of capital income at the 75th percentile of the distribution in the SCF (about $2000). In effect, we assume that the underreporting of income in the CEX, though it gets the value of capital income wrong, preserves the household's place in the capital income distribution.

---

[12] The CEX is available at http://www.bls.gov/cex/, and the SCF is available at http://www.federalreserve.gov/pubs/oss/oss2/scfindex.html.

[13] We drop households in the CEX with negative reported total income (1.6% of observations). We drop those in the SCF with negative reported capital income (2.5% of those observations). The mean value of reported capital gains income in the SCF is $5358, while this category is missing in the CEX. The mean value of business income (including farms) in the SCF is $11316, while it is only $3252 in the CEX. Our use of the SCF resembles that of Metcalf et al. (2010).

Fullerton and Heutel: Effects of Energy Policy

Table 2 summarizes the distribution of income and expenditures by annual income decile. Columns two through four present the distribution of income between wage, capital, and transfer income. Wage and salary income are directly reported in the CEX. Capital income from the SCF is the sum of income from interest, dividends, capital gains, and farm and other business income (but not any capital income within retirement accounts).[14] Wage and capital income sum to less than 100% because of transfer income sources: Social Security, unemployment and workers' compensation, and other public assistance. We omit the category "other income," which accounts for less than 1% of total income.

(1) Annual Income Decile	(2) % of Income from Wages	(3) % of Income from Capital	(4) % of Income from Transfers	(5) Capital-Wage Ratio (%)	(6) Dirty Good Expenditure as % of Income	(7) Clean Good Expenditure as % of Income
All	69.1	24.6	6.3	35.6	6.6	58.7
1	35.8	5.7	58.5	16.0	47.4	361.0
2	33.9	4.1	62.1	12.0	20.3	141.9
3	55.1	6.5	38.4	11.8	16.7	116.5
4	68.1	7.4	24.5	10.9	13.5	97.3
5	79.9	7.8	12.2	9.8	11.1	84.0
6	83.4	8.8	7.8	10.6	9.6	74.8
7	86.6	9.1	4.3	10.5	8.3	68.0
8	86.8	10.6	2.6	12.2	7.2	62.9
9	84.9	13.2	1.9	15.6	5.9	58.1
10	53.5	45.6	0.9	85.3	2.5	32.6

Table 2: Sources and Uses of Income for each Annual Income Group

Overall, about 69% of consumer income is from wages, 25% from capital, and 6% from transfers. These fractions vary by income group. The fraction of income coming from transfers is declining over income deciles (with the exception that the lowest income decile has a slightly lower fraction than the next decile), and the fraction of income coming from capital is increasing (with the same lone exception). Column 5 presents the capital-wage income ratio for each group, excluding any income from transfers. This ratio is U-shaped, with a much higher value for the richest decile.

The fact that the lowest annual income decile has the second-highest capital-wage ratio (16%) indicates one major problem with using annual income to categorize families from rich to poor. The lowest decile includes a lot of

---

[14] The omission of capital income from retirement accounts understates total capital income. While the SCF includes retirement account withdrawals and balances, it does not show income.

The B.E. Journal of Economic Analysis & Policy, Vol. 10 [2010], Iss. 2 (Symposium), Art. 15

retired individuals who have no labor income and are living off their retirement savings. These individuals may not really be "poor" on a lifetime basis. In other words, though we may want to classify households by the *stock* of lifetime wealth, we instead are classifying them by a *flow* of annual income. If individuals smooth consumption over their lifetime, as pointed out by Poterba (1989), then total annual consumption might be a good proxy for lifetime income (or at least, a better proxy than is annual income). We investigate this alternative below.

The final two columns in Table 2 present a distribution of expenditures between the clean and dirty outputs.[15]   Each value is a ratio of that expenditure to annual income, not to total expenditures, so these two values do not add to 100% in each row. The poorest deciles spend more than they earn, and the richest deciles spend less than they earn.[16]

Our earlier distinction between clean and dirty *production* sectors does not present us with an immediate mapping into clean and dirty *consumption* goods. Some of the outputs of the industries defined as dirty are used as inputs to industries defined as clean. A complete analysis would account for these inputs, for example by using Input-Output matrices as in Hassett et al. (2009). Here, we simply assign final consumption goods into either a clean or dirty category. Four categories of expenditures (out of the 74 total) are labeled as dirty: electricity, natural gas, fuel oil and other fuels, and gasoline. These are the goods whose consumption directly involves the combustion of fossil fuels (save for electricity, some of which is generated by nuclear or renewable sources). This choice is justified by a more complete analysis considering the pass-through of costs through intermediate goods (Hassett et al. 2009). For a CO_2 tax of $15 per metric ton, they find that the prices of these four categories of goods increase by 8-13%, while no other category of goods sees a price increase of greater than 1%.[17]

Overall, in Table 2, about 7% of income goes toward these dirty goods, and about nine times as much goes toward clean goods. The pattern of expenditures for these annual income groups is smoother than is the pattern for

---

[15] Only 65.3% of total income is spent (see the top row of Table 2). This ratio is low, compared to the 85% ratio in data from the National Income and Product Accounts (NIPA) of the Bureau of Economic Analysis (BEA) at http://www.bea.gov/national/nipaweb/SelectTable.asp?Selected=Y (see Table 2.1). Using the CEX data alone, however, the overall ratio of expenditure to income is only 78.7%. We then add some imputed capital income from the SCF, which reduces the overall spending/income ratio from 78.7% to 65.3%.

[16] One reasonable approach would scale all household expenditures upward so that their sum is 85% of income as in the NIPA accounts, but we wish to avoid unnecessary manipulation of the data. A proportional scaling would not change our relative burden results in any case.

[17] The exception is air transportation, whose price increases by 1.86%. The CEX tables do not list expenditures on air transportation separately (they are lumped with public transportation).

Fullerton and Heutel: Effects of Energy Policy

income sources. Higher income households spend a lower fraction of their total income on dirty goods than do lower income households.

Table 3: Sources and Uses of Income for each Annual Expenditure Group						
(1) Expend. Decile	(2) Wage Income as % of Expend.	(3) Capital Income as % of Expend.	(4) Transfer Income as % of Expend.	(5) Capital- Wage Ratio (%)	(6) % of Expend. on Dirty Good	(7) % of Expend. on Clean Good
All	105.8	37.6	9.7	35.6	10.1	89.9
1	42.8	13.5	63.5	31.6	14.5	85.5
2	74.5	13.8	36.6	18.5	15.2	84.8
3	86.3	16.2	26.8	18.7	14.6	85.4
4	103.5	18.0	17.7	17.4	13.9	86.1
5	108.8	20.4	13.8	18.7	13.2	86.8
6	114.4	29.4	10.0	25.7	12.3	87.7
7	118.8	31.2	7.3	26.2	11.5	88.5
8	120.0	38.4	5.7	32.0	10.8	89.2
9	124.6	45.1	3.9	36.2	9.3	90.7
10	93.4	54.7	2.4	58.6	5.9	94.1

Because of the issues discussed earlier with measuring incidence across annual income groups, Table 3 presents the same decompositions across deciles defined by a different measure of "income", namely total annual expenditure (which serves as a proxy for lifetime income). Yet we do not have lifetime breakdowns of wages and capital income. Therefore, in Table 3, all annual income sources do not sum to this measure of income, while all annual expenditures do sum to this measure of income. On average, the sum of all sources of annual income is higher than total annual expenditure. In Table 3, the pattern of spending across clean and dirty goods is qualitatively the same as in Table 2; richer households have a lower ratio for expenditures on dirty goods. In fact, using consumption deciles rather than annual income decile reduces the variance in the fraction spent on the dirty good. The gap between the richest and poorest groups' percentage spent on the dirty good is 9 percentage points in Table 3, versus 45 percentage points in Table 2.

The implications of this phenomenon will be seen below in the simulation results. Briefly, when a CO_2 tax hike increases the relative price of the dirty good, then the tax hike appears less regressive when households are divided into annual consumption groups than when households are divided into annual income groups. This corroborates previous research on the uses-side incidence of energy policy (Hassett et al. 2009) and more generally of consumption taxes (Lyon and Schwab 1995).

The B.E. Journal of Economic Analysis & Policy, Vol. 10 [2010], Iss. 2 (Symposium), Art. 15

Capital income's share is monotonically increasing across the expenditure deciles, and transfer income's share declines across the expenditure deciles. The capital-wage ratio is high for the poorest group compared to the second group, but then shows a roughly increasing pattern through the remaining nine deciles. For all groups, the sum of annual income sources exceeds total spending.

III. Numerical Results

We consider the effects of doubling the CO_2 tax from \$15 to \$30 per ton, that is, a 100% increase in the tax rate ($\hat{\tau}_Z = 1$). The base case results for changes in goods prices and factor prices are presented in the first column of Table 4. Other columns present results from sensitivity analyses, discussed later. In all columns the price of the dirty good changes by more than 7%, while factor prices change by less than one percent. But these results do not imply that effects on the uses side outweigh effects on the sources side, because the 7% output price change applies to only the 6.6% of income spent on the dirty good, while a "small" factor price change may apply to more than half of a group's income. Later we will see that uses effects usually outweigh sources effects, but not always.

Table 4: Simulation Results: Effect on Factor and Output Prices (%)

Change in Price of:	(1) Base Case	(2) Capital a Better Substitute	(3) Labor a Better Substitute	(4) Low Substitution in Utility	(5) High Substitution in Utility
Dirty good, $\hat{p}_Y$	7.20	7.26	7.07	7.23	7.16
Wage rate, $\hat{w}$	0.0718	−0.067	0.35	0.00100	0.14
Return, $\hat{r}$	−0.12	0.11	−0.58	−0.00166	−0.24

In the base case, the change in the relative output price $\hat{p}_Y$ (0.0720) is very close to $\theta_{YZ}\hat{\tau}_Z$ (0.0723, see Table 1), which we called the "direct" effect from passing through the tax increase. The relative changes in the wage and the capital rental rate are small, but we expect them to be small. They come from doubling the price of an input that represents 7% of a sector, which itself comprises about 7% of the economy. The change in the capital rental rate $\hat{r}$ is negative, and the change in the wage $\hat{w}$ is positive, so capital bears a higher than proportional share of the burden of the tax increase. Using our base case parameters, capital is a better substitute for pollution than is labor ($e_{KZ} > e_{LZ}$), so the substitution effect pushes more of the burden onto labor. However, the dirty sector is capital-intensive, so the output effect pushes more of the burden on capital. Here, the output effect dominates the substitution effect.

We then use Table 2 to translate these price changes into relative uses-side and sources-side burdens for different annual income groups. For each income group, we first calculate $\hat{p}_Y$ times expenditures on the dirty good, plus $\hat{p}_X$ times expenditures spent on the clean good (all divided by the group's income). Because our numeraire used in solving the system sets $\hat{p}_X = 0$, these burdens will be positive for every group. Yet none of these results should imply anything about how much of the burden is on the uses side compared to the sources side; that comparison depends entirely on the choice of numeraire (or equivalently, on whether monetary policy accommodates the increase in output prices or forces the burden to be felt by falling factor prices). Since the choice of numeraire does not affect the real incidence of a tax, the discussion of burdens on the uses side should focus only on who spends relatively more on each good (not on how much of the burden is on the uses side). Similarly, the discussion of sources side should focus only on who earns relatively more from each factor.

For this reason, we normalize the calculated uses side burden for each group by subtracting from it a uses side calculation based on the entire sample. Those groups with a positive value see the ratio of their expenditures to income increase more than the average, and those groups with a negative value see their ratio of expenditures to income increase less than the average. The calculation is similar for the sources-side incidence: $\hat{w}$ times income from wages plus $\hat{r}$ times income from capital, all divided by total income, minus this ratio for the entire sample. Using this procedure, results do not depend on the choice of numeraire. We change the sign of the sources side calculation, however, so that those income groups whose income decreases more than the average have a positive "burden", while those groups whose income decreases less than the average have a negative burden (a relative gain). Finally, we calculate each group's normalized overall burden by summing the uses side and sources side burdens.

Table 5: Incidence with Base Case Parameters for Annual Income Deciles

Annual Income Decile	Relative Burden from Output Price Changes (%)	Relative Burden from Factor Price Changes (%)	Relative Overall Burden (%)
1	2.936	0.001	2.937
2	0.986	0.001	0.986
3	0.724	-0.012	0.713
4	0.496	-0.020	0.476
5	0.323	-0.028	0.295
6	0.216	-0.029	0.186
7	0.123	-0.031	0.092
8	0.045	-0.029	0.015
9	-0.051	-0.025	-0.076
10	-0.297	0.036	-0.261

The B.E. Journal of Economic Analysis & Policy, Vol. 10 [2010], Iss. 2 (Symposium), Art. 15

These incidence results are presented in Table 5. The pattern of uses-side burdens in the first column is clear: the highest income groups (deciles 9 and 10) suffer a smaller than average share of this burden. Their cost of goods decreases relative to the average, because they spend a lower than average fraction on the dirty good. With our choice of numeraire, the average increase in overall price is about 0.48% (a 7.2% increase in the price of the good that constitutes 6.6% of total annual income). Thus, Table 5 tells us that the highest income group's price increase under this normalization overall is only about 0.18%, whereas the lowest income group sees an overall price increase of about 3.4%. These results are consistent with those in Hassett et al. (2009), who examine the uses-side incidence of a CO_2 tax. They find that the relative burden is monotonically decreasing across the income deciles. Burtraw et al. (2009) find the same result of uses side regressivity for a cap-and-trade policy.

The sources-side burden in the second column of Table 5 is felt most by the highest and lowest income deciles; the positive burdens for the lowest deciles indicate that their incomes fall proportionally more than average. In the base case simulation, $\hat{w} > 0$ and $\hat{r} < 0$, so earning a higher fraction of income from capital tends to increase overall real burdens. Table 2 shows that the capital-wage income ratio is U-shaped across the ten deciles, and so the sources side burden in Table 5 is U-shaped. This effect is muted in the lowest decile, however, because of their high share of income from transfers. Because the sources side burdens are all small relative to the uses side burdens, the overall pattern in the last column mimics the regressive burdens on the uses side.

Table 6: Incidence with Base Case Parameters for Expenditure Deciles			
Annual Expenditure Decile	Relative Burden from Output Price Changes (%)	Relative Burden from Factor Price Changes (%)	Relative Overall Burden (%)
1	0.316	0.016	0.333
2	0.366	-0.006	0.360
3	0.319	-0.012	0.307
4	0.273	-0.022	0.251
5	0.218	-0.023	0.195
6	0.157	-0.016	0.141
7	0.099	-0.017	0.082
8	0.046	-0.009	0.036
9	-0.063	-0.005	-0.068
10	-0.303	0.029	-0.273

Table 6 presents the same calculations for income groups defined by annual consumption rather than by annual income. The uses-side incidence in the first column is again regressive, and the sources-side incidence in the second column is again U-shaped. And because the uses side burden again dominates the sources side, the overall burden is still regressive. When groups are defined by annual consumption, the uses-side incidence is *less* regressive than when defined by annual income. The lowest expenditure decile's relative price increase in Table 6 (0.32%) is smaller than the lowest annual income decile's price increase in Table 5 (2.94%). This pattern can be seen in Figure 1, which plots the relative uses-side burdens by income and expenditure deciles. This result occurs because the between-decile variance in the fraction of spending on the dirty good is lower across consumption deciles than across annual income deciles.

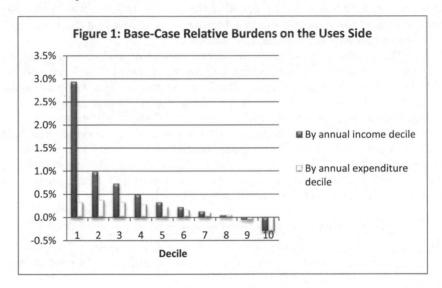

Figure 1: Base-Case Relative Burdens on the Uses Side

Sources-side incidence is U-shaped in Table 6 due to the U-shaped pattern of capital-wage income ratios in Table 3, because the wage is rising ($\hat{w} = +0.07\%$), while the return to capital is falling ($\hat{r} = -0.12\%$). Both sets of relative sources-side burdens are plotted in Figure 2.

Sensitivity Analysis

The results in Tables 5-6 are calculated using our base case parameter values. Some of these parameters are based on solid information about factor shares or consumption shares, but some of the parameters are known with little precision. Thus, sensitivity analysis is in order. In particular, the elasticities of

The B.E. Journal of Economic Analysis & Policy, Vol. 10 [2010], Iss. 2 (Symposium), Art. 15

substitution in production for the dirty sector have not been directly estimated.[18] Next, we present alternative incidence calculations for different sets of parameter values. The changes in prices under these alternative parameter values are presented in columns 2 through 5 of Table 4. In columns 2 and 3, all of the parameters are identical to their base case values except for the dirty sector substitution elasticities. In column 2, we set $e_{KL} = 0.1$, $e_{KZ} = 0.5$, and $e_{LZ} = -0.5$. In this column, capital is a much better substitute for pollution than is labor; in fact, labor is a complement for pollution rather than a substitute. As we expect, under these parameters, labor ends up relatively worse off with a pollution tax increase. The signs of the price changes in w and r switch from the base case. The second set of results (in column 3) are based on parameters where labor is a much better substitute for pollution than is capital: $e_{KL} = 0.1$, $e_{KZ} = -0.5$, and $e_{LZ} = 0.5$. Under these parameter values, capital bears a larger share of the tax burden than even in the base case, since the fact that labor is a substitute for pollution enables it to avoid more of the burden.

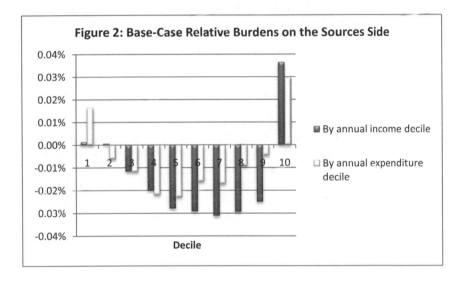

Figure 2: Base-Case Relative Burdens on the Sources Side

Table 7 presents the resulting incidence calculations across annual expenditure groups. Columns 2 and 3 present the relative burdens from the uses

---

[18] We calibrated elasticities for our model based on estimates in Jin and Jorgenson (2010) that were based on a somewhat different model, with more sectors, and where firms use labor, capital, and energy. Our dirty sector uses labor, capital, and pollution (which is not the same as energy, because firms can reduce their pollution per unit of energy).

Fullerton and Heutel: Effects of Energy Policy

side and sources side for the first set of alternative parameter values, where capital is a better substitute for pollution than is labor. The uses-side incidence results are not affected very much (relative to Table 6), even with this large change of production substitution elasticities. The households with the lowest expenditures tend to see higher than average increases in their costs, indicating a regressive uses-side incidence. However, under these parameter values the sources-side incidence results are starkly different from the U-shaped burdens associated with the base case parameters in Table 6 (where the reduction in r hurts capital owners in the highest and lowest income deciles). In column 3 of Table 7, the return to capital rises (as shown in column 2 of Table 4), so the richest and poorest deciles see their incomes rise relative to the average, while the middle deciles see their incomes fall relative to the average. Under these parameters, capital is a better substitute for pollution than is labor, so the pollution tax increase means that labor is made relatively worse off. The highest annual expenditure group has the highest capital/wage income ratio, and thus it gains the most on the sources side under this parameterization. Here, instead of offsetting the uses side, the sources side exacerbates the regressivity of the uses side.

	Capital a better substitute for pollution		Labor a better substitute for pollution	
(1) Annual Expenditure Decile	(2) Relative Output Price Burden (%)	(3) Relative Factor Price Burden (%)	(4) Relative Output Price Burden (%)	(5) Relative Factor Price Burden (%)
1	0.319	-0.016	0.311	0.081
2	0.369	0.005	0.360	-0.029
3	0.321	0.011	0.313	-0.056
4	0.276	0.020	0.268	-0.106
5	0.220	0.021	0.214	-0.111
6	0.158	0.015	0.154	-0.078
7	0.099	0.016	0.097	-0.083
8	0.046	0.009	0.045	-0.045
9	-0.064	0.004	-0.062	-0.023
10	-0.305	-0.027	-0.297	0.143

Table 7: Incidence for Expenditure Deciles, Sensitivity Analysis on Production Parameters

Columns 4 and 5 present incidence calculations under the parameter values that make labor a much better substitute for pollution than capital. Again, the uses-side incidence results are virtually no different than in the base case; the burden is regressive. However, the sources-side incidence results in column 5 are opposite to those in column 3 and are in the same direction as in the base case.

The B.E. Journal of Economic Analysis & Policy, Vol. 10 [2010], Iss. 2 (Symposium), Art. 15

Here, because the wage rate rises, the poorest and richest deciles have more burden relative to the average, and the middle deciles have less burden relative to the average. The highest income group has the most capital and is therefore burdened the most, so the sources-side incidence is more progressive than in the base case results. The degree of progressivity is higher than in the base case, since the magnitudes of the factor price changes are higher in this simulation than in the base case.

An additional sensitivity analysis we perform involves varying the elasticity of substitution in utility, σ_u. In the base case this value is one. The analytical solutions of the model show that the value of this parameter affects the strength of the output effect. As with the elasticities of substitution in production for the dirty sector, the true value of this parameter is not known. We choose two alternate values for σ_u: 0.5 and 1.5. These results are presented in columns 4 and 5 of Table 4. The substitution elasticities in production are kept at the base case values. When $\sigma_u = 0.5$, the relative price changes in w and r are very close to zero, indicating a proportionally-shared burden between labor and capital. When $\sigma_u = 1.5$, capital bears a higher share of the burden than in the base case. In all of these cases the dirty sector is capital-intensive, and so the output effect makes capital worse off. When $\sigma_u = 0.5$, the output effect burden on capital is small and completely offset by the substitution effect (which helps capital, since $e_{KZ} > e_{LZ}$). When $\sigma_u = 1.5$, the output effect is large and dominates the substitution effect, so capital bears relatively more of the burden of the CO_2 tax increase. The uses-side incidence $\hat{p}_Y$ does not vary much with σ_u.

Table 8 presents incidence calculations for the alternative parameter values of σ_u. This parameter does not affect the uses-side incidence, which is regressive for both alternate values of σ_u. It does, however, affect the sources-side incidence. When σ_u is low, as in columns 2 and 3 of Table 8, the output effect hurting capital is small and dominated by the substitution effect, so the burden on capital is roughly proportional (to three decimal places). On the other hand, when σ_u is high, the burden on capital owners is increased since the output effect dominates. Thus, the richest and poorest households bear relatively more of the burden.[19]

Middle-Aged Heads of Household

Annual income is a poor proxy for lifetime income. Annual consumption may be a better proxy, but even this case leaves us with only one year's capital-wage income ratio. This measure may fail to capture the desired long-term capital-wage income ratio for each permanent income group. A large part of the

---

[19] All of the findings from Table 7 and 8 hold for annual income deciles as well as annual expenditure deciles, though those results are not presented.

problem is that individuals have different income patterns at different stages of their lives. Retirees have low annual income but high a capital share, while college students have low annual income but high borrowing. Some evidence for this pattern appears in the CEX data. The ten annual income deciles have average ages that range from 45.3 years to 58.9 years, whereas the ten annual consumption deciles have average ages that range only from 47.2 years to 53.1 years.

	Low substitution in utility		High substitution in utility	
(1)	(2)	(3)	(4)	(5)
Annual Expenditure Decile	Relative Output Price Burden (%)	Relative Factor Price Burden (%)	Relative Output Price Burden (%)	Relative Factor Price Burden (%)
1	0.318	0.000	0.315	0.030
2	0.368	0.000	0.364	-0.013
3	0.320	0.000	0.317	-0.024
4	0.274	0.000	0.272	-0.044
5	0.219	0.000	0.217	-0.046
6	0.158	0.000	0.156	-0.032
7	0.099	0.000	0.098	-0.034
8	0.046	0.000	0.045	-0.018
9	-0.063	0.000	-0.063	-0.009
10	-0.304	0.000	-0.301	0.058

Table 8: Incidence for Expenditure Deciles, Sensitivity Analysis on Substitution in Utility

An alternative method of overcoming this life-cycle problem is to focus on only one age group for head of household. We choose households whose heads are 41-50 years old.[20] Table 9 summarizes the income sources and expenditure data across the ten annual expenditure deciles of these households. Overall, these households have a lower capital-wage ratio (25.1%) than do all households (35.6%, in Table 3). The fraction of expenditures on dirty goods (10.2%) is virtually identical to that for all households (10.1%, Table 3). Across expenditure deciles, the decreasing fraction of income from transfers is again seen. Here, though, this fraction drops to single-digit percentages by the second decile, and overall, transfers are only 2.6% rather than 9.7% of expenditure (Table 3). These 41-50 year old household heads are not receiving nearly as much Social Security retirement income as all other households in Table 3.

The big difference in Table 9 compared to Table 3 is that the capital-wage income ratio is clearly rising with income (in an almost monotonic fashion). In

---

[20] Another approach, which we do not pursue here, is to attempt to create a synthetic cohort of households using multiple years of the CEX, as in Jorgenson and Slesnick (2008).

The B.E. Journal of Economic Analysis & Policy, Vol. 10 [2010], Iss. 2 (Symposium), Art. 15

Table 3, it was U-shaped. While using expenditure deciles in Table 3 may capture permanent income, each decile still contains young and old with very different income sources. The 41-50 year olds in Table 9 may have income sources that better reflect their long run income sources.

Table 9: Sources and Uses of Income for each Annual Expenditure Group, Households with Heads aged 41-50 only						
(1) Annual Expend. Decile	(2) Wage Income as % of Expend.	(3) Capital Income as % of Expend.	(4) Transfer Income as % of Expend.	(5) Capital-Wage Ratio (%)	(6) % of Expend. on Dirty Good	(7) % of Expend. on Clean Good
All	125.5	31.5	2.6	25.1	10.2	89.8
1	94.9	7.7	26.7	8.1	15.2	84.8
2	125.2	11.3	9.9	9.0	15.4	84.6
3	131.8	12.6	8.0	9.6	14.6	85.4
4	137.2	22.3	3.3	16.2	14.0	86.0
5	141.3	16.9	2.5	12.0	13.2	86.8
6	140.3	29.7	2.0	21.2	12.1	87.9
7	133.1	35.4	1.5	26.6	11.7	88.3
8	142.7	34.2	1.4	23.9	10.8	89.2
9	131.4	31.3	0.8	23.8	8.9	91.1
10	101.5	44.3	0.5	43.6	5.9	94.1

Table 10 presents the incidence calculations for these 41-50 year old household heads. Columns 2 and 3 present results using the base case parameters, columns 4 and 5 are from the alternative substitution elasticity values where capital is a much better substitute for pollution than is labor, and columns 6 and 7 are from the alternative substitution elasticity values where labor is a much better substitute for pollution. As before, the uses-side burden is regressive and consistent across parameter values. In the base case, the sources-side incidence is progressive, in contrast to the U-shaped result from the base case for all households (Table 6). Under the alternate substitution parameters, the sources-side burden is regressive when capital is a better substitute for pollution, and progressive when labor is a better substitute for pollution.

Whereas the capital-wage ratio for all households in Table 3 varies from 0.185 to 0.586 (a factor of three), the ratio for 41-50 year olds in Table 9 varies from 0.081 to 0.436 (a factor of more than five). If capital income is an important lifetime source of income for the well-to-do, and if the return falls as much as 0.58% (column 3 of Table 4 and last column of Table 10), then the sources side could be particularly progressive. In fact, the sources-side burden here is sufficiently progressive that it comes close to offsetting the regressivity of the

20

uses side burden: the *overall* burden is much less regressive than in any other simulation. If the return rises 0.11% (column 2 of Table 4 and column 5 of Table 10), then the sources side could be regressive – exacerbating the regressive effects of carbon pricing on the uses side. We conclude that general equilibrium effects are potentially important.

Table 10: Incidence for Households with Heads aged 41-50 only						
	Base Case		Capital better substitute for pollution		Labor better substitute for pollution	
(1)	(2)	(3)	(4)	(5)	(6)	(7)
Annual Expend Decile	Relative Output Price Burden (%)	Relative Factor Price Burden (%)	Relative Output Price Burden (%)	Relative Factor Price Burden (%)	Relative Output Price Burden (%)	Relative Factor Price Burden (%)
1	0.358	-0.007	0.361	0.006	0.351	-0.031
2	0.374	-0.024	0.377	0.022	0.367	-0.116
3	0.314	-0.027	0.317	0.025	0.308	-0.132
4	0.274	-0.019	0.277	0.018	0.269	-0.094
5	0.211	-0.029	0.213	0.027	0.207	-0.140
6	0.135	-0.013	0.137	0.012	0.133	-0.062
7	0.106	-0.001	0.107	0.001	0.104	-0.003
8	0.041	-0.009	0.041	0.009	0.040	-0.045
9	-0.098	-0.004	-0.099	0.004	-0.096	-0.022
10	-0.312	0.033	-0.315	-0.030	-0.306	0.158

Regional Incidence

Incidence can be defined across groups defined in ways other than annual income or annual expenditure. We look also at incidence across regions. The CEX data along with imputed capital income from the SCF data are used to tabulate expenditure and income data by the four census regions. Results are summarized here. Households in the West region have a substantially higher capital-wage ratio (42% vs. an average of 35.6%). Households in the West spend a lower fraction of their expenditures on dirty goods (8.4%), and households in the South spend a higher fraction on the dirty good (11.3%), compared to the average (10.1%, Table 3). A reason is that households in the South spend more than elsewhere on electricity for their air conditioners. The incidence results follow from these facts. On the uses side, the West faces a lower burden (0.125% less than average) and the South faces a higher burden (0.087% higher than average). The deviations from a proportional burden on the sources side are small, but the West's burden is somewhat higher than the average (0.009%).

IV. Conclusion

We use an analytical general equilibrium tax incidence model to examine the uses-side and sources-side distribution of burdens from a carbon tax. In general,

The B.E. Journal of Economic Analysis & Policy, Vol. 10 [2010], Iss. 2 (Symposium), Art. 15

the uses-side costs are relatively more burdensome on low income households, who spend more than average on dirty goods (electricity, natural gas, gasoline, heating oil).   This reinforces previous findings that the uses-side incidence is regressive (Hassett et al. 2009, or Burtraw et al. 2009).   The base case results suggest that the sources-side costs are relatively more burdensome on those who earn a higher than average fraction of their income from capital (because carbon-intensive industries tend to be relatively capital-intensive).   This implies a U-shaped burden when households are divided by annual expenditure or by annual income.   This result is sensitive to chosen parameter values for substitution elasticities that are not known.   The burden on the sources side can even be regressive if the wage falls relative to the rental rate, such as when capital is better than labor as a substitute for pollution.

Many extensions to the model are possible, including more sectors, more final goods, intermediate goods, market power, or other refinements.[21]   In particular, consideration of imperfect factor mobility could significantly affect the results; the transition costs for both capital and labor are likely to be large components of the overall burden of any policy.   The effect of market power or industry regulation may be of particular relevance to a carbon tax, since electric utilities are large emitters and are often highly regulated.   The particular policy could be modeled more carefully, rather than just looking at a simple tax.[22]   A more complex CGE model may allow more specific results, but at the expense of analytical solutions made possible by our simple two-sector model.[23]   Finally, this model does not incorporate the benefits of pollution reduction, which themselves may be progressive or regressive.

References

Allen, R.G.D. *Mathematical Analysis for Economists*. New York: St. Martin's (1938)

Burtraw, Dallas, Margaret Walls, and Joshua Blonz. "Distributional Impacts of Carbon Pricing Policies in the Electricity Sector." In Gilbert E. Metcalf, ed., *U.S. Energy Tax Policy*, Cambridge: Cambridge University Press (2010): 10-40.

---

[21] Capital may bear none of the burden, for example, in a dynamic model with capital accumulation, or in an open economy with international mobility where the world-wide rate of return is fixed (though see Gravelle and Smetters, 2006).

[22] See Burtraw et al. (2010) for an analysis of how the choice of allocation of carbon permits to the electricity sector affects the distribution of costs across income groups and regions.

[23] For example, Rausch et al. (2010) use a CGE model with a detailed structure of the U.S. energy sector to investigate the distributional impacts of a carbon tax.

Burtraw, Dallas, Richard Sweeney, and Margaret Walls. "The Incidence of U.S. Climate Policy: Alternative Uses of Revenues from a Cap-and-Trade Auction." *National Tax Journal* 62, no. 3 (September 2009): 497-518.

de Mooij, Ruud, and Lans Bovenberg. "Environmental Taxes, International Capital Mobility and Inefficient Tax Systems: Tax Burden vs. Tax Shifting." *International Tax and Public Finance* 5, no. 1 (February 1998): 7-39.

Fullerton, Don. "Introduction." In Don Fullerton, ed., *Distributional Effects of Environmental and Energy Policy*, Aldershot, UK: Ashgate (2009).

Fullerton, Don, and Garth Heutel. "The General Equilibrium Incidence of Environmental Taxes." *Journal of Public Economics* 91, no. 3-4 (April 2007): 571-591.

Gravelle, Jane, and Kent Smetters. "Does the Open Economy Assumption Really Mean that Labor Bears the Burden of a Capital Income Tax?" *Advances in Economic Analysis & Policy* 6, no. 1, article 3 (2006): 1-42.

Harberger, Arnold. "The Incidence of the Corporation Income Tax." *Journal of Political Economy* 70, no. 3 (June 1962): 215-240.

Hassett, Kevin, Aparna Mathur, and Gilbert Metcalf. "The Incidence of a U.S. Carbon Tax: A Lifetime and Regional Analysis." *Energy Journal* 30, 2 (2009): 155-177.

Metcalf, Gilbert E., Aparna Mathur, Kevin A. Hassett, "Distributional Impacts in a Comprehensive Climate Policy Package", NBER Working Paper No. 16101, Cambridge, MA (June 2010).

Jin, Hui and Dale Jorgenson. "Econometric Modeling of Technical Change," *Journal of Econometrics*, 157, no. 2 (August 2010): 205-219.

Jorgenson, Dale, Mun Ho, and Kevin Stiroh. "A Retrospective Look at the U.S. Productivity Growth Resurgence." *Journal of Economic Perspectives* 22, no. 1 (Winter 2008): 3-24.

Jorgenson, Dale, and Daniel Slesnick. "Consumption and Labor Supply." *Journal of Econometrics* 147, no. 2 (December 2008): 326-335.

Lyon, Andrew, and Robert Schwab. "Consumption Taxes in a Life-Cycle Framework: Are Sin Taxes Regressive?" *The Review of Economics and Statistics* 77, no. 3 (August 1995): 389-406.

Mieszkowski, Peter. "The Property Tax: An Excise Tax or a Profits Tax?" *Journal of Public Economics* 1, no. 1 (April 1972): 73-96.

The B.E. Journal of Economic Analysis & Policy, Vol. 10 [2010], Iss. 2 (Symposium), Art. 15

Poterba, James M. "Lifetime Incidence and the Distributional Burden of Excise Taxes." *American Economic Review* 79 (May 1989): 325-30.

Rausch, Sebastian, Gilbert E. Metcalf, John M. Reilly, and Sergey Paltsev. "Distributional Implications of Alternative U.S. Greenhouse Gas Control Measures." *The B.E. Journal of Economic Analysis & Policy* 10, no. 2, Symposium, article 1 (2010).

[6]

The B.E. Journal of Economic Analysis & Policy

Symposium

Volume 10, Issue 2	2010	Article 16

DISTRIBUTIONAL ASPECTS OF ENERGY AND CLIMATE POLICY

Comment on "Analytical General Equilibrium Effects of Energy Policy on Output and Factor Prices"

Samuel Kortum*

*University of Chicago, kortum@uchicago.edu

Kortum: Comment on "Analytical General Equilibrium"

While economists view a carbon tax as a natural solution to the perceived externality of climate change, they tend to gloss over any distributional impact of such a policy. Yet, the distributional impact may be crucial in determining what policies might be chosen. Don Fullerton and Garth Heutel (2010) approach this distributional issue using a simple quantitative general-equilibrium analysis. The general equilibrium part means that the authors consider the effect of a carbon tax on both output prices and factor prices. The former have distributional consequences due to differences in what we consume while the latter due to differences in our sources of income.

Progress in quantifying distribution effects of carbon policy will likely come from using large computable general equilibrium (CGE) models that incorporate substantial sectoral, demographic, and policy detail. Yet such models would necessarily be quite complicated, making it difficult to interpret the results. We will need simple versions of CGE models if we ever hope to clearly communicate the results from detailed CGE models.

The Fullerton and Heutel paper implicitly takes this view, proceeding in two steps. First, it uses a stripped-down model to make transparent the general-equilibrium changes in output and factor prices resulting from a carbon tax. Second, it uses household data on consumption patterns and sources of income to calculate the distributional impact of these general-equilibrium price changes.

I will comment on each step. On the first, I relate the analysis to a text-book model of international trade. While this analysis does not change the results in the paper, it provides a simple way to see the general equilibrium effects. On the second, I point out an analytical shortcoming of treating the distributional issues separately from the general-equilibrium analysis. I briefly conclude with a discussion of the quantitative results.

An Analogy to Trade Theory

With one simplification, the model becomes identical to a standard Heckscher-Ohlin analysis with two factors of production [capital (K) and labor (L)] and two goods [the clean good (X) and the dirty good (Y)]. We can focus on just a single country with preferences $U = U(X, Y)$. Technologies are constant returns to scale, factors are mobile across sectors, and markets are competitive.

To get the model into this form requires a restriction on the technology for production of the dirty good. With a slight abuse of notation:

$$Y = Y(K_Y, L_Y, Z) = \min \{aZ, Y(K_Y, L_Y)\}. \tag{1}$$

The B.E. Journal of Economic Analysis & Policy, Vol. 10 [2010], Iss. 2 (Symposium), Art. 16

With this restriction carbon emissions are necessarily proportional to output, $Z = Y/a$, so that production of the dirty good can be expressed as

$$Y = Y(K_Y, L_Y).$$

The price of Y paid by consumers is still P_Y, while the price received by producers, net of the carbon tax $\tau_Z \geq 0$, is

$$\tilde{P}_Y = P_Y - \tau_Z/a.$$

As in the paper, the clean good serves as the numeraire, with $P_X = 1$. Assume, in line with the calibration in the paper, that production of the Y good is capital intensive.

Consider a small open economy facing world prices P_Y. An equilibrium, in which the economy produces both goods, can be depicted using the Lerner diagram, as in Figure 1.[1] With K on the vertical axis and L on the horizontal axis plot the isoquants for 1 unit of the X good and $1/\tilde{P}_Y$ units of the Y good. Since the Y good is capital intensive the isoquant for the Y good is flatter than that for the X good along any ray from the origin. Producers in both industries must break even while minimizing costs. Hence these unit-revenue isoquants must both be tangent to a line with slope w/r, representing combinations of factors that can be purchased for one unit of the X good. This unit-cost line crosses the vertical axis at $1/r$ and the horizontal axis at $1/w$.

Now consider an increase in the carbon tax. The increase in τ_Z lowers $\tilde{P}_Y$ given P_Y. The unit-revenue isoquant for the Y good therefore shifts up. Assuming both goods continue to be produced, the unit-cost line must get steeper, crossing the vertical axis at a higher value of K and crossing the horizontal axis at a lower value of L. Thus, as in the Stolper-Samuelson Theorem, the wage must rise to $w' > w$ and the rental price of capital must fall to $r' < r$, illustrated in Figure 1. In the case of a small open economy producing both goods, an increase in the carbon tax has no consequences for the prices faced by consumers. But, it does shift factor prices in favor of labor, the factor used intensively by the industry that doesn't experience the tax increase.

How about a closed economy? Conditional on $\tilde{P}_Y$ the analysis above goes through, yet now P_Y is endogenous. At one extreme, the analysis is unchanged if consumers view the goods as perfect substitutes, one unit of Y always trading for P_Y units of X. At the other extreme, with Leontief preferences, an increase in the carbon tax would raise P_Y exactly to the point at which $\tilde{P}_Y$ and factor prices remain unchanged. For preferences U intermediate between these two extremes

---

[1]See, for example, chapter 5 of Bhagwati, Panagariya, and Srinivasan (1998). I am indebted to Kelsey Moser for producing Figure 1.

we expect both P_Y and $\tilde{P}_Y$ to rise. The increase in P_Y means a change in prices faced by consumers while the increase in $\tilde{P}_Y$ leads to a rise in the wage and a fall in the rental price. These changes in factor prices are the same as those predicted for the small open-economy case, although with a closed economy it is much less restrictive to assume that both goods are being produced. This analysis provides a simple illustration of the general-equilibrium effects derived by Fullerton and Heutel, and is consistent with the numerical results in their Table 4 [except for column (2) which is ruled out by our restriction (1)].

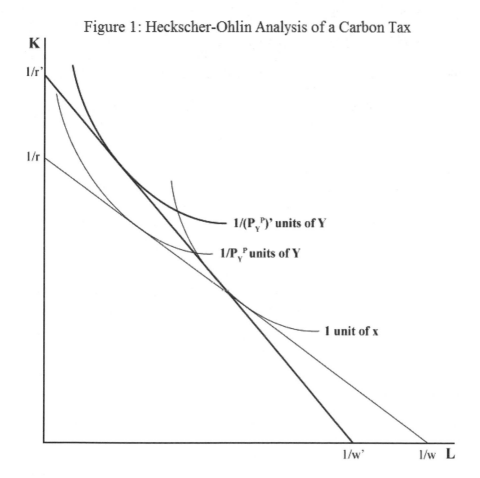

Figure 1: Heckscher-Ohlin Analysis of a Carbon Tax

The B.E. Journal of Economic Analysis & Policy, Vol. 10 [2010], Iss. 2 (Symposium), Art. 16

A Conceptual Difficulty

The authors retain the simplicity of the analysis by proceeding in two steps, first calculating general-equilibrium effects of a carbon tax on output and factor prices and then computing distributional effects of these price changes. That two-step approach is perfectly justified if the only concern is to compute the consequences of heterogeneity in individuals' sources of income. It is more problematic, at least conceptually, when one focusses on the consequences of heterogeneity in individuals' pattern of consumption.

Suppose individual i in the economy works an amount l_i and owns capital k_i so that his income is $y_i = wl_i + rk_i$. If k_i is small enough relative to l_i he may well gain from the income effects of the carbon tax. This heterogeneity at the individual level is no problem for the analysis, as we can aggregate over all individuals and get $Y = wL + rK$, without putting any restrictions on the distribution of the ownership of capital.

Unfortunately, we can't make a similar argument for heterogeneity in preferences. To represent preferences in the aggregate by $U(X,Y)$ we need to assume identical homothetic preferences for each individual. If not, aggregate preferences will also depend on the distribution of income. Thus, acknowledging heterogeneity in tastes undercuts the representative consumer framework that is used to calculate the general equilibrium effects on output and factor prices. There is a fundamental inconsistency here. It may not be a first-order issue quantitatively, but it would require a fairly intricate CGE analysis to make sure that's true.

The Results

My discussion so far has been about the modeling. A major contribution of the Fullerton and Heutel paper, however, is that it applies the analysis to real data. That's never easy. The paper does a nice job of explaining the new issues that arise.

The first step is to compute the general-equilibrium changes in output and factor prices that result from doubling a carbon tax (from a hypothetical baseline of $15 per ton of CO_2). The results are in Table 4. The price of the dirty good rises by 7.2 percent relative to the clean good in the baseline case. Changes in factor prices are much smaller. The wage increases by less than 0.1 percent and the the rental price of capital falls by somewhat more than 0.1 percent relative to the price of the clean good.

Moving to the second step requires measuring how sources of income (wage vs. capital income) and uses of income (spending on dirty vs. clean goods)

Kortum: Comment on "Analytical General Equilibrium"

vary across the income distribution. The authors carry out this measurement in several ways. I find Table 9 most compelling. Motivated by the permanent income hypothesis, it locates positions in the income distribution using total household expenditures. Motivated by the life-cycle hypothesis, it fixes the age of household head (at age 41-50). The ratio of capital income to wage income rises by a factor of five as we move from the lowest to highest permanent income decile. The percentage of expenditure on the dirty good moves in the opposite direction, falling by almost a factor of three.

Table 10 shows the bottom line of the analysis. The poor bear the greater burden of the carbon tax via their relatively higher expenditures on the dirty good. This burden is offset, but to only a small extent, by changes in factor prices (an increase in the wage and decrease in the rental price of capital), which favor the poor. The message is that the price effects dominate, and that these changes will make a carbon tax regressive.

The income effects in this analysis are small. Future work should explore a model in which capital is immobile. In that case there could be much larger income effects on households whose income derives from capital that is specific to the dirty-good sector. I could no longer make the analogy to the Heckscher-Ohlin model, but instead to the specific-factors model of international trade.

References:

Bhagwati, Jagdish N., Arvind Panagariya, and T.N. Srinivasan (1998), *Lectures on International Trade*, second edition, MIT Press.

Fullerton, Don and Garth Heutel (2010), "Analytical General Equilibrium Effects of Energy Policy on Output and Factor Prices," *The B.E. Journal of Economic Analysis & Policy*, 10.2.

[7]

The B.E. Journal of Economic Analysis & Policy

Symposium

Volume 10, *Issue* 2	2010	*Article* 5

DISTRIBUTIONAL ASPECTS OF ENERGY AND CLIMATE POLICY

Climate Policy's Uncertain Outcomes for Households: The Role of Complex Allocation Schemes in Cap-and-Trade

Joshua Blonz*   Dallas Burtraw[†]

Margaret A. Walls[‡]

*Resources for the Future, blonz@rff.org
[†]Resources for the Future, burtraw@rff.org
[‡]Resources for the Future, walls@rff.org

Recommended Citation
Joshua Blonz, Dallas Burtraw, and Margaret A. Walls (2010) "Climate Policy's Uncertain Outcomes for Households: The Role of Complex Allocation Schemes in Cap-and-Trade," *The B.E. Journal of Economic Analysis & Policy*: Vol. 10: Iss. 2 (Symposium), Article 5.
Available at: http://www.bepress.com/bejeap/vol10/iss2/art5

Climate Policy's Uncertain Outcomes for Households: The Role of Complex Allocation Schemes in Cap-and-Trade*

Joshua Blonz, Dallas Burtraw, and Margaret A. Walls

Abstract

The design and implementation of the allocation of CO_2 emissions allowances in recent bills in the U.S. Congress introduces a new source of uncertainty to the climate policy debate. We examine the Waxman-Markey bill (H.R. 2454) with scenarios that vary outcomes associated with allocations to electricity local distribution companies, investments in energy efficiency and technology development. The average net household burden in 2016 ranges from \$133, with a CO_2 allowance price of \$13.19, to \$418, with an allowance price of \$23.41. The uncertainty about average burdens does not carry over to the distribution of those burdens; both scenarios impose the greatest burden as a percentage of income on middle-income households. A third scenario that allocates a substantial portion of allowance value as lump-sum payments imposes an average net household burden of \$206, with a price of \$17.37 and with highly progressive distributional impacts.

KEYWORDS: cap-and-trade, allocation, distributional effects, cost, burden, regulated entities

---

*Dallas Burtraw and Margaret Walls are Senior Fellows and Joshua Blonz is a Research Assistant at Resources for the Future, Washington, DC. This research was supported by grants from the National Commission for Energy Policy, the Doris Duke Charitable Foundation and Mistra's Climate Policy Research Program. We are indebted to Karen Palmer, Anthony Paul, and especially Matt Woerman for assistance, and to Arik Levinson for comments.

Introduction

As the U.S. draws nearer to adopting a national climate policy, the allocation of emissions allowances in a cap-and-trade program has become a front and center issue in the policy debate. With estimates of allowance value at approximately $100 billion per year initially, rising to $230 billion according to the Energy Information Agency (EIA) (2009a) and even higher according to other studies, the stakes are high. Many vested interests have lobbied for a dedicated slice of the allowance pie, and this has led to detailed and complex allocation schemes in current bills in Congress. The front-runners are H.R. 2454, the Waxman–Markey bill, which came out of the House of Representatives in June 2009; S. 1733, the Kerry–Boxer bill, introduced in the Senate in September 2009; and S. 2877, the Cantwell–Collins bill introduced in the Senate in December 2009. These bills have many provisions for allocating allowances among various programs, industry groups, and consumers.

These proposals differ not only in the allocation of emissions allowances or their value, but they differ also in a qualitative way. The allocation schemes introduce an additional level of uncertainty over the outcome of climate policy for households, because the schemes are often complex and therefore hard to understand, their implementation is not completely specified, and the effects of the allocation are in fact uncertain. This paper does not focus on scientific or economic uncertainty, and only partially on technological uncertainty. Instead, we focus on uncertainty introduced in the design and implementation of legislative formulas for the allocation of emissions allowances in recent bills in the U.S. Congress. We examine how that policy-induced uncertainty affects the level and distribution of the burden on households.

Studies by economists have shown that the allocation of allowances can be a key determinant of both the welfare costs of a cap-and-trade program and the distribution of impacts across households. A number of studies have emphasized, for example, that using the revenues generated from an allowance auction (or a carbon tax) to reduce pre-existing distortionary taxes in the economy, such as income and payroll taxes, can improve the overall efficiency of the program (Parry et al. 1999; Goulder et al. 1999). On the other hand, several studies have shown that reducing payroll or income taxes in combination with a carbon price is likely to be regressive, benefiting high-income households more than low-income households (Burtraw et al. 2009). Giving allowances away for free to existing emitters, so-called "grandfathering" of allowances, has also been shown to be regressive, because the benefit flows to shareholders who are predominantly in higher income groups (Parry, 2004; Dinan and Rogers 2002; Burtraw et al. 2009). In a recent paper (Burtraw et al. 2009), we highlighted the impacts of giving allowances away for free to electricity local distribution companies (LDCs), the

The B.E. Journal of Economic Analysis & Policy, Vol. 10 [2010], Iss. 2 (Symposium), Art. 5

regulated entities that provide electricity to end users. This approach is a feature of H.R. 2454, accounting for 30 percent of allowances, with another 9 percent allocated to natural gas LDCs; S. 1733 has similar features. Our results suggest this approach has implications for both the welfare costs of the policy and the distribution of impacts across households.[1]

This extensive cap-and-trade literature is instructive in understanding the implications of various allocation provisions. However, the current bills in the U.S. Congress are far more complicated than the arrangements analyzed by existing studies. H.R. 2454 has over 20 provisions for allocating allowances. The bill has money flowing directly to industry, such as to domestic refineries and so-called "energy-intensive, trade-exposed industries". Money goes to consumers, via direct rebates for low-income households and via electricity and natural gas LDCs. Money is also targeted to support a variety of funds to address adaptation to climate change, including extensive technology development, renewable energy, energy efficiency measures, and some international programs. It is unclear how all this money will eventually make its way into household benefits. Language in the bill is vague, for one thing, and even for those less ambiguous provisions, the degree of uncertainty in implementation and in market outcomes is substantial.

In this paper, we analyze the allocation provisions of H.R. 2454 and attempt to determine how the various allocation schemes will flow through to households. We focus on 2016 as the year when all of the main provisions of the bill come into effect, and we construct an optimistic and a pessimistic scenario for the LDC provisions and for the various energy efficiency and technology development provisions in the bill. These three sets of provisions account for approximately 43 percent of allowances in 2016, introducing a wide range of possible outcomes for the burden that households will face. Some of this uncertainty stems from the familiar uncertainty associated with technology development, but most of it arises from questions about how the policy would be implemented.

Additionally, we assume that government's own energy costs will increase, although no revenue is set aside in H.R. 2454 to pay for it. Without such an allocation, this increase in government costs would constitute a hidden tax, which we assume to be 14 percent of total allowance revenue based on estimates from the Congressional Budget Office (2009a). The magnitude of government costs changes with the allowance price, which varies across scenarios.

---

[1] In a recent paper, Hahn and Stavins (2010) enumerate conditions under which the initial allocation of allowances is independent of the efficiency properties of a cap-and-trade system. These conditions center on general issues such as transaction costs, market power, uncertainty, and the like. We show a specific case in which the initial allocation, through its effect on energy prices, is not independent of efficiency.

Our optimistic and pessimistic scenarios are distinguished in the degree to which they affect allowance prices and the efficiency of the program. In the optimistic scenario, we assume LDC allocations end up reducing fixed charges on monthly electricity bills for some customers. In contrast, in the pessimistic scenario, the LDC allocations reduce variable charges for all customers. The efficiency consequences under these scenarios depend on the degree to which the policy results in a weaker price signal to consumers. If consumers receive a weak price signal, there will be less of a behavioral change by end users of energy. This will lead to more emissions in this sector and will require greater emissions reductions from other sectors of the economy, which raises the allowance price and the overall cost of the program. This is the sense in which the efficiency implications of the policy may depend on the allocation of allowances. Reducing fixed charges, as in our optimistic scenario, preserves the incentive features of pricing carbon and is thus more efficient.

Our pessimistic scenario also assumes the money spent on energy efficiency and technology development provisions is simply wasted, i.e., that no additional emissions reductions are obtained with the money dedicated to these efforts. In our optimistic case, we accept the opinion of some experts that society can benefit more from government investments in various energy efficiency and clean technology development programs than from the efforts of the private market (McKinsey and Company 2007). Among the expected benefits would be additional reductions in CO_2 emissions, a diminished burden on other sectors of the economy, and lowered allowance prices. Again, the allocation of allowances can affect overall efficiency of the policy. In addition, household electricity bills would be reduced as electricity consumption falls.

These two sets of extreme assumptions make a big difference in the costs of climate policy. We find that the net household burden – the consumer surplus loss minus the refunded allowance value– for an average household in 2016 is three times higher in our pessimistic scenario than in our optimistic one. Under the pessimistic assumptions, the average household incurs a net loss of $418; under optimistic assumptions, the loss is only $133. The CO_2 allowance price in the optimistic scenario is only $13.19/ton, while it is more than $10 higher, $23.43/ton, in the pessimistic scenario. The difference between these two cases is demonstrated by calculating Shapley values to show which of the optimistic assumptions are responsible for what portion of the cost savings to households (Roth 1988).

While uncertainty characterizes the difference between the magnitudes of the costs in the Waxman-Markey scenarios, the shape of the distribution is similar across the two H.R. 2454 scenarios. The allocation scheme in the H.R. 2454 bill is progressive over the bottom four-fifths of the income distribution under either set of assumptions. The average household in the lowest income quintile enjoys a net

The B.E. Journal of Economic Analysis & Policy, Vol. 10 [2010], Iss. 2 (Symposium), Art. 5

gain (negative burden) in both scenarios, and households in the higher quintiles incur relatively more of the burden of the policy through the fourth quintile. The fifth quintile incurs a slightly smaller burden than the fourth as a percentage of income because of how some of the provisions pass through to shareholders (who are predominantly in the highest income quintile). Additional analysis is then conducted using consumption quintiles, which are often used as proxies for lifetime income quintiles. The results are similar to those with annual income quintiles, but with all three policies being progressive across the entire consumption distribution. This finding highlights some of the complicated ways in which the provisions impact households.

For comparison purposes, we also assess a simple allocation scheme in which 75 percent of allowances are auctioned and returned to households as a lump-sum payment per person, i.e., a so-called cap-and-dividend approach.[2] The average net household burden in this case is $206, and the allowance price is $17.37/ton, so the results lie in between our optimistic and pessimistic scenarios for H.R. 2454. With no LDC allocation, there is no distortion in electricity markets in the cap-and-dividend scenario, and this helps to reduce costs and the allowance price. On the other hand, with no allocation to energy efficiency and technology programs, there is no possibility of reaping those benefits that we obtained in our optimistic scenario. This is balanced against the possibility that the energy efficiency and technology programs might not benefit households, and the revenue might essentially be lost. The cap-and-dividend approach is progressive across the entire income distribution, because the lump-sum return of revenue on a per capita basis benefits low and middle-income households relatively more.

Our results are not meant to be precise representations of the outcomes under proposed climate legislation. Rather, they are illustrative of the range of possibilities in a cap-and-trade program that has a complex, multi-faceted allocation scheme. Because of the uncertainty in how the separate provisions affect energy use, emissions, and household welfare, it is difficult to say exactly what the impacts of the legislation will be. Complex allocations and uncertain outcomes also make it difficult to protect certain vulnerable groups, because allowance value and initial welfare losses remain undetermined. In contrast, a simple scheme such as embodied in the cap-and-dividend approach has more predictable and straightforward impacts on average burden and the distribution of that burden across income groups.

---

[2] The optimistic and pessimistic scenarios are designed around specific provisions of H.R. 2454, but bear some similarity to provisions of S. 1733 because of the similarity of the allocations in the legislation. In contrast, the cap with 75 percent dividend explored in our third scenario bears some resemblance to S. 2877, the Cantwell–Collins bill.

In the next section, we provide a review of the literature on the economic impacts of climate policy on households. We then describe our data and methodology, lay out the specific provisions of the Waxman–Markey bill, explain how we account for these provisions in our model, and present our results for both average net household burdens and the distribution across quintiles.

Literature Review

Several studies have evaluated the impacts of alternative allocation schemes in a carbon cap-and-trade program. Dinan and Rogers (2002) find that distributional effects hinge crucially on whether allowances are initially distributed free of charge to incumbent emitters (grandfathered) or auctioned, and whether revenues from allowance auctions, or from indirect taxation of allowance rents, are used to cut payroll or corporate taxes or to provide lump-sum transfers to households. They find grandfathering to be very regressive, as a result of the value flowing through to shareholders who are primarily in the upper-income groups. Parry (2004) also obtained this result in a calibrated analytical model. In contrast, if allowances are auctioned, Dinan and Rogers (2002) find that returning revenues in equal lump-sum rebates for all households reverses the regressively finding. Using auction revenues to cut payroll or corporate taxes is found to be regressive, though less so than grandfathering.

Metcalf (2009) also analyzes reductions in payroll taxes. Specifically, he looks at a policy where revenues from a CO_2 tax are used to give each worker in a household a tax credit equal to the first $560 of payroll taxes; this would be equivalent to exempting from the payroll tax the first $3,660 of wages per worker. He finds that this option leads to approximately equal net impacts, as a percentage of income, across income quintiles. An option that couples this rebate with an adjustment to Social Security payments that benefits the lowest-income households makes the CO_2 policy more progressive. Finally, Metcalf compares these options to a lump-sum redistribution of the CO_2 tax revenues and finds that this last option is the most progressive of all.

Burtraw et al. (2009) assess the effects of a cap-and-trade program. They examine two cap-and dividend scenarios, one in which dividends are taxed and one in which they are not taxed. They also explore three other scenarios: reducing the payroll tax, reducing the personal income tax, and expanding the Earned Income Tax Credit (EITC), which is available for low-income households. The authors find that the cap-and-dividend options and the EITC alternative reverse

The B.E. Journal of Economic Analysis & Policy, Vol. 10 [2010], Iss. 2 (Symposium), Art. 5

the regressivity of carbon pricing. Reducing payroll or income taxes, however, exacerbates the regressivity.[3]

Most of these studies use data from the Bureau of Labor Statistics Consumer Expenditure Survey (CES), including direct energy expenditures as well as indirect expenditures through the purchase of goods and services. Most take a partial equilibrium approach, and focus on expenditures as a fraction of annual income. Some of the studies compare results using a measure of lifetime income, in addition to current income. The measure used in most studies as a proxy for lifetime income is consumption expenditures. Most consumption taxes, including CO_2 taxes or a cap-and-trade program, look more regressive on the basis of annual income than on the basis of consumption.[4] We focus primarily on annual income in this analysis, but also conduct sensitivity analysis using a consumption measure. Rausch et al. (2009) use a regionally disaggregated general equilibrium model to trace through the impacts of carbon pricing to changes in wages and returns on capital, issues that are ignored in the partial equilibrium studies.[5] The authors estimate the distributional impacts of a carbon pricing policy with seven alternatives for return of revenue to households. Results are generally similar to the partial equilibrium findings: lump-sum payments make for a progressive policy, while reduction in personal income taxes, capital income taxes, or payroll taxes is regressive. Payroll tax reductions are the least regressive of the three. A large literature in environmental economics has argued the efficiency merits of reductions in these kinds of taxes (Parry et al. 1999; Goulder et al. 1999), but Rausch et al. (2009) find the efficiency gains to be relatively modest. The authors argue that this is likely a result of the revenue neutrality requirement in their model. They fix government revenue relative to GDP at the same level as in the no-policy scenario, which means that not all carbon pricing revenue is available for recycling purposes. In some cases, only about one-half of the revenue is available.

---

[3] This study also focused on regional impacts, finding that whereas net impacts for average households in each region are quite similar for any given allocation alternative, differences for low-income households are more pronounced. Other studies that look at regional incidence of various cap-and-trade schemes (though not by income group and region together) are Boyce and Riddle (2009), Hassett et al. (2009), and Pizer et al. (2009). We ignore regional issues in the results we present in this paper, but regional variation is an underlying characteristic of the model, as we will explain below.

[4] Grainger and Kolstad (2009) discuss income measures in more detail and calculate the incidence of carbon pricing using alternative measures. However, this study does not deal with allocation of allowances or use of allowance revenue.

[5] The model is a regional version of the MIT Emissions Prediction and Policy Analysis (EPPA) model (Paltsev et al. 2005). For more discussion of the various key parameters and relationships in a general equilibrium framework that determine the ultimate burden of a tax on a pollutant, see Fullerton and Heutel (2007).

All of the above studies, with the exception of Rausch et al. (2009), separate the distributional impacts from the efficiency costs of the cap-and-trade or carbon tax policy. Thus, in each case, the price of carbon is the same across alternative allocation schemes.[6] The resource cost of the policy and the amount of allowance value created are also the same, and all that changes is the distribution of that value across sectors of the economy and households. Some allocation schemes affect the allowance price, however. The LDC allocation in H.R. 2454 is a case in point. If it leads to lower electricity prices than the prices in a full auction scenario, then it alters the costs of the program and the size of the allowance pie. Our earlier paper focused on this issue, arguing that reductions in variable electricity prices are a likely outcome. We then assessed the additional burden on society, and the distribution of that burden across income deciles and regions, compared with a full auction coupled with a per capita dividend of allowance value (Burtraw et al. 2009). We calculated that the annual net household burden for an average household would be $157 higher under this approach than under cap-and-dividend.[7] However, we did not evaluate the full impacts of the LDC approach in the context of the entire H.R. 2454 allocation scheme—the exercise that is the focus of this paper.[8]

Three government studies have analyzed the costs of H.R. 2454 to the U.S. economy, the U.S. Department of Energy's Energy Information Administration (EIA 2009a), the U.S. Congressional Budget Office (CBO 2009b), and the U.S. Environmental Protection Agency (EPA 2009). These analyses differ from the aforementioned academic studies in that they analyze only the specific provisions detailed in the legislation.

CBO (2009b) estimates the loss of purchasing power that households will experience in 2020 to be $160, or 0.2 percent of income, at projected 2010 income levels. The cost, however, varies greatly over household income levels. They estimate an average household in the lowest income quintile will see a net benefit of $125, or 0.7 percent of income, while an average household in the highest quintile will experience a cost of $165, or 0.1 percent of income. The

---

[6] Rausch et al. (2009) do not change the allowance price across scenarios; essentially, their policy is more like a carbon tax with changes in emissions and the size of the allowance pie as a result, but no change in the tax rate or carbon price.

[7] This assumes that allowance value outside the electricity sector is not returned to and does not benefit households. Under the other extreme assumption that it all goes to the benefit of households, the difference in net burden for an average houshold was estimated to be $36.

[8] Rausch et al. (2009) assess two versions of the LDC allocation scheme, one of which is a subsidy to electricity consumption. However, the ultimate impact of the extra electricity consumption in their framework is simply higher electricity costs, not additional abatement from other sectors (with concomitant increase in the allowance price as in our model).

The B.E. Journal of Economic Analysis & Policy, Vol. 10 [2010], Iss. 2 (Symposium), Art. 5

middle-income quintile will incur the most significant cost of this policy, with an average cost of $310, or 0.6 percent of income, per household.

EPA (2009) also estimates the cost of H.R.2454 to households. Rather than purchasing power, however, they estimate the average loss of consumption under the policy. Using their ADAGE computable general equilibrium model, they find an average household will incur a cost of $105 in 2020 (2005$).[9] This cost is equivalent to 0.11 percent of total household consumption in 2020. This analysis corresponds to the ADAGE model's allowance price in 2020 of $16 per metric ton CO_2. EPA does not look at the distribution of costs across different income groups.

EIA (2009) estimates the cost of H.R. 2454 as the average loss of consumption per household. Based on the results of their National Energy Modeling System (NEMS) energy–economy model, the cost of the policy will be $134 (2007 $) for an average household in 2020. Like EPA, the EIA also does not calculate how this cost would be distributed among households of different income levels. The cost estimate corresponds to the NEMS allowance price in 2020 of $32 per metric ton CO_2. The EIA also allocates allowances in the proportions specified by H.R. 2454, such as those given to LDCs and trade-exposed industries. They do not specify how this allowance value is then passed on to households and shareholders, either within the NEMS model or outside of the modeling framework.

Academic and governmental analyses provide various measures of the household burden of climate policy using different data and types of models and slightly different welfare measures. Each study has its strengths and weaknesses, but together they add to the continually growing body of literature on the incidence of climate policy. Our analysis seeks to further the literature by analyzing detailed and specific allowance allocation mechanisms such as exist in proposed legislation. Unlike other studies, we provide a range of estimates that are intended to capture the uncertainty associated with these complex allocation schemes. The simple cap-and-dividend approach is provided as a point of comparison.

Data and Methodology

We base our analysis on CES data from 2004 through 2008. The population sampled in the CES includes 133,421 observations for 51,694 households; an observation equals one household in one quarter. We use these observations to construct national after-tax income quintiles. Our sample for examining regional effects includes 112,306 observations for 42,828 households in 43 states plus the

---

[9] For more information on ADAGE see Ross (2008).

District of Columbia.[10] We aggregate the observations into 11 regions. Although we do not use observations with missing state identifiers in our regional-level calculations, we do include them in our calculations at the national level.

We account for direct energy expenditures and indirect expenditures through the purchase of goods and services.[11] We focus the analysis on 2016, assuming that the distribution of consumption across regions and income groups would be the same as in our data period (2004-2008) in the absence of climate policy. The consumption data is combined with the average carbon contents of goods from Hassett et al. (2009) to estimate the CO_2 content of every household's consumption bundle. The average CO_2 content of household goods is scaled to reflect changes in production and consumption that are predicted by EIA baseline forecast for 2016 outside the electricity sector and the forecast of RFF's Haiku electricity market model in the electricity sector. This procedure ensures that our baseline emissions estimates for 2016 match those of EIA and the Haiku model.

We use Haiku to model the electricity sector more accurately. The model solves for electricity market equilibria in 21 regions of the country that are mapped into the 11 regions we use for the distributional analysis. The electricity model accounts for price-sensitive demand, electricity transmission between regions, system operation for three seasons of the year (spring and fall are combined) and four times of day, and changes in demand and supply-side investment and retirement over a 25-year horizon (Paul et al. 2009). The Haiku model also captures differences in the regulatory environment across regions and allows us to model different behavioral assumptions corresponding to fixed and variable charges for residential, commercial, and industrial customers, as we explained in the introduction. Table 1 illustrates the electricity sector results for the 11 regions of the country that we model, with an indication of how we aggregate states into these regions.[12] The model calculates a national baseline emissions rate of 0.602 tons of CO_2 per megawatt-hour (MWh) for 2016 in the absence of any climate policy. In this example, the introduction of an emissions cap (and scenario characteristics that correspond to the "pessimistic case" described below) lead to an emissions allowance price of \$20.86/metric ton CO_2 (mtCO_2) (2006 dollars). Emissions fall to 0.516 tons/MWh. Table 1 also reports the percentage change in electricity price on a regional basis and the percentage

---

[10] Five states (Iowa, New Mexico, North Dakota, Vermont, and Wyoming) are dropped because of missing information.

[11] Indirect consumption accounts for approximately 49 percent of an average household's carbon emissions.

[12] The 48 contiguous states and the District of Columbia are included in the electricity modeling, but as noted, five states (Iowa, New Mexico, North Dakota, Vermont, and Wyoming) are dropped when calculating effects on households at the regional level. However, national estimates always include these five states.

The B.E. Journal of Economic Analysis & Policy, Vol. 10 [2010], Iss. 2 (Symposium), Art. 5

change in consumption that is expected to result from the introduction of the price on CO_2 emissions.

Table 1. Illustrative Results for the Haiku Electricity Sector Model

		Haiku Modeling Results - Waxman Markey Basic Case			
Region	States	Baseline CO2 Emissions Per MWh of Generation	Post-Cap CO2 Emissions Per MWh of Generation	Price Change	Change in Consumption
Southeast	AL, AR, DC, GA, LA, MS, NC, SC,	0.602	0.516	3%	-1%
California	CA	0.190	0.215	3%	-1%
Texas	TX	0.619	0.514	5%	-1%
Florida	FL	0.557	0.493	3%	0%
Ohio Valley	IL, IN, KY, MI, MO, OH, WV, WI	0.782	0.660	7%	-3%
Mid-Atlantic	DE, MD, NJ, PA	0.573	0.500	8%	-1%
Northeast	CT, ME, MA, NH, RI	0.371	0.336	6%	-2%
Northwest	ID, MT, OR, UT, WA	0.406	0.234	1%	0%
New York	NY	0.318	0.293	8%	-1%
Plains	KS, MN, NE, OK, SD	0.844	0.809	9%	-4%
Mountains	AZ, CO, NV	0.628	0.609	4%	-2%
National		0.608	0.523	5%	-2%

Figure 1 illustrates the mechanism of placing a price on CO_2 emissions through the introduction of a cap-and-trade policy. The horizontal axis in the graph represents the reduction in emissions (moving to the right implies lower emissions), and the upward-sloping curve is the incremental resource cost of a schedule of measures to reduce emissions; thus, it sketches out the marginal abatement cost (MAC) curve. The electricity sector MAC is generated from the

Haiku model, whereas abatement behavior for the rest of the economy, excluding the electricity sector, is taken from the Energy Information Administration (2009) analysis of Waxman–Markey (H.R. 2454). The EIA MAC curve is combined with the Haiku electricity MAC curve to model economy-wide abatement behavior. Using this economy-wide curve, an endogenous allowance price (represented by the height of the rectangle) is calculated such that emissions in our incidence model match capped 2016 EIA levels (17 mtCO2 equivalent).[13]

Figure 1. Resource Cost and Allowance Value

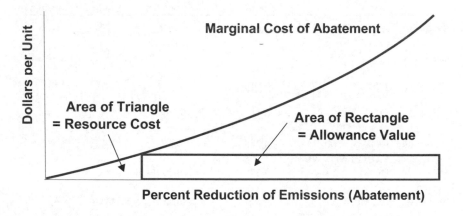

The triangular area under the marginal cost curve up to the emissions abatement target is the cost of resources used to achieve emission reductions. The rectangle represents the value of emissions allowances generated (number of allowances times price per allowance) by the trading program. EIA's analysis of H.R. 2454 provides an estimate of the aggregate burden, i.e., the sum of these two areas shown on the graph, along with a breakdown of this burden among sectors. We treat the electricity sector separately, using the Haiku model to obtain changes in emissions due to the CO_2 price. All other sectors' reductions and costs are assumed to match EIA.

To calculate the distribution of losses across regions and households, we use baseline emissions intensities and own-price elasticities along with consumption expenditure and price increases. These data provide a first-order indication of the relative change in burden across various consumption categories resulting from the introduction of a price on CO_2. This distribution of losses is then scaled proportionately across categories of consumption to match the

---

[13] http://www.eia.doe.gov/oiaf/servicerpt/hr2454/excel/hr2454cap.xls

The B.E. Journal of Economic Analysis & Policy, Vol. 10 [2010], Iss. 2 (Symposium), Art. 5

changes predicted by EIA and Haiku to generate an initial household burden. This explicitly assumes the initial change in household welfare before accounting for allowance value equals the sum of the resource cost and the allowance value estimated by EIA and Haiku. This approach rests on the implicit assumption that producers pass all costs through to consumers and bear none of the costs themselves, which is approximately true in the short run, when demand is relatively inelastic. As a result, our estimate of the initial household burden outside the electricity sector for the average household matches EIA's estimate of abatement cost (including allowance cost).[14]

After initial household burdens are calculated, allowance value is distributed based on scenario-dependent allocation schemes, which introduces the source of uncertainty that is the focus of this paper. This refunded allowance value is subtracted from the initial household burden to calculate a net household burden.

As mentioned above, for electricity we use the Haiku electricity market model in place of EIA's forecast for the sector. The Haiku model reports price changes that are somewhat different across regions and slightly lower on average than EIA. Using Haiku in place of the EIA electricity model preserves substantial detail with respect to regions and customer classes and allows for the various treatments of customer classes and allocation scenarios that we discuss below.

We use the cost estimate that is predicted by EIA outside the electricity sector and by Haiku within the electricity sector to estimate the total burden on households to achieve an emissions reduction target for 2016. EIA's 2016 emissions level provides our emissions target, and we hold it constant across scenarios.[15] Across regions and income groups, these effects are distributed according to the expenditure patterns revealed in the Haiku model and the CES data.

---

[14] This exercise does not affect our distributional findings. It does, however, provide estimates of the overall level of household burden consistant with EIA estimates of abatement cost.

[15] Both the EIA and Haiku models solve for aggregate intertemporal emissions targets and an intertemporal equilibrium that include potential changes in banking and the purchase of offsets. The models vary to a small degree with respect to emissions obtained in 2016. To hold constant the emissions in that year, we scale the results across scenarios. This introduces a small inconsistency in the aggregate level of intertemporal emissions reductions achieved over the modeling horizon, but this has only a slight effect on the level of burden for the average net household because most of the burden is determined by the scenario dependent allocation of allowances. It does not affect the distributional issues that are our focus in this paper.

Description of H.R. 2454 Cap-and-Trade Program

Title III of H.R. 2454 sets an annual cap on greenhouse gas emissions covering approximately 85 percent of all emissions in the U.S. including, among many others, those from oil refineries, natural gas suppliers, electricity generators, and industries such as cement, paper, iron, steel, and chemicals. The cap becomes gradually more stringent over time. In 2012, the first year the law would go into effect, covered emissions are required to be 3 percent below 2005 levels; by 2020, 2030, and 2050, this figure rises to 17, 42, and 83 percent, respectively. The cap is met by allocating emissions allowances among the regulated entities; those allowances can then be traded. The bill has approximately 22 separate provisions dealing with allowance allocation. Table 2 shows the breakdown for 2016, the year of our analysis.

We have divided the provisions into categories. A significant percentage of total emissions allowances go toward relieving the burden of higher energy prices on households. Thus, both electricity and natural gas LDCs receive allowances. For electricity LDCs, which receive 30 percent of total allowances, distribution is based one-half on emissions and one-half on electricity output. Natural gas LDCs receive 9 percent of allowances. LDCs are regulated or publicly owned entities in all 50 states, so they can be expected to act as trustees on behalf of consumers, passing the value of allowances on to customers through lower rates (or equivalently funding investments) rather than retaining it as profits. How exactly they will do this is the subject of much debate and we return to this issue below. Other allocations are designed for a variety of purposes – to reduce the burden on low-income households, to provide money for adaptation to climate change, to fund supplemental international reductions, to lessen the burden on industries that are thought to be particularly hard hit by climate policy, and to enhance energy efficiency, technology development, and the development of renewable energy.

Many of the allocations will be handled by federal and state government to further climate related goals. These allocations, however, do not address the impact of a carbon price on government. We return to this issue and explain how we handle it in the next section of the paper.

How money will be spent is unclear in many cases. For example, money flows into several funds, such as the Natural Resource Climate Change Adaptation Fund, but no guidance is provided on how those funds are to be spent. The same problem exists for many of the energy efficiency, renewables, and technology development programs. Much of the money is allocated to state administered programs, but with no strict requirements as to how the funds are spent.

Table 2. Allowance Allocation in 2016 as Specified in H.R. 2454 and Distribution of Allowances in Model

	Allocation Section	Program Description Section	Description	Percent	Optimistic Case	Pessimistic Case
Household Energy Consumption	782(a)(1)	783(b)	LDCs - Electricity	30.00	Allocated through fixed charge on electricity bills	Allocated through variable charge on electricity bills
	782(b)	784	LDCs - Natural gas	9.00	Captured in MAC curve	Captured in MAC curve
	782(c)	785	Home heating oil, propane	1.50	Captured in MAC curve	Captured in MAC curve
Low Income	782(d)	Title IV (C)	Energy refunds for low-income consumers	15.00	Per-capita dividend to low income households	
Industry	782(a)(1)	783(c)(d)	Merchant coal, long-term contracts	5.01	Shareholders	Shareholders
	782(e)	765	Trade-vulnerable, energy-intensive industries	14.44	Captured in MAC curve	Captured in MAC curve
	782(j)	787	Refineries	2.25	Shareholders	Shareholders
Adaptation & Adjustment	782(k)	Title IV (B2)	Climate Change Worker Adj. Assistance Fund	0.50	Per-capita dividend to all households	Per-capita dividend to all households
	782(l)	453, 467	Domestic adaptation	1.00	Per-capita dividend to all households	Per-capita dividend to all households
	782(m)	480(a)(b)	Wildlife/natural resources	1.00	Per-capita dividend to all households	Per-capita dividend to all households
	782(u)	788(b)	Supplemental agriculture	0.14	Per-capita dividend to all households	Per-capita dividend to all households
International	782(n)	Title IV (E2)	International adaptation	1.00	Lost revenue overseas	Lost revenue overseas
	782(o)	Title IV (D)	International clean technology development	1.00	Lost revenue overseas	Lost revenue overseas
	781(a)	Title III (E)	International forestry (REDD)	5.00	Lost revenue overseas	Lost revenue overseas
Energy Efficiency, Energy R&D, Technology Development, Renewables	782(i)	124	Investment in clean vehicle technology	3.00	Abatement at $75/ton[*]	Revenue lost
	782(f)	786	Development of CCS Technology	1.75	Abatement at $50/ton[*]	Revenue lost
	782(u)	788(c)	Supplemental Renewable Energy	0.14	Abatement at $34/ton[*]	Revenue lost
	782(g)(1)	132	Investment in energy efficiency and renewable energy	3.25	Abatement at $34/ton[*]	Revenue lost
	782(a)(2)	783(e)	Efficiency etc through small electricity LDCs	3.25	Electricity consumption reduced at 2.8 cents/kwh[*]	Revenue lost
	782(a)(2)	783(e)	Efficiency etc through small electricity LDCs	0.50		Revenue lost
	782(g)(3)	132	State renewable, energy efficiency programs	0.05		Revenue lost
	782(h)	171, 172	Energy innovation hubs, advanced research	1.50		Revenue lost
	782(g)(2)	201	Efficiency in building codes	0.50	Captured in MAC curve	Revenue lost
TOTAL				100.78		
Cap in 2016			5,482 MMT CO2e			

[*] Affects households through lower allowance price or through lower electricity bills.

Even in those provisions that contain more specifics, outcomes and impacts on households are still highly uncertain. The LDC provisions are a case in point. Because they are regulated, LDCs are required to pass on the allowance value to consumers, but it is not obvious exactly how they will do that. H.R. 2454 has language suggesting that LDCs reduce fixed charges on monthly electricity bills for residential and commercial class customers "to the maximum extent feasible," but such a change is thought to be infeasible in practice and the more likely outcome is a reduction in variable electricity rates (Burtraw 2009). The outcome has serious implications for the cost of the policy. If variable rates are reduced, electricity consumption remains higher than it would be if the price of electricity reflected the price of CO_2 allowances used in generation. Consequently, other sectors of the economy must work harder to reduce emissions in order to meet the economy-wide cap. This increases the allowance price and the overall cost of the program (Burtraw et al. 2009). How households are affected, on net, is unclear, but we expect on average that households are made worse off for the effort to subsidize their electricity consumption. On the one hand, electricity prices are lower as are electricity expenditures. On the other hand, a higher allowance price results in increased prices for expenditures on goods and services other than electricity. Because the overall value of allowances is greater, how these allowances flow into various categories will also affect the outcome.

The energy efficiency, renewables, and technology development provisions also generate highly uncertain outcomes. Some observers and efficiency advocates have argued that a great deal of "low-hanging fruit" is available for reducing energy use and CO_2 emissions. In fact, an iconic image of the last decade, in energy policy circles, is the McKinsey efficiency curve that indicates substantial opportunities to reduce energy use at no cost or significant negative cost (McKinsey and Company 2007). That study and several replicas, for example Sweeney and Weyant (2008), identify marginal abatement cost curves for reducing CO_2 emissions. These curves show engineering cost estimates for achieving the same or a comparable level of energy services through a variety of energy-saving options in production and consumption. From an economic perspective, the technical costs of energy-saving options only address half the problem. There may be significant behavioral, informational or social barriers to realizing these potential technical gains. In any case, the ultimate impact of these provisions on energy use, emissions, and household welfare are unclear; allocating money to energy efficiency programs, technology development, renewable energy, and the like may improve upon private market outcomes or it might simply lead to wasted expenditures.

The B.E. Journal of Economic Analysis & Policy, Vol. 10 [2010], Iss. 2 (Symposium), Art. 5

Modeling Strategy

To illustrate the range of potential impacts on the overall costs of the policy and the distribution of those costs across households, we model two bookend scenarios with respect to the LDC and energy efficiency provisions of the Waxman–Markey bill. One scenario is optimistic and the other pessimistic, distinguished by the possibility that uncertain provisions of the bill lead to efficiency-enhancing outcomes or not.

With respect to the LDC allocations, in the pessimistic scenario, we assume the allowance value flows to all classes of customers—residential, commercial, and industrial—via a reduction in the variable electricity rates on monthly bills. As we explained above, consumption is higher as a consequence of the subsidy to electricity prices, so the allowance price is higher in this case and the overall costs of the policy are higher as a result; this is the sense in which the assumption is pessimistic. In the optimistic scenario, we assume that fixed charges are reduced for industrial and commercial customers. However, we assume this remains infeasible for residential customers. As explained in Burtraw et al. (2009), a review of current billing practices and state public utility commission behavior suggests that significant hurdles exist to implementing this kind of pricing (a reduction in the fixed portion of the electricity bill) by 2016. One prominent reason for this is that in almost no case does the fixed portion of costs appear as a separate line item on bills for residential class customers, and even when it does appear separately, the cost is recovered almost entirely through volumetric charges. Furthermore, even if it were possible to return allowance value to residential customers through fixed payments it is unclear how residential customers would respond. Many observers have suggested that residential customers are unlikely to understand and respond in an economically rational way to an increase in the variable rate (marginal cost) by reducing consumption if their overall bill were reduced.[16] Industrial and commercial electricity consumers, however, may have a more sophisticated understanding of the difference between the fixed and variable parts of their bill, and thus we evaluate this possibility in our optimistic scenario with respect to the allocation to LDCs.

Another of the significant sources of uncertainty affecting the future development of energy and climate policy is the ability to harvest potential low cost opportunities for improvements in the way producers and consumers use energy. Indeed, how well previous investments of this nature have performed remains controversial (Arimura et al. 2009). The Waxman–Markey bill would

---

[16] Borenstein (2009) illustrates that models of consumer response that have been used in many previous studies of increasing-block pricing are not realistic models of the information consumers have at the time they make consumption decisions.

expand the previous programs to a national scale, and how well this approach would perform in the future in regions that have no experience with such programs is even more uncertain. These so-called "opportunity regions" may have substantial undeveloped potential, but they also lack the infrastructure, expertise and regulatory rules such as decoupling of cost recovery from energy sales that have developed over many years in other regions. In addition, these regions historically have lacked the will (perhaps because of historically low prices) to implement such programs. On the other hand, intuition and some evidence indicate that the greatest efficiency improvements at the least cost may be technically possible in these regions. Arimura et al. (2009) speculate that incremental spending by utilities that had low previous levels of spending could achieve savings at one-half the incremental cost of previous programs.

In the case of allowance value directed to energy efficiency and technology development, our optimistic scenario uses selective estimates in the literature for the cost of reducing electricity consumption through end-use efficiency improvements. We based the optimistic scenario on the premise that these options would not be adopted in response to the price signal introduced from the cap-and-trade program alone. The advocates of spending resources through LDCs on energy efficiency programs argue that this spending can overcome institutional or market barriers and, in effect, accomplish cost-effective investments (of the type illustrated by McKinsey 2007) that households cannot or do not make for themselves (Cowart 2008). Hence, in the optimistic scenario, spending by government or LDCs on energy efficiency provides a net benefit to society.

In a large survey of the energy efficiency literature, Gillingham et al. (2004) estimate that investments in efficiency have reduced electricity use at a payoff of about 2.8 cents per kWh of electricity saved (2002$). However, subsequent studies characterize this as a very optimistic estimate. Using improved statistical methods, Arimura et al. (2009) estimate an average cost of 6.2 cents per kWh (2007$) saved in previous programs. Accounting for the possibility of improvement in program design and the expansion to new regions of the country that provide low-cost opportunities, we use the Gillingham et al. estimate of 2.8 cents per kWh as an optimistic forecast of the cost-effectiveness of future investments in efficiency on average across the nation. We apply an average emissions intensity of electricity generation in the country of 0.000523 tons CO_2 per kWh of generation as forecast by the Haiku model in the baseline to arrive at an optimistic estimate of $53.50 as the cost per ton of avoided emissions resulting from energy efficiency investments.[17]

---

[17] For comparison, EIA (2009b) indicates average emissions intensity in 2010 is 0.000558 tons CO_2 per kWh.

The B.E. Journal of Economic Analysis & Policy, Vol. 10 [2010], Iss. 2 (Symposium), Art. 5

Compared to other estimates of the cost of emissions reductions and our estimated allowance prices, the cost per ton for reductions that could be achieved through energy efficiency are relatively expensive. However, these investments not only reduce emissions but they also reduce spending on electricity, providing a direct savings and, thus, an additional benefit to households. We also capture this impact in the model.

The optimistic estimates for the technology development provisions—those provisions targeting renewable energy, clean vehicles, and carbon capture and storage (CCS)—come from McKinsey and Company (2007), Kammen et al. (2009), and Al-Juaied and Whitmore (2009). These studies provide estimates of the cost per ton of CO_2 emissions reduced when particular technologies or alternative fuels are used in place of the conventional option. Al-Juaied and Whitmore (2009), for example, estimate the cost of abating CO_2 emissions through CCS, both for a first-of-a-kind plant and a mature technology plant in 2030, using a range of cost estimates from several previous studies. They conclude a first-of-a-kind plant is likely to have an abatement cost of $100–150 per metric ton CO_2 avoided, while a mature technology plant is likely to have an abatement cost of $30–50 per metric ton CO_2 avoided. Kammen et al. (2009) estimate the cost per ton of emissions reduced in a range of scenarios for plug-in hybrid and electric vehicles compared with conventional gasoline vehicles. As mentioned above, McKinsey and Company (2007) has cost and effectiveness estimates for a range of technologies and scenarios.

Based on findings in these studies, we assume in the optimistic case that CCS, renewable energy, and clean vehicle technology provide CO_2 emissions reductions at costs of $50/ton, $34/ton, and $75/ton, respectively.[18] In the pessimistic scenario, we assume that these activities and investments have no benefit and thus all of the allowance value devoted to them is lost.

Using these estimates along with the percentages devoted to these activities as specified in the bill, we calculate the total amount of money available, and then compute total emissions reductions. This means that fewer reductions are needed elsewhere in the economy, thus lowering the allowance price and providing a benefit to households. The pessimistic scenario simply assumes that the money going to all of the efficiency and technology provisions is lost.

Table 2 summarizes the assumptions that we use in our two scenarios for the Waxman–Markey bill. About 57 percent of allowances are allocated in the same way across the two scenarios; these are shown in the bottom portion of the

---

[18] We do not solve with any specific introduction date for any of the technology development provisions, and they may not be deployed by 2016. Instead, we assume that some of these technologies will work to reduce cumulative emissions targets over the lifetime of the cap-and-trade policy, and consequently 2016 allowance prices, through the banking mechanism.

table. The first four rows of that section show allocations that are captured in EIA's marginal abatement cost curve and that are taken into account in both the optimistic and pessimistic representation of H.R. 2454.[19] The provision dealing with low-income consumers is for households with annual incomes less than 150 percent of the poverty line. The EIA is directed to estimate loss in purchasing power for this group, and to refund it via direct cash transfers. We assume the allocations to refineries and to merchant (independently owned, unregulated) coal generation plants do not affect the variable costs of production, and hence are earned as industry profits that flow back to shareholders. Assigning the value of the domestic adaptation provisions to households is difficult. This money goes to dedicated trust funds; the money from those funds is to be spent on programs that offset the impacts of climate change. For lack of a better alternative, we simply distribute this money back to households as a lump-sum payment per person. Finally, we assume that all of the value of allowances going to international efforts — adaptation, forestry, and clean technology development — is lost to the U.S. economy.

The remaining allowances, or 43 percent of the total, are examined under the alternative optimistic and pessimistic scenarios. These allowances go to electricity LDCs and the variety of energy efficiency and technology development provisions discussed above.

We emphasize that the optimistic and pessimistic scenarios, particularly in the case of the energy efficiency and technology provisions, are truly bookends, i.e., they reflect the range of plausible outcomes and the truth may lie somewhere in between. Nonetheless, they provide a sense of the uncertainty in the outcomes with such a large and complex climate bill.

For a third and final comparison, we model the impacts of a cap-and-trade program with 100 percent auction of allowances, with 75 percent of the revenue returned as a lump-sum per capita payment, i.e., a cap-and-dividend program. This formulation roughly corresponds to the Cantwell–Collins proposal (S. 2877), which reserves 25 percent of allowance value for a set of unspecified spending priorities. To model this, we assume that the funds that are not returned as dividends include money directed to lowering consumer costs for home heating from natural gas and fuel oil (6.75 percent) and expenditures on related energy efficiency programs and building codes (4.25 percent). In addition, we assume the maintenance of allowance cost rebates to energy-intensive, trade-exposed industries (14.44 percent). These priorities are embodied in the marginal cost curves we adopt from the EIA.

---

[19] This includes 3.75 percent of allocation directed to LDCs and home heating that EIA models explicitly as investments in energy efficiency. This is incorporated in the EIA marginal cost curves, and is held constant across our scenarios. Allocations to natural gas LDCs and home heating also are captured in the EIA model and are constant across our scenarios.

The B.E. Journal of Economic Analysis & Policy, Vol. 10 [2010], Iss. 2 (Symposium), Art. 5

Finally, across all three scenarios we maintain the assumption that government incurs additional costs for its own direct expenditures on energy and other goods and services that are equal to 14 percent of allowance value.[20] It is not known if governments would meet these costs through tax increases or spending decreases, so the incidence of the costs is uncertain. Previous studies have frequently omitted this change in government cost, thereby implicitly introducing a tax that is hidden from their analyses, which will be paid ultimately by households. CBO (2009b) does not account for this cost directly, but nets it out against the 18 percent of allowances to government for spending on projects such as renewable energy and energy efficiency, and thus does not distribute to households these increases in government costs and spending. EPA (2009) predicts that increased prices of goods and services will increase government spending and assumes the increase is made up by adjusting taxes in a lump-sum manner. EIA (2009) analysis of H.R. 2454 does not specify if government costs increase under the policy, or if so, how the NEMS model accounts for the increased costs.

To account for the effect this has on household well being, we assume an increase in government costs equal to 14 percent of allowance value. Although the share is constant across scenarios, the absolute value of government's change in costs differs with the allowance price. We assume budget neutrality in this regard, as does EPA, but we assume this is achieved through an increase in the average personal income tax for each income group. The increase in personal income taxes is accounted for in calculating the net burden per household.

Results

Table 3 shows the average net household burden by income quintile and for all households for the three scenarios: the two H.R. 2454 scenarios and the 75 percent cap-and-dividend scenario. We first discuss the national average household burden and compare the different scenarios; in the following section, we explain the distributional results.

---

[20] CBO (2009a) predicts this cost for federal, state and local government combined to be 14 percent of allowance value; CBO (2009b) predicts it to be 13 percent.

Table 3. Average Net Household Burden of Three Cap-and-Trade Scenarios, by Income Quintile (in 2006$)

Household Income Quintile	Average Income	Waxman-Markey Optimistic Case	Waxman-Markey Pessimistic Case	75 Percent Cap-and-Dividend
1	11,276	-110	-122	-65
2	26,804	21	111	7
3	44,060	181	408	106
4	68,609	314	667	255
5	140,283	258	1,029	725
All	58,207	133	418	206
Allowance Price		$13.19 mt/CO2	$23.41 mt/CO2	$17.37 mt/CO2

Impacts on Average Households

The average net household burden varies widely across the three scenarios, from $133/household in the optimistic H.R. 2454 scenario to $418/household in the pessimistic scenario, with the cap-and-dividend option in between, at $206/household. There are a number of factors driving the differences. We focus first on the comparison of our optimistic and pessimistic scenarios.

Three categories of assumptions distinguish the optimistic and pessimistic scenarios: the assumptions about how LDCs will pass on the value of allowances to customers, in fixed or variable charges, and the assumptions about whether the value of allowances devoted to energy efficiency and technology programs yield benefits or are simply wasted resources. Separating the effects of these three policies on the difference between the optimistic and pessimistic scenarios is not straightforward. Each policy affects the allowance price, which in turn has a direct effect on household expenditures and on the revenue that flows to the other program areas. For example, a change in the LDC allocation from optimistic to pessimistic scenario causes the allowance price to increase, making more allowance revenue available as compensation for low-income families and for both technology and energy efficiency programs.

To understand the effects of these three policies we use the Shapley value, which is an axiomatic approach to sharing of costs or benefits (Roth 1988). The Shapley value is based on the calculation of the marginal contribution of every possible combination of policies that distinguish the optimistic and pessimistic scenarios, resulting in a unique estimate of the share of that difference that can be attributed to each policy. Table 4 shows the results for the pessimistic and

The B.E. Journal of Economic Analysis & Policy, Vol. 10 [2010], Iss. 2 (Symposium), Art. 5

optimistic cases, along with the Shapley values that describe the percent of the reduction in cost and allowance price that can be attributed to each of the policies.

Table 4. Shapley Values by Income Quintile per Household for LDC, Technology Provisions and Electricity Energy Efficiency Programs

Household Income Quintile	Waxman-Markey Pessimistic Case	Percentage Reduction in Household Cost Due to**			Waxman-Markey Optimistic Case
		LDC Billing Behavior	Technology Programs	Electricity Energy Efficiency	
1*	-122	485%	284%	-669%	-110
2	111	-77%	15%	162%	21
3	408	-23%	34%	89%	181
4	667	-9%	37%	72%	314
5	1,029	40%	21%	39%	258
All	418	6%	25%	69%	133
		Percent Reduction in Allowance Price			
Allowance Price $23.41 mt/CO2		21%	51%	28%	$13.19 mt/CO2

* The first income quintile experiences net losses under the optimistic assumptions. Consequently negative Shapley values signify gains from a given assumption and positive Shapley values signify losses.
**Percentage reductions for each quintile do not average to the value for all households.

The differences between the pessimistic and optimistic cases that are due to the LDC billing behavior are minimal, only 6 percent of the difference in the household burden and 21 percent of the difference in allowance price is due to LDC assumptions. This is because the difference between the pessimistic and optimistic LDC assumptions is not as extreme as it is for the other two policies. Changing from pessimistic to optimistic assumptions for LDCs is a reallocation from one method of spending the revenue to a more efficient one. For the other two programs, the switch from pessimistic to optimistic constitutes a shift from an unproductive loss of the revenue to productive investments that benefit households.

The savings to households due to changes in LDC assumptions are dwarfed by those of the electricity energy efficiency program, which accounts for 69 percent of the reduction in average household burden.[21] Efficiency policies provide savings through two separate mechanisms. Because of improved efficiency, households purchase less electricity than they would otherwise, and the lower quantity of consumption leads to a lower equilibrium price in electricity

---

[21] This percentage does not include efficiency measures totaling 4.25 percent that are included in the EIA marginal abatement cost curves and held constant across scenarios.

markets that further benefits households. In addition, emissions decline because less electricity is generated, meaning that other sectors of the economy do not need to reduce emissions as much; this lowers the allowance price and reduces overall costs. The efficiency investments account for cost savings of $99 due to avoided electricity expenditures for the average household compared to the pessimistic cases where no such investments exist.[22] This effect, in combination with the 28 percent allowance price reduction caused by the program, yields the full $197 (69 percent) savings due to the optimistic assumptions.

Although the payoff from electricity efficiency initiatives can be, and often is, sharply debated, we feel that the technology development initiatives in H.R. 2454 may be even more speculative. Over 8 percent of allowances are directed to CCS, clean vehicle technology, and renewable energy, and in our optimistic scenario, we assume these three investments provide CO_2 abatement at $50, $75 and $34 per ton respectively. These costs are greater than the value attained for emissions reductions in other components of the model, so they raise program costs at least in the near term (2016). Nonetheless, in the optimistic scenario they provide a total reduction in emissions of 127 million tons. This lessens the burden on other sectors and lowers the allowance price compared to the pessimistic scenario. We hasten to point out that because of their relatively high abatement costs, these initiatives do not lower the burden in comparison with an alternative where the money would be returned to households directly or spent on more cost-effective CO_2 reduction options.

Even considering the high abatement costs of the technology programs, they account for 51 percent, or $5.22, of the reduction in the allowance price between the pessimistic and optimistic cases. This large decrease is responsible for 25 percent of the savings ($71 per household). While these savings are significant, it is important to note that they are much smaller than those of the electricity energy efficiency program, although technology programs receive 8.1 percent of allowance value, compared to only 5.3 percent dedicated to the efficiency program.

As shown in Table 2, all of the revenue directed at energy efficiency and abatement programs is assumed to be lost in the pessimistic scenario. This money constitutes 13.44 percent of allowance value, approximately $17.4 billion ($134/household) of revenue, which does not find its way back to households. Households would experience a decreased burden of $286 if this money were rebated directly, and they experience even greater savings as it is spent on energy

---

[22] The government benefits from the electricity energy efficiency programs in the same way that households realize savings from decreased utility bills. The energy efficiency programs reduce the revenue collected by the government to pay for their increased energy costs under cap-and-trade. The electricity energy efficiency programs save households $16/year in increased tax revenue in the optimistic case.

The B.E. Journal of Economic Analysis & Policy, Vol. 10 [2010], Iss. 2 (Symposium), Art. 5

efficiency and direct abatement programs. The cap-and-dividend case does not share the complexities of the H.R. 2454 scenarios and is hence much easier to understand. There is no uncertainty as to the potential effectiveness of energy efficiency and technology development programs or the outcomes of electricity LDC allocations.[23] Instead, 75 percent of auction revenue is directed as a per capita dividend back to households that, as a result, experience a modest burden of $206 per year (last column of Table 3). The allowance price in this scenario is $17.37—substantially lower than in the pessimistic H.R. 2454 case but higher than in the optimistic case.

One reason the price and the net household burden are lower than in the pessimistic case is because of the 7 percent of allowance value going overseas to adaptation and supplemental reduction purposes in H.R. 2454 but not in the cap-and-dividend case. If this program were included in the cap-and-dividend scenario, directing a comparable 7 percent share of allowance value overseas, it would raise the average cost per household from $206 to $257. Furthermore, in the cap-and-dividend case, there is no cost from LDC allocations; but even in our optimistic scenario, we assume that LDC allocation leads to lower electricity prices for residential households, which increases the cost of the policy. However, most importantly, the uncertainty is eliminated. The wide range in burdens under the optimistic and pessimistic scenarios is a result of the uncertainty in how the LDC allocation will play out, as well as how the energy efficiency and technology provisions will work in practice. These sources of uncertainty do not exist with cap-and-dividend.

Distributional Consequences across Income Groups

In addition to differences for average households, our scenarios also reveal potential differences by income quintiles. However, the overall distributional consequence appears of secondary importance, compared to the overall changes in the level of the policy, which are reflected in the cost for the average household. The outcomes are reported in Table 3.

The H.R. 2454 optimistic scenario exhibits comparatively low costs for most income quintiles compared to the other scenarios, which is driven by the low allowance price and electricity savings due to energy efficiency programs. In the optimistic H.R. 2454 scenario, we assume that LDCs are able to pass the allowance value through the fixed portion of electricity bills for commercial and

---

[23] As in the optimistic and pessimistic Waxman–Markey cases, a small portion of allowance value directed to energy efficiency programs, as well as to subsidize home heating and to support trade-exposed industries is built into EIA modeling of the policy. Concequently, the outcome from these measures does not vary across our scenarios. In the cap-and-dividend case, this constitutes the 25.44 percent of allowance value that is not returned as dividends to households.

industrial customers; in the pessimistic scenario, the value ends up reducing variable electricity prices. These differences have significant distributional consequences. In the optimistic case, the transfer ultimately finds its way to shareholders, the vast majority of which belong to the top income quintile. In the pessimistic case, commercial and industrial customers pass through the reduced electricity price to lower prices of goods and services; this benefits all consumers. Consumption is less concentrated in the upper income quintiles than is the share of share ownership, which explains the fact that the average household in the fourth quintile bears a higher burden than the average household in the top quintile, in the optimistic scenario.

This effect can be seen in the Shapley values in Table 4. The fifth income quintile sees significant benefits from the optimistic LDC assumption due to transfers to shareholders. Forty percent of the reduction in high income household burdens moving from the pessimistic to optimistic case is due to the LDC assumption. The bottom four income quintiles, however, actually see their burdens rise from the LDC assumptions. They lose out because they do not receive the LDC subsidy through lower electricity and goods prices nor through an increase in shareholder value. Households in quintiles 1-4 gain relatively more from the energy efficiency assumptions. While high-income households may receive more absolute savings from the energy efficiency provisions, savings to households in the lower quintiles are a larger share of their overall savings that result from the optimistic scenario.

Households in the lowest income quintile have almost the same net burden in the optimistic and pessimistic cases. This results from the fact that low-income households receive 15 percent of total allowance value in both cases. The lower allowance price in the optimistic case leads to a smaller amount of money going to these households and this offsets some of the gains from the optimistic assumptions.

Figure 2 displays the net household burdens of the three scenarios as a percent of income. Both of the H.R. 2454 scenarios display a similarly shaped distribution in which costs increase as a percent of income for the 1st through 4th income quintile, but decline for the 5th quintile. This is most striking in the optimistic case, where the 5th quintile has a lower burden as a percentage of income than the 3rd quintile. As discussed above, these results are mostly due to LDC allocations ultimately finding their way to shareholders. By contrast, the cap-and-dividend scenario exhibits progressivity across the full income distribution.

The B.E. Journal of Economic Analysis & Policy, Vol. 10 [2010], Iss. 2 (Symposium), Art. 5

Figure 2. Average Net Household Burden as a Percent of Income for Three Cap-and-Trade Scenarios, by Income Quintile

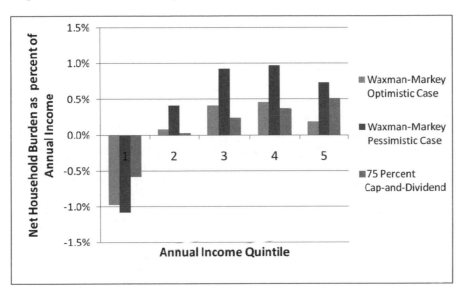

The results suggest that the uncertainty in outcomes in the Waxman-Markey bill does not lead to sharp differences in the *relative* impacts on different income groups, just differences in the absolute impacts. For example, the second quintile's burden as a percent of total burden across all quintiles is equal to 3 percent and 5 percent in the optimistic and pessimistic cases respectively; the fifth quintile's burden is 39 and 49 percent of the total in each case. These are small differences considering the absolute differences in burdens between the optimistic and pessimistic scenarios.

The first income quintile is a slight exception, however. As explained above, it has roughly the same burden as a percent of income in either the optimistic and pessimistic case. This is a result of our assumption about how the bill's provisions for low-income consumers flow to households (15 percent of total allowance value), which we hold constant across the optimistic and pessimistic scenarios.

Economists have long understood that current consumption is often based more on lifetime or permanent income than on current income. Some households with low current earnings, for example, might actually be consuming at a higher level due to their position in their career path (beginning or retirement) or other factors. For this reason, an analysis of the distributional impacts of a tax or other policy that groups households by annual income may misrepresent the true distributional impacts (Fullerton and Rogers 1993). Other problems such as inaccuracy in the reporting of annual incomes in a survey could contribute to the problem.[24]

Unfortunately, in cross-section data such as the CES, a measure of lifetime income is unavailable. As in other cross-section studies, however, we can use current consumption as a proxy for lifetime income (Poterba 1989; Hassett, Marthur, and Metcalf 2009; Burtraw et al. 2009). Figure 3 displays net household burdens as a percent of consumption for all three policy scenarios. The results are quite similar to those using annual income. Minor differences include the cap-and-dividend scenario being slightly more progressive than when annual income is used and the two Waxman-Markey scenarios looking progressive across the full distribution rather than just through the fourth quintile.

Figure 3. Average Net Household Burden as a Percent of Consumption for Three Cap-and-Trade Scenarios, by Consumption Quintile

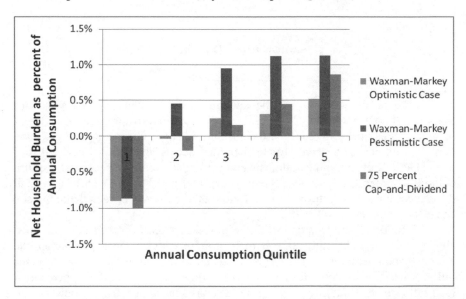

Examining the carbon pricing component alone, ignoring the distribution of allowance value, shows the policy to be regressive on the basis of annual income but approximately proportional on the basis of consumption. This is a

---

[24] It is important to note that starting in 2004 (our first year of data) the consumer expenditure survey started imputing annual income. This fills in nonresponess and improves the overall qualtiy of CES annual income data (Fisher 2006).

The B.E. Journal of Economic Analysis & Policy, Vol. 10 [2010], Iss. 2 (Symposium), Art. 5

typical finding in studies of energy taxes, general sales taxes, and other consumption-based taxes.[25] But with the complex distribution of allowance value to households, these differences across the two different income measures are muted and results for the two metrics are roughly comparable. It is important to note also that a large share of the rebate programs in Waxman-Markey (and as represented in this analysis) are means tested based on annual income. Allocation to shareholders is also based on annual income, not consumption. These factors help explain why the income and consumption measures of burden are similar, despite differences in the effects of carbon pricing before accounting for allowance value.

Conclusion

Climate change poses an extraordinarily challenging coordination problem for the international community and for the domestic body politic. This difficulty is exacerbated because the science of climate change, the underlying characterization of opportunities to reduce emissions, and the prospects for technological change are all extremely uncertain. Although the uncertainty about effects and outcomes may be seen by the scientific community as the motivation for the implementation of policy to reduce emissions, the same sources of uncertainty have emerged among critics as fundamental reasons to object to climate policy in political debate.

This paper focuses not on scientific or economic uncertainty, and only partially on technological uncertainty. Instead, we focus on a new source of uncertainty that is introduced in the design and implementation of policy intended to address this challenge. Specifically, we focus narrowly on uncertainty about the level and distribution of the burden on households that result from the formula for the allocation of emissions allowances in recent bills in the U.S. Congress. Even with this narrow focus, we find uncertainty about the implementation of the program and its costs on households to be substantial. The finding that policy design may exacerbate the fundamental characteristic that makes climate policy such a daunting challenge in the first place may be important to making progress on the policy debate.

We examine the allocation under H.R. 2454 (the Waxman-Markey bill) under bookend scenarios that we label optimistic and pessimistic. In the scenarios, we explore the possible outcome with respect to three sets of issues. The first two

---

[25] The Poterba (1989) study showed this, as have several others since then. Indeed, in our earlier study, we showed that the effects of carbon pricing alone were regressive on the basis of annual income but roughly proportional on the basis of annual consumption (Burtraw et al. 2009). However, the *net* effect, incorporating the distribution of allowance value (in various alternative schemes) looked quite similar across the two income measures.

scenarios explore the effect of allocations to electricity LDCs (30 percent of allowances) and the effect of allocations to promote efficiency (5.3 percent) and energy technology (8.14 percent). The remaining allowance allocations are held constant across the scenarios. In the optimistic scenario, we assume that the allocation to LDCs introduces relatively minimal distortions away from efficient pricing of electricity, although distortions remain important for the residential class of customers, and we assume favorable outcomes with respect to expenditures on efficiency and technology development. The pessimistic scenario has contrary assumptions; in fact, in the pessimistic case, we assume that efficiency and technology investments are wasted. For comparison, we consider a third scenario that directs 75 percent of allowance value directly back to households, keeping the remaining 25 percent directed to some of the categories that are held constant in the other two scenarios; this "cap-and-dividend" scenario is similar to the Cantwell-Collins bill (S. 2877).

Although the assumptions underlying our optimistic and pessimistic scenarios are bookends, they highlight the wide range of potential outcomes under a complex allocation scheme in some current climate bills, and they represent assessments that are commonly found as part of the contemporary political debate. We find that if investments in electricity energy efficiency programs and technology development pay off, and if, for industrial and commercial class customers, allocations to electricity LDCs do not dampen the rise in electricity prices brought about by climate policy, then the average net household burden experienced by households is only $136 in 2016 and the allowance price is as low as $13.20/ton. On the other hand, if the energy efficiency and technology investments are wasted and LDC allocations do dampen price rises for all electricity customers, the net household burden rises to $420, with an allowance price of $23.43/ton.

Shapley values help untangle the contribution of each policy in contributing to difference in burden under the optimistic assumptions. The LDC assumption is only responsible for 7 percent of the difference, with a large part of the benefits accruing to the highest income quintile at the expense of the lower quintiles. The electricity energy efficiency programs, on the other hand, are responsible for 69 percent of the average difference in household burden through a combination of reduced electricity expenditures and emissions. These results point to the potential gains to households from energy efficiency spending if the program produces results along the lines of the optimistic assumption.

There are important questions about and limitations to our analysis, but we conclude by drawing attention to just two. One important limitation is the use of partial equilibrium modeling for what is fundamentally a general equilibrium problem. The partial equilibrium approach allows us greater flexibility in manipulating institutional representation in the model, but important feedbacks within the economy are lost to the analysis. These include changes in factor prices, investment returns, and housing prices, as well as international trade effects and a host of other impacts associated with introducing new regulatory costs to the economy. Our partial equilibrium approach also does not take the benefits of mitigation into account. A second limitation is the actual schedule of opportunities for energy efficiency and technological innovation. Under each of our scenarios that assume optimistic or pessimistic outcomes, we assume a constant return to scale for these endeavors. In fact, efficiency or technology could exhibit increasing cost, but through learning by doing and other factors, it could alternatively exhibit decreasing costs. Moreover, the examination of these technological possibilities on a piecewise basis rather than in an integrated technological model introduces the opportunity for double counting of emissions reductions opportunities. While concerns about returns to scale arguably introduce bias in any direction, the possibility for double counting is likely to bias our cost estimates downward. In any event, this illustrates one more way in which the cost of climate policy to households is uncertain.

References

Al-Juaied, Mohammed, and Adam Whitmore. 2009. Realistic Costs of Carbon Capture. Discussion Paper 2009-08. Cambridge, MA: Energy Technology Innovation Research Group, Belfer Center for Science and International Affairs, Harvard Kennedy School.

Arimura, Toshi, Richard G. Newell, and Karen L. Palmer. 2009. Cost-Effectiveness of Electricity Energy Efficiency Programs. Discussion paper 09-48. Washington, DC: Resources for the Future.

Borenstein, Severin 2009. To What Electricity Price Do Consumers Respond? Residential Demand Elasticity under Increasing-Block Pricing. Paper presented at NBER., Summer Institute. July 2009, Cambridge, MA. http://www.nber.org/~confer/2009/SI2009/EEE/Borenstein.pdf

Boyce, James, and Matthew Riddle. 2009. Cap and Dividend: A State-by-State Analysis. Working paper August 5, 2009. Amherst, MA: Political Economy Research Institute University of Massachusetts.

Burtraw, Dallas. 2009. Climate Change Legislation: Allowance and Revenue Distribution. Testimony prepared for the U.S. Senate Committee on Finance. August 2009, Washington, DC. http://www.rff.org/rff/ documents/ 09- 08_testimony_burtraw_climate_legislation.pdf (accessed March 1, 2010).

Burtraw, Dallas, Richard Sweeney, and Margaret Walls. 2009. The Incidence of U.S. Climate Policy: Alternative Uses of Revenues from a Cap-and-Trade Auction. *National Tax Journal*, 62 (3): 497–518.

Burtraw, Dallas, Margaret Walls, and Joshua Blonz (Forthcoming), "Distributional Impacts of Carbon Pricing Policies in the Electricity Sector," in G.E. Metcalf, ed., Energy Taxation , Cambridge: Cambridge University Press.

Cowart, Richard. 2008. Carbon Caps and Efficiency Resources. *Vermont Law Review* 33: 201–223.

Dinan, Terry M., and Diane L. Rogers. 2002. Distributional Effects of Carbon Allowance Trading: How Government Decisions Determine Winners and Losers. *National Tax Journal* 55(2): 199–221.

Fisher, Jonathan. 2006. Income Imputation and the Analysis of Expenditure Data in the Consumer Expenditure Survey. Working Paper 394, Washington DC: U.S. Bureau of Labor Statistics.

Fullerton, Don and Diane Lim Rogers. Who Bears the Lifetime Tax Burden? Washington, DC: Brookings Institution, 1993.

Fullerton, Don and Garth Heutel. 2007. The general equilibrium incidence of environmental taxes" *Journal of Public Economics*, 91(3-4): 571-591.

Gillingham, Kenneth, Richard Newell, and Karen Palmer. 2004. Retrospective Examination of Demand-Side Energy Efficiency Policies. Discussion Paper 04-19. Washington, DC: Resources for the Future.

Goulder, Lawrence H., Ian W.H. Parry, Roberton C. Williams III, and Dallas Burtraw. 1999. The Cost-Effectiveness of Alternative Instruments for Environmental Protection in a Second-Best Setting. *Journal of Public Economics* 72(3): 329–360.

Grainger, Corbett A., and Charles D. Kolstad. 2009. Who Pays a Price on Carbon? NBER Working Paper 15239. Cambridge, MA.

Hahn, Robert and Robert Stavins. 2010. The Effect of Allowance Allocations on Cap-and-Trade System Performance. NBER Working Paper 15854. Cambridge, MA.

Hassett, Kevin A., Aparna Marthur, and Gilbert E. Metcalf. 2009. The Incidence of a U.S. Carbon Tax: A Lifetime and Regional Analysis. *The Energy Journal*, 30(2): 157–179.

Kammen, Daniel M., Samuel M. Arons, Derek M. Lemoine, and Holmes Hummel. 2009. Cost-Effectiveness of Greenhouse Gas Emission Reductions from Plug-in Hybrid Electric Vehicles. In Plug-In Electric Vehicles: What Role for Washington? David B. Sandalow, ed., Brookings Institution Press, 170–191.

The B.E. Journal of Economic Analysis & Policy, Vol. 10 [2010], Iss. 2 (Symposium), Art. 5

McKinsey and Company. 2007. Reducing U.S. Greenhouse Gas Emissions: How Much and at What Cost? U.S. Greenhouse Gas Abatement Mapping Initiative Executive Report. Chicago: McKinsey & Company, Inc. http://www.mckinsey.com/clientservice/sustainability/pdf/US_ghg_final_r eport.pdf (accessed January 15, 2010).

Metcalf, Gilbert E. 2009. Designing a Carbon Tax to Reduce U.S. Greenhouse Gas Emissions. *Review of Environmental Economics and Policy* 3(1): 63–83.

Paltsev, Sergey, John M. Reilly, Henry D. Jacoby, Angelo C. Gurgel, Gilbert E. Metcalf, Andrei P. Sokolov, and Jennifer F. Holak.2007. NBER Working paper 13176. Cambridge, MA.

Parry, Ian W.H. 2004. Are Emissions Permits Regressive? *Journal of Environmental Economics and Management* 47(2): 364–87.

Parry, Ian W.H., Roberton C. Williams, and Lawrence H. Goulder. 1999. When Can Carbon Abatement Policies Increase Welfare? The Fundamental Role of Distorted Factor Markets. *Journal of Environmental Economics and Management* 37(1): 52–84.

Paul, Anthony, Dallas Burtraw, and Karen Palmer. 2009. Haiku Documentation: RFF's Electricity Market Model Version 2.0. Washington, DC: Resources for the Future.

Pizer, William A., James N. Sanchirico, and Michael B. Batz. 2009. Regional Patterns of U.S. Household Carbon Emissions. *Climatic Change.* 99(1-2): 47–63.

Poterba, J.M. 1989. Lifetime Incidence and the Distributional Burden of Excise Taxes. *American Economic Review* 79(2): 325–30.

Rausch, Sebastian, Gilbert E. Metcalf, John M. Reilly, and Sergey Paltsev. 2009. Distributional Impacts of a U.S. Greenhouse Gas Policy: A General Equilibrium Analysis of Carbon Pricing Joint Program Report Series, Report 182. Cambridge, MA: Massachusetts Institute of Technology.

Ross, Martin T. 2008. Documentation of the Applied Dynamic Analysis of the Global Economy (ADAGE) model. RTI International Working paper 08_01. http://www.rti.org/pubs/adage-model-doc_ross_sep08.pdf

Roth, Alvin E. 1988. *The Shapley Value—Essays in Honor of Lloyd S. Shapley.* Cambridge, U.K.: Cambridge University Press.

Sweeney, Jim, and John Weyant. 2008. Analysis of Measures to Meet the Requirements of California's Assembly Bill 32. Discussion draft, September 2008. Stanford, CA: Stanford University Precourt Institute for Energy Efficiency.

U.S. (Congressional Budget Office). CBO. 2009a. The Estimated Costs to Households from the Cap-and-Trade Provisions of H.R. 2454. June 19, 2009. Washington, DC: U.S. CBO.

————. 2009b. The Economic Effects of Legislation to Reduce Greenhouse-Gas Emissions. Washington, DC: CBO. http://www.cbo.gov/ftpdocs/105xx /doc10573/09-17-Greenhouse-Gas.pdf (accessed January 13, 2010).

U.S. (Environmental Protection Agency). EPA. 2009. EPA Analysis of the American Clean Energy and Security Act of 2009—H.R. 2454 in the 111th Congress. Washington, DC: U.S. EPA. http://www.epa.gov/climatechange/economics/pdfs/HR2454_Analysis.pdf (accessed January 13, 2010).

U.S. (Energy Information Administration). EIA. 2005. Production Tax Credit for Renewable Electricity Generation. *Annual Energy Outlook*, 2005. Washington, DC: U.S. EIA.

————. 2009a. Energy Market and Economic Impacts of H.R. 2454, the American Clean Energy and Security Act of 2009. SR/OIAF/2009-05. Washington, DC: U.S. EIA. http://www.eia.doe.gov/oiaf/servicerpt/hr2454/pdf/sroiaf(2009)05.pdf (accessed January 13, 2010).

————. 2009b, *Annual Energy Outlook*, 2010. Washington, DC: U.S. EIA.

The B.E. Journal of Economic Analysis & Policy

Symposium

Volume 10, Issue 2	2010	Article 6

DISTRIBUTIONAL ASPECTS OF ENERGY AND CLIMATE POLICY

Comment on "Climate Policy's Uncertain Outcomes for Households: The Role of Complex Allocation Schemes in Cap-and-Trade"

Arik M. Levinson*

*Georgetown University, aml6@georgetown.edu

Recommended Citation

Arik M. Levinson (2010) "Comment on "Climate Policy's Uncertain Outcomes for Households: The Role of Complex Allocation Schemes in Cap-and-Trade"," *The B.E. Journal of Economic Analysis & Policy*: Vol. 10: Iss. 2 (Symposium), Article 6.
Available at: http://www.bepress.com/bejeap/vol10/iss2/art6

Levinson: Comment on "Climate Policy's Uncertain Outcomes"

Policy uncertainty has become a hot topic in Washington DC. The National Association of Manufacturers (NAM) cites uncertainty about "expiration of the 2001 and 2003 tax cuts" as a primary cause of slack labor demand in 2010 (NAM, 2010). With regard to climate change, commenters on both sides of the debate lament the current uncertainty surrounding U.S. climate policy. The right contends that climate policy uncertainty stifles economic growth and promoted California's Proposition 23, which would have indefinitely delayed enforcement of the state's far-reaching climate legislation. The left contends that climate policy uncertainty suppresses investment in clean energy and campaigned against Proposition 23.

Into this fray step Blonz, *et al.* (2010), with some actual evidence on policy uncertainty. They model one particular aspect of climate policy uncertainty, the potential distribution of carbon emission allowances under the Waxman-Markey cap-and-trade bill passed by the U.S. House of Representatives in 2009. How large is the resulting uncertainty U.S. households face? As it turns out, not so large.

Table 1 displays Blonz, *et al.*'s lowest and highest estimates of the cost per household. The low estimate comes from their "optimistic" case, where allocations to electricity local distribution companies (LDCs) result in lower *fixed* charges on some customers' utility bills. This gives those customers the marginal incentive to conserve, but offsets the average costs with a subsidy. The high estimate comes from the Blonz, *et al.* "pessimistic" case, where LDC allocations result in lower *variable* electricity charges. This reduces those customers' marginal incentives to conserve, making it harder to reduce emissions from the electricity sector, and requiring more costly abatement from other sectors. As line 1 of Table 1 shows, the low and high estimates differ by $285 per year per household, which is plus or minus 52 percent of the average of the two estimates.

Taken out of context, a swing of 52 percent seems highly uncertain. In fact, it is not – even when compared to other models of climate change. The Energy Information Agency (EIA) scoring of the Waxman-Markey bill focused on two different sources of policy uncertainty: the amount of electricity generated with carbon capture and sequestration and nuclear power, and the treatment of offsets (what type would be approved and limits on their use). The cases examined by EIA vary from $100 to $746 in increased household energy expenditures (in year 2020, in 2006 dollars), a range of plus or minus 76 percent. Measured as a decrease in household consumption, the EIA cases range from $29 to $352, plus or minus 85 percent.

Similarly, a study organized by the American Council for Capital Formation (ACCF) and NAM (ACCF/NAM, 2009) used the same underlying model as the EIA study, but with different high and low cost scenarios. They

The B.E. Journal of Economic Analysis & Policy, Vol. 10 [2010], Iss. 2 (Symposium), Art. 6

report decreases in household disposable income ranging from $115 to $243 per year, which is plus or minus 36 percent from the mean of those two numbers.

Table 1: Uncertainty in Household Expenditures ($ per year per household)

Panel A: Uncertainty in Waxman-Markey

Study	Source of uncertainty	Measure of welfare	Low	High	Gap	+/- above or below the mean
Blonz, et al. (2010)	Allowance allocations	Net household burden = consumer surplus loss minus refunded allowance value	133	418	285	+/- 52%
EIA (2009)	Offset and banking treatment	Increased household energy expenditures (2020)	100	746	646	+/- 76%
EIA (2009)	Offset and banking treatment	Reduced household consumption (2020)	29	352	323	+/- 85%
ACCF/ NAM (2009)	Low-cost and high- cost cases	Decrease in household disposable income (2020)	115	243	128	+/- 36%

Panel B: Uncertainty in other household outcomes

Source of uncertainty	Low	High	Gap	
Gasoline expenditures (20,000 miles at 20 mpg)	1,464	3,652	2,188	+/- 43%
$10,000 invested in the S&P 500 in 2005	5,350	12,806	7,456	+/- 41%

All values in $2006. The +/- % are relative to the simple averages of the high and low figures.

Uncertainty about LDC allowance allocations would likely not be the greatest source of climate policy uncertainty facing households, but nor would it be the greatest source of energy-related uncertainty. Consider the cost of gasoline. Average weekly national retail prices varied from $1.56 to $3.90, just during the three years from 2007-2009. A typical household driving 20,000 miles per year in a vehicle getting 20 miles per gallon could see annual gas expenditures fluctuating from $1,464 to $3,652 based on those weekly prices, a swing of $2,188 or 43 percent.

For even more uncertainty, consider financial wealth. A household with $10,000 invested in an S&P 500 index fund in 2005 would have seen its real

Levinson: Comment on "Climate Policy's Uncertain Outcomes"

wealth rise to $12,806 in October of 2007 and fall to $5,350 in March of 2009, a swing of $7,456 or 41 percent.

Two results jump out from this. First, despite the rhetoric, the annual cost of climate policy is not crippling. Even the ACCF and NAM – not exactly climate policy boosters – estimate the annual cost per household at $243. And that's their high cost estimate.

Second, while Blonz *et al.* show that the *distribution* of outcomes across household income quintiles is not dramatically affected by LDC allocations, I would argue that nor is the *level* of outcomes. Sure, the pessimistic case costs households $285 per year more than the optimistic case. But that's not large relative to uncertainty about other features of the bill, such as the treatment of banking and offsets, uncertainty about the pace of technology change and the adoption of nuclear and CCS energy generation, and uncertainty about whether or not any bill will be passed at all. Moreover, all of those sources of household uncertainty combined are still tiny relative to the everyday uncertainty households face in their annual energy costs and financial wealth.

References

American Council for Capital Formation / National Association of Manufacturers (ACCF/NAM). 2009. "Economic Impact of the Waxman-Markey American Clean Energy and Security Act." August 12, 2009. (http://www.accf.org/media/docs/nam/2009/National.pdf)

Blonz, Joshua. Dallas Burtraw, and Margaret A. Walls. 2010. "Climate Policy's Uncertain Outcomes for Households: The Role of Complex Allocation Schemes in Cap-and-Trade" *The B.E. Journal of Economic Analysis & Policy* Vol.10, Iss.2, Article 5.

Energy Information Administration (EIA). 2009. "Energy Market and Economic Impacts of H.R. 2454, The American Clean Energy and Security Act of 2009" U.S. Department of Energy, Washington DC. (http://www.eia.doe.gov/oiaf/servicerpt/hr2454/index.html)

National Association of Manufacturers (NAM). 2010. "Labor Day 2010: The Impact of Anti-Labor Policies on Working Men and Women." (http://www.nam.org/Communications/Articles/2010/09/NAM-Report-Examines-Impact-of-Anti-Labor-Policies-on-Working-Men-and-Women.aspx)

The B.E. Journal of Economic Analysis & Policy

Symposium

| *Volume* 10, *Issue* 2 | 2010 | *Article* 9 |

DISTRIBUTIONAL ASPECTS OF ENERGY AND CLIMATE POLICY

What are the Costs of Meeting Distributional Objectives for Climate Policy?

Ian W. H. Parry*   Roberton C. Williams[†]

*Resources for the Future, parry@rff.org

[†]University of Maryland, Resources for the Future, and NBER, rwilliams@arec.umd.edu

What are the Costs of Meeting Distributional Objectives for Climate Policy?*

Ian W. H. Parry and Roberton C. Williams III

Abstract

This paper develops an analytical model to quantify the costs and distributional effects of various fiscal options for allocating the large rents created under proposed cap-and-trade programs to reduce domestic, energy-related CO_2 emissions. The trade-off between cost effectiveness and distribution is striking. The welfare costs of different policies, accounting for linkages with the broader fiscal system, range from negative $6 billion/year to a positive $53 billion/year in 2020 (or from -$12 to almost $100 per ton of CO_2 reductions). The least costly policy involves auctioning all allowances with revenues used to cut income taxes, while the most costly policies involve recycling revenues in lump-sum dividends or grandfathering emissions allowances. The least costly policy is regressive, however, while the dividend policy is progressive. Grandfathering permits is both costly and regressive. A distribution-neutral policy entails costs of $18 to $42 per ton.

KEYWORDS: cap-and-trade, welfare cost, distributional incidence, revenue-recycling

*We are grateful to Don Fullerton, Bill Randolph, an anonymous referee, and participants in the Chicago-RFF-Illinois Symposium on Distributional Aspects of Energy and Climate Policy for helpful comments and suggestions. And we thank Fanqing Ye for her excellent research assistance.

1. Introduction

One of the most important issues in the design of domestic, market-based climate policy is what to do with the rents or revenues created under cap-and-trade or emissions tax systems.[1] How these policy rents are allocated and used critically affects not only the distributional impacts of the policy, but also its overall cost-effectiveness.

There are many ways to allocate policy rents. One is to grandfather allowances in a cap-and-trade system, that is, give them away free to existing sources, typically based on their historical emission rates. The main motivation for this approach is political—providing compensation for producers affected by the regulation may make it easier to move legislation forward. Almost all of the allowances in the US program to cap SO_2 emissions from power plants were given away for free and similarly in the initial phases of the European Union's CO_2 trading program. However, more recent European and US federal climate initiatives have moved away from full grandfathering, as it is widely appreciated that this approach substantially overcompensates emitters for their compliance costs.[2]

More recently, the distributional concern in the United States has been on the household side, particularly insulating low-income households from the prospective increase in energy prices from climate policy. Some proposals involve granting free allowances to local distribution companies with the expectation that the value of these allowances will be rebated to households in the form of lower electricity bills. Another approach that has recently gained traction is known as "cap-and-dividend", which involves a cap-and-trade program with full allowance auctions with all the revenue returned in equal lump-sum transfers for all individuals. An argument for this approach is that all individuals have equal ownership rights to the environment and therefore proceeds from charging for use of the environment should be shared equally.

A key drawback of both of these approaches is that they forgo the potentially large efficiency gains from using allowance auction revenues to cut

---

[1] Other key issues include the overall stringency of the policy, its sectoral coverage (including provisions for emission offsets), whether cap-and-trade should include price stabilization mechanisms, and to what extent supplementary instruments are warranted to address other possible market failures (e.g., technology spillovers). For a general discussion of these issues see, for example, Aldy et al. (2010).

[2] Power companies in the European Union earned large windfall profits when the CO_2 cap was first introduced (e.g., Sijm et al., 2006). At least for a moderately scaled CO_2 permit system, only about 15-20 percent of allowances are needed to compensate energy intensive industries for their loss of producer surplus, so the huge bulk of the allowances could still be auctioned (e.g., Bovenberg and Goulder, 2001, Smith et al., 2002).

The B.E. Journal of Economic Analysis & Policy, Vol. 10 [2010], Iss. 2 (Symposium), Art. 9

other distortionary taxes. Personal income and other taxes distort capital and labor markets by depressing net factor returns. They also distort household consumption decisions by allowing exemptions or deductions for particular types of spending (e.g., employer medical insurance). Therefore, using revenues to lower the rates of these distortionary taxes produces gains in economic efficiency that can substantially lower the overall welfare costs of climate policy (though this approach may do little to help with distributional objectives). In fact, without this counteracting revenue-recycling effect, the welfare costs of market-based climate policies are substantially higher in the presence of distortionary taxes, because those distortions raise production costs and product prices, and thereby lower real factor returns and factor supply (e.g., Goulder et al., 1999, Fullerton and Metcalf, 2001).

To date, no climate bills have been introduced in the US Congress that would use policy revenues to reduce marginal income tax rates, though some include uses that would have similar effects.[3] Moreover, considering this case helps clarify the quite striking trade-offs involved in allocating climate policy rents.

In this paper, we develop and parameterize an analytical general-equilibrium model that synthesizes two different strands in the literature on domestic climate policy, one focusing on the cost-effectiveness of alternative market-based policies in a homogeneous agent framework and the other that looks at distributional effects in multi-agent models but with no attention to efficiency. While some of our results—such as the low cost-effectiveness of cap-and-dividend and grandfathered permits—have been recognized for some time, our analysis offers a more complete picture of efficiency/distributional trade-offs than can be gleaned from prior literature.[4]

---

[3] The Waxman-Markey, bill for example, devotes a portion of revenue to deficit reduction. If that deficit reduction would otherwise have been accomplished by raising marginal rates, then this would have the same effect as lowering marginal rates. In addition, a bill sponsored by Rep. Larson (HR 1337) would have used revenues to exempt a portion of income from payroll taxes. While this form of recycling would increase the return to labor force participation it would be less efficient than cutting marginal income tax rates. The latter would also increase effort on the job and reduce the bias towards tax-preferred spending. CO_2 taxes which are largely revenue neutral have been implemented in British Columbia and in Scandinavian countries, though a tax proposal for France (with very limited coverage) was recently ruled unconstitutional.

[4] Prior analyses by Burtraw et al. (2009), Dinan and Rogers (2002), and Parry (2004) have studied the distributional impacts of various options for recycling climate policy revenues through the tax system. And many papers have been written on the linkages between environmental policies and the broader tax system and how these affect the overall costs of policy (see, for example, Goulder et al., 1999, Fullerton and Metcalf, 2001, Bovenberg and Goulder, 2002, and Schöb, 2006, among many others).

Our analysis is more than just the sum of findings from these two literatures, however. Most importantly, the efficiency and distributional results come from the same model, and thus are

Our analysis considers three bounding cases for rent allocation and use under cap-and-trade systems. These include full grandfathering of allowances; full auction of allowances with revenues returned in lump-sum-transfers; and full auctions with revenues recycled to cut income taxes, such that all household groups get the same percentage-point reduction in tax rates (i.e., the tax cut is the same proportion of pre-tax income for all income groups). We also consider a "distribution neutral" policy where the tax and benefit system is adjusted to neutralize any distributional effects of the cap-and-trade policy on households, leaving the overall policy change neither progressive nor regressive. Policymakers are more likely to choose combinations of the bounding cases studied here, rather than using 100 percent of rents for one purpose alone, however the implications of these combinations are easily inferred by taking the appropriate weighted averages of our bounding cases. Furthermore, although our discussion is couched in terms of cap-and-trade, each cap-and-trade variant has an emissions tax counterpart in our analysis.[5]

New revenue sources might be used in many ways other than those studied here. These run the gamut from program enhancing measures like clean technology programs, incentives for energy efficiency, and capacity investments to facilitate adaptation to climate change, to general increases in public spending and federal deficit reduction. However, it is difficult to make general statements about the potential efficiency implications from these broader alternatives for revenue use, without specific evidence on the benefit and costs of the spending measures, accounting for possible market failures that they might address. Even the effects of deficit reduction are opaque, as it is unclear whether lower debt burdens for future generations will lead to lower taxes or more public spending.

We find that the allocation of rents under cap-and-trade (or use of revenues under carbon taxes) can have huge efficiency and distributional

---

directly comparable. This model has several additional advantages. Unlike in most prior literature we account for broader distortions of the tax system beyond those in factor markets, so we can integrate empirical findings from recent public finance literature on the taxable income elasticity (see below). Furthermore, our distributional analysis takes into account implicit burdens on households from the changes in deadweight loss from the broader fiscal system. Moreover, we account for inflation indexing of the tax/benefit system, which provides some (albeit limited) automatic adjustment to higher energy prices. In addition, the type of multi-agent model used here is needed to avoid aggregation bias when households face different tax rates, benefit differently from revenue-recycling schemes, and have different behavioral responses to tax changes. Finally, the distribution-neutral approach described below has not previously been analyzed in the context of carbon policy.

[5] For example, a carbon tax with revenues returned in lump-sum transfers to firms in proportion to their historical emissions would be equivalent to cap-and-trade with grandfathered allowances. The tax and cap-and-trade approaches would differ if our analysis were extended to incorporate uncertainty over future emissions abatement costs (e.g., Pizer, 2003).

The B.E. Journal of Economic Analysis & Policy, Vol. 10 [2010], Iss. 2 (Symposium), Art. 9

consequences for carbon policy, effects that can be much larger than the direct effects of the carbon policy itself.

Within the range of options we consider, the direct cost of reducing CO_2 emissions to 9% below business-as-usual levels in the year 2020 is $9 billion (all figures are in year 2007 $). However, the overall welfare costs of different policies, taking into account linkages with the broader fiscal system, ranges from *negative* $6 billion/year to $53 billion/year. In terms of average cost of CO_2 reductions this is a huge range—from minus $12 to almost $100 per ton.

The distribution of that cost ranges from highly progressive (with the bottom two income quintiles bearing a negative burden) to highly regressive (with the top two quintiles bearing a negative burden). There are stark trade-offs between cost-effectiveness and distribution in the design of market-based climate control policies. A cap-and-dividend approach makes carbon policy more progressive at the expense of dramatically raising the overall cost, and conversely, using revenue to fund a proportional (i.e., equal-percentage-point) income tax cut lowers overall cost but leads to a regressive distribution of those costs. A distribution-neutralizing tax change represents a middle ground for both efficiency and distributional effects. Cap-and-trade with grandfathered allowances performs poorly on both cost and distributional grounds, though it may reduce industry opposition and thus be more politically feasible.

The rest of the paper is organized as follows. Section 2 develops our analytical framework and key formulas for the efficiency cost and distributional burden of different policies. Section 3 describes the model parameterization. Section 4 presents the main quantitative findings and sensitivity analysis. Section 5 offers concluding remarks.

2. Analytical Framework

A. Model Assumptions

We use a long run, comparative static model with multiple agents, each representing an average over all households within a particular income class. Households choose consumption of general goods, products that are energy intensive, products that receive favorable treatment from the tax system, and labor supply. The government prices carbon dioxide (CO_2) emissions, through a cap-and-trade system, which in turn drives up energy costs and product prices in general. Rents created by climate policy may accrue to the private sector (e.g., through free allowance allocation) or to the government. In the latter case, revenues are used to adjust the income tax and transfer system.

(i) Household utility. We divide households into 5 equal-sized income groups, indexed by i, where $i = 1$ and 5 denote the lowest and highest income quintiles,

respectively. Each group is modeled as a single representative household, with the utility function:

(1) $u_i(X_{iE}, X_{iF}, X_{iC}, L_i, G^{PUB})$

X_{iE} denotes consumption of an aggregate of goods whose production or use is energy intensive (e.g., electricity-using durables, auto travel, home heating fuels). X_{iF} is consumption of goods that are favored through the broader tax system, such as employer-provided medical insurance or owner-occupied housing. X_{iC} is an aggregate of all other (non-energy-intensive, non-tax-favored) goods. L_i is work effort, implicitly combining labor force participation rates and average hours worked on the job by households in that group. G^{PUB} is government spending on public goods, which is fixed, and is included only to scale income tax rates to their observed levels. u_i is quasi-concave, twice differentiable, decreasing in labor supply, and increasing in all other arguments.

(ii) Household budget constraint. The household budget constraint is given by:

(2) $\sum_j p_j X_{ij} = I_i$

where $j = E, F, C$ indexes the three goods and p_j is the market price of good j.
    The tax and transfer system, reflecting state and federal personal income taxes, employer and employee payroll taxes, and sales taxes, minus government transfers, is represented as a piecewise-linear function with five segments and each of the five income groups falling on a different segment. For simplicity, we assume that any policy-induced changes in taxable income will be sufficiently small that no group will move to a different segment of the tax schedule. Under this assumption, the tax schedule group i faces is equivalent to a linear tax with rate t_i and intercept $-G_i$ (which can be found by extending the relevant segment of the piecewise-linear tax schedule out to a point with zero pre-tax income). Disposable income (I_i) consists of taxable income (Γ_i), net of taxes paid on that income. This gives

(3a) $I_i = (1 - t_i)\Gamma_i + G_i$

    Taxable income consists of the wage w_i times labor supply, plus lump-sum net-of-tax profit income π_i, less spending on tax-favored goods. Profit income accrues to households via their ownership of firms and reflects any producer rents created by cap-and-trade policy.

(3b) $\Gamma_i = w_i L_i + \pi_i - p_F X_{iF}$

The B.E. Journal of Economic Analysis & Policy, Vol. 10 [2010], Iss. 2 (Symposium), Art. 9

The (pre-existing) tax system causes two sources of distortion. First, it distorts labor supply by depressing the returns to work effort. Second, it creates a bias towards tax-preferred spending. The combined effect of carbon policies on these two distortions can be summarized by changes in taxable income.[6]

(iii) Production, CO_2 emissions, prices and rents. All goods are produced under constant returns to scale under perfect competition (hence there are no pure profits in the absence of carbon policy). Market prices of goods, which implicitly incorporate any energy costs associated with their use, are determined by:

(4) $p_j = \rho_j p_H + \beta_j$

Here p_H is the price of energy, β_j is a parameter representing non-energy costs, and ρ_j is the energy intensity of good j, where $\rho_E > \rho_C$ and $\rho_E > \rho_F$.[7]

The price of energy is determined as follows

(5) $p_H = c(z^0 - z) + \tau z + p_H^0$

where z is CO_2 emissions per unit of energy, $c(.)$ is a positive, convex abatement cost function ($c(0) = 0$) and a superscript 0 denotes the initial value of a variable (prior to the introduction of carbon policy). A reduction in z represents a switch to lower carbon but more expensive fuels in energy production (e.g., a switch from coal to renewables or nuclear power). Emissions are priced at rate τ, which reflects the allowance price. Our assumptions also imply that emissions prices and abatement costs will be fully passed through into energy prices. As discussed below, this assumption is debatable, although it does not affect the key focus of our paper, which is the trade-off between costs and distributional goals in allowance allocation.

Economy-wide CO_2 emissions, Z, are the product of the emissions intensity of energy and energy use aggregated over all products and households:

(6) $Z = z \sum_j \rho_j \sum_i X_{ij}$

For simplicity, we focus on a policy covering CO_2 only, which accounts for about 80 percent of total US greenhouse gases. In addition, the (energy-related

---

[6] We implicitly assume that that tax preferences do not address any market failures, and thus are purely distortionary.

[7] For simplicity ρ_j is taken as fixed though relaxing this would not affect the results, given our parameterization below.

domestic) CO_2 reductions, and the emissions price, are approximately consistent with those projected under recent climate legislation.[8]

Profit income to household group i is given by

(7) $\pi_i = \theta_i \phi \tau Z$

where θ_i is the fraction of (economy-wide) energy capital owned by household group i (implicitly, this includes both retirement and non-retirement bond and stockholdings), where $\Sigma_i \theta_i = 1$ and θ_i is larger for higher income groups. ϕ is the portion of policy rents that are left in the private sector, as opposed to accruing to the government in revenue. $\phi = 0$ represents a cap-and-trade system with full allowance auctions. $\phi = 1$ represents a system with 100 percent free allowance allocation to energy firms, where entire policy rents τZ are reflected in higher firm equity values and stockholder wealth, and such income is not taxed. Intermediate cases might represent partial taxation of allowance rents or partial auctioning of allowances. Outcomes for these cases are easily inferred by taking weighted averages of the results below.

(iv) Government constraints and policy. The government is subject to the budget constraint

(8) $G^{PUB} = \sum_i t_i \cdot \Gamma_i - \sum_i G_i + (1 - \phi) \tau Z$

This constraint requires that spending on public goods equals revenue from the income tax system, plus revenue from the carbon policy.

Following the introduction of a carbon policy, the intercept terms in the tax schedule are adjusted as follows:

(9) $G_i = G_i^0 \bar{p} + \hat{G}_i,$ $\bar{p} = \sum_j \mu_j p_j,$ $\mu_j = \dfrac{\sum_i X_{ij}^0}{\sum_{ij} X_{ij}^0}$

---

[8] Although under current legislation firms may purchase offsets in other domestic sectors (e.g., forestry and agriculture) and in other countries we do not count those as part of the emission reduction, nor do we include the deadweight losses (in other sectors or countries) associated with those broader reductions. A key concern with offsets is the difficulty of verifying whether or not these broader emission reductions would have occurred anyway, and whether they might be negated through increased emissions elsewhere (e.g., slowed deforestation in one country might accelerate deforestation in other countries through a rise in global timber prices).

The B.E. Journal of Economic Analysis & Policy, Vol. 10 [2010], Iss. 2 (Symposium), Art. 9

where we have normalized all initial product prices to unity. $\overline{P}$ is the general price level, a weighted average of market prices where the weight μ_j is the (initial) share of economy-wide spending (or consumption) on good j.

According to (9), the nominal tax schedule is automatically "inflation-indexed" to reflect increases in the general price level (this implies that marginal and average tax rates depend on real income). In addition, some schemes devote a portion of the new revenues to making the policy more progressive, which implies an additional adjustment $\hat{G}_i$ to transfer payments.

More specifically, we consider four bounding cases for the allocation/use of policy rents (each defined for the same emissions reduction/emissions price):

Proportional income tax cut ($\phi = 0$, $\hat{G}_i = 0$). In this case, all policy rents go to the government and, after indexing the nominal tax schedule, all other revenue is used to finance a tax cut that is proportional to pre-tax income (i.e., equal percentage-point cuts in all tax brackets), such that $dt_i / d\tau = dt / d\tau < 0, \forall i$.[9]

Cap and dividend ($\phi = 0$, $\hat{G}_i = \hat{G} > 0$). For this policy, allowances under a cap-and-trade policy are fully auctioned with revenues (after indexing) returned in equal lump-sum transfers to households.

Grandfathered permits ($\phi = 1$, $\hat{G}_i = 0$). Here all allowances are given away for free and all policy rents accrue to households through their ownership of firms. (Again, we abstract from taxation of rents).

Distribution-neutral ($\phi = 0$, $\hat{G}_i \neq 0$). Whereas the tax change in the proportional tax cut case entails the same change to all marginal tax rates, the distribution-neutral case requires changing marginal income tax rates by different amounts at different points in the income distribution, in order to equalize the net burden as a percentage of income across all quintiles.[10] The efficiency gains from revenue recycling are smaller in the distribution-neutral case than in the case with proportional tax rate reductions. This occurs primarily because marginal rates fall by less in this case than in the proportional tax cut case. Cutting tax rates on low income groups leads to a relatively large drop in tax revenue, because all higher income groups also benefit from the rate reductions for the lower income

---

[9] This allocation of tax rate changes implies that the tax cut, taken by itself, is neither progressive nor regressive.

[10] This is similar to the approach taken in Williams (2009a and 2009b) when considering a distribution-neutralizing tax change in other policy contexts, but while those papers assume a continuous income distribution, this paper works with a discrete distribution made up of five quintiles. The approach in those papers in turn draws on earlier work by Kaplow (2004).

Parry and Williams: The Costs of Meeting Distributional Objectives for Climate Policy

brackets. Consequently, a tax cut disproportionately targeted at lower income groups will yield a smaller reduction in marginal rates than would a proportional tax cut.

In all four cases we assume that any indirect revenue losses are made up through equal, proportionate adjustments to all marginal income tax rates.[11] These losses stem from reductions in taxable income as households, for example, reduce labor supply in response to lower real wages as higher energy prices drive up the general price level.

One noteworthy limitation of our analysis is that the model does not capture distortions from taxes on capital income. In this regard we mischaracterize, though perhaps only moderately, the efficiency gains from, and the incidence of, proportional cuts in marginal income taxes. More generally, incorporating taxes on capital income would admit a wider range of possibilities for changes in the taxation of personal and corporate income in response to recycling revenue from allowance sales.

B. Formulas for Efficiency Costs and Distributional Burdens of Carbon Policies

Here we go straight to the main formulas of interest. These are general expressions for the components of the efficiency costs and distributional burdens of carbon policies, and how those components vary under the allocation/recycling options just described. The formulas below are (reasonable) approximations given the scale of emission reductions considered and would be exact if demand and marginal abatement cost curves were linear over the relevant range. Derivations for the formulas are provided in the Appendix.

(i) Efficiency costs. The (approximate) efficiency cost of a carbon policy that reduces aggregate CO_2 emissions by an amount $\Delta Z = Z^0 - Z$, with associated emissions price τ, can be decomposed into the following four components:

(10a) $\quad WC^{HT} - WG^{RR} + WC^{TI} + WC^{INC}$

$$WC^{HT} = \frac{1}{2} \cdot \tau \cdot \Delta Z, \quad WG^{RR} = MEB \cdot \left\{ (1-\phi)\tau E - \sum_i \Delta G_i \right\},$$

---

[11] Alternatively, we could assume that indirect revenue losses are deducted from the amount of revenue returned in transfers under cap-and-trade, or the size of the allowance giveaway under grandfathered permits, with significant implications for distributional incidence. However, our assumption is more in keeping with actual policy proposals (not least, given the difficulty of accurately projecting indirect revenue losses).

$$WC^{TI} = (1 + MEB) \cdot \sum_i t_i \frac{\partial \Gamma_i}{\partial \bar{p}} \Delta \bar{p} = (1 + MEB) \cdot \sum_i \alpha_i' t_i \varepsilon_i^{\bar{p}} \tau \left(Z + \frac{\Delta Z}{2} \right),$$

$$WC^{INC} = -(1 + MEB) \cdot \sum_i t_i \frac{\partial \Gamma_i}{\partial G_i} (\Delta G_i + \pi_i) = -(1 + MEB) \cdot \sum_i \alpha_i' t_i \varepsilon_i' \frac{\Gamma_i}{I_i} \Delta G_i$$

where

$$(10b) \quad MEB = \frac{-\sum_i t_i \frac{\partial \Gamma_i}{\partial t_i}}{\sum_i \Gamma_i + \sum_i t_i \frac{\partial \Gamma_i}{\partial t_i}} = \frac{\sum_i \alpha_i^\Gamma \varepsilon_i' \frac{t_i}{1 - t_i}}{1 - \sum_i \alpha_i^\Gamma \varepsilon_i' \frac{t_i}{1 - t_i}},$$

$$\varepsilon_i' = \frac{\partial \Gamma_i}{\partial (1 - t_i)} \frac{1 - t_i}{\Gamma_i}, \quad \varepsilon_i^{\bar{p}} = \frac{\partial \Gamma_i}{\partial \bar{p}} \frac{\bar{p}}{\Gamma_i}, \quad \varepsilon_i' = \frac{\partial \Gamma_i}{\partial I_i} \frac{I_i}{\Gamma_i}, \quad \alpha_i^\Gamma = \frac{\Gamma_i}{\sum_i \Gamma_i},$$

$$\alpha_i' = \frac{\Gamma_i}{\sum_i I_i}$$

Beginning with (10b), ε_i', $\varepsilon_i^{\bar{p}}$, and ε_i' denote three different taxable income elasticities for household i defined with respect to changes in marginal tax rates, changes in the general price level, and changes in (taxable) household income, respectively. α_i^Γ is the share of household group i's taxable income in economy-wide income.

MEB is the marginal excess burden of taxation, the efficiency cost of raising an extra dollar of revenue through a proportionate increase in marginal income taxes. The numerator of *MEB* is the efficiency loss from an incremental increase in the marginal income tax for group i, aggregated over all households. The efficiency loss for group i is the induced reduction in taxable income (reflecting both reductions in labor supply and increases in tax-preferred spending) times the marginal tax rate. The denominator of the *MEB* is the extra revenue from the tax increase, summed over all households. Alternatively, the *MEB* can be expressed as a function of the weighted sum of taxable income elasticities (with respect to tax rates) for different household groups, with weights equal to the households' share in economy-wide taxable income. Note that, according to the way we have defined it in (10b), behavioral responses underlying the *MEB* are uncompensated.

In equation (10a), the first component of the welfare cost, WC^{HT}, corresponds to the Harberger triangle under the economy-wide marginal abatement cost curve. This curve represents the envelope of other marginal abatement cost curves for each margin of behavior for reducing emissions—reducing emissions per unit of, and reducing overall consumption of, energy-intensive products.[12] The Harberger triangle is the same under all four policy scenarios.

The second component in (10a), WG^{RR}, termed the revenue-recycling effect (e.g., Goulder 1995), is the efficiency gain that results to the extent that revenues are used to reduce marginal income taxes. This component is the product of the *MEB* and the amount of revenue recycled in this fashion, where the latter is the policy rents retained by the government less what is spent on transfer payments and indexing.

The third welfare cost component, WC^{TI}, is the tax-interaction effect. In most prior work this referred to the welfare loss from the reduction in labor supply as policy-induced increases in the general price level reduced the real return to work effort (e.g., Goulder 1995). The welfare loss was the change in labor earnings, multiplied by the labor tax distortion, and multiplied by 1+*MEB*, as lost labor tax revenues must be made up through higher (distortionary) taxes to maintain budget balance. In these models, the change in labor earnings depends on both the responsiveness of labor supply to higher prices, and the loss of worker surplus from the price increase. The latter is (approximately) equal to allowance rent plus the Harberger triangle under the marginal abatement cost curve, and corresponds to $\tau(Z + \Delta Z / 2)$ in our formula for WC^{TI}. However, in our case the tax-interaction effect is measured with respect to changes in taxable income, in response to higher product prices, rather than labor earnings, and hence it is the taxable income, rather than labor supply, elasticity that appears in our formula. The tax-interaction effect (as defined here) is also constant across different policies.

Our expression for WC^{TI} embeds a number of simplifying assumptions that might, to some degree, be open to question (see the Appendix). One is that we assume all goods are equal substitutes for leisure and therefore not deserving of any Ramsey tax or subsidy from an optimal tax perspective. In the absence of evidence to the contrary however, this seems a reasonable, neutral assumption. Another simplification is that compliance costs are fully passed forward into higher prices. To the extent, for example, that compliance costs come at the expense of infra-marginal rent earned on base load power generation, the price effect and tax-interaction effect will be weaker.[13] In short, the reader should bear

---

[12] These marginal cost curves all come out of the origin given that our model abstracts from other distortions affecting the production and use of energy, such as fuel taxes, automobile congestion, and non-competitive pricing of electricity.

[13] The price pass through may also be imperfect in states that retain cost-of-service regulation for power generation—in this case, utilities receiving free allowance allocations cannot pass forward

The B.E. Journal of Economic Analysis & Policy, Vol. 10 [2010], Iss. 2 (Symposium), Art. 9

in mind that the absolute value of the tax-interaction effect is difficult to pin down accurately, though this is not relevant for the cost/distributional tradeoffs involved in allocating cap-and-trade rents, which is the main focus of our paper.

The final component of welfare cost, WC^{INC}, reflects efficiency losses from the reduction in taxable income in response to higher lump-sum income—through higher government transfers and/or profit income. This largely reflects a reduction in labor supply as households take more leisure, which is a normal good.

(ii) Distributional burden. The distributional burden of the emissions pricing policy on household group i, relative to income, is fairly straightforward and can be approximated by (see the Appendix for derivations):

$$(11) \quad \sum_j \frac{\Delta p_j^0 X_{ij}^0}{I_i^0} + \frac{\Delta G_i + \pi_i}{I_i^0} + \frac{\Delta t_i \Gamma_i}{I_i^0}$$

The first terms reflects the (first order) consumer surplus loss from the induced increase in energy prices. The second term picks up possible benefits from dividends and rent income through stock ownership under grandfathered permits. And the third term reflects gains from reductions in marginal tax rates. The efficiency consequences from the revenue-recycling, tax-interaction, and income effect on labor supply are all included in the distributional analysis, through overall changes in marginal income tax rates, in the third term. We omit the Harberger triangle from the distributional analysis, as this is small relative to the first order loss of surplus to households from higher energy prices.

3. Parameterization of the Model

Our baseline parameter assumptions are summarized in Table 1. We focus on projections for year 2020. All monetary figures are expressed in year 2008 dollars (or thereabouts).

---

the opportunity cost of such allowances in higher generation prices. See Parry (2005) for more discussion of these issues.

Parry and Williams: The Costs of Meeting Distributional Objectives for Climate Policy

Table 1. Summary of Baseline Data

(projections for year 2020 with monetary data in year 2008 dollars)

Household data	Income quintile					Average
	1	2	3	4	5	
Income, $	11,307	22,882	46,638	81,467	181,626	68,784
Consumption, $	22,566	34,187	47,835	66,202	108,573	55,873
Burden of carbon tax						
as percentage of income	6.03%	4.27%	2.71%	1.87%	1.16%	1.59%
as percentage of consumption	3.02%	2.86%	2.64%	2.31%	1.94%	1.96%
Fraction of stockholder wealth owned	2.63%	8.92%	8.63%	19.99%	59.84%	
Lump-sum transfer component of tax, $	4,048	6,762	9,404	11,909	7,949	8,014

	Consumption quintile					Average
	1	2	3	4	5	
Income, $	18,204	35,476	53,203	82,719	145,079	66,936
Consumption, $	19,253	34,605	49,644	70,428	126,173	60,020
Burden of carbon tax						
as percentage of income	3.27%	2.74%	2.27%	1.87%	1.54%	1.64%
as percentage of consumption	3.09%	2.81%	2.43%	2.20%	1.77%	1.83%
Fraction of stockholder wealth owned	1.60%	3.30%	12.85%	24.63%	57.62%	
Lump-sum transfer component of tax, $	3,453	7,716	11,085	13,743	10,314	9,262

	Income or Consumption quintile					Average
	1	2	3	4	5	
Income tax rate	17%	33%	40%	44%	41%	40%
Taxable income elasticities						
with respect to taxes	0.20	0.22	0.25	0.28	0.35	0.31
with respect to price level	0.15	0.15	0.15	0.15	0.15	0.15
with respect to income	-0.20	-0.20	-0.20	-0.20	-0.20	-0.20

Other data

Marginal excess burden of taxation	0.27
BAU CO_2 emissions, billion tons	6.0
Emissions price, $/ton	33
Reduction in emissions from BAU	9.0%
Policy rents/revenues, $ billion	180

Source. See text.

Note that averages differ slightly between the income and consumption quintile cases,
due to dropping households with consumption under $7,500

The B.E. Journal of Economic Analysis & Policy, Vol. 10 [2010], Iss. 2 (Symposium), Art. 9

Income distribution. To obtain the income distribution we use data from the Consumer Expenditure Survey (CEX) for years 2007 and 2008, for households that completed their year in the survey panel during 2008.[14] As is standard in incidence analysis we divide households into quintiles in two different ways: based on their pre-tax annual income and based on their total consumption.[15] To the extent that households are able to smooth consumption over time, the consumption-based measure will more accurately reflect lifetime income. Distributional differences are more muted when measured based on consumption quintiles than based on income quintiles.

Under either measure, quintile 1 is the lowest 20 percent of income earners, quintile 2 the next lowest 20 percent, and so on. For simplicity, we assume the real income and real consumption of each household group grows at the same 1.5 percent annual rate out to 2020. In 2020, income per household varies from \$11,307 for the lowest income quintile to \$181,626 for the top quintile, and consumption varies from \$19,253 for the lowest consumption quintile to \$126,173 for the highest (Table 1).

Budget shares for energy-intensive goods. In the analytical model, goods are either energy-intensive or non-energy intensive. In the real world and in the data, goods vary widely in their energy (and thus carbon) intensity. One could simply divide goods into energy-intensive and non-energy-intensive categories and assign all energy-intensive goods the same embodied carbon content and all non-energy-intensive goods a lower or zero embodied carbon content. But given the wide variation in embodied carbon across goods, this could introduce substantial inaccuracy when measuring the distributional burden of a carbon tax.

Instead, we estimate the total carbon embodied in all of the goods a household consumes, and then define the consumption of energy-intensive goods in the model to be proportional to that total. In effect, this treats consuming a particular good from the CEX as consuming some of the energy-intensive good and some of the non-energy-intensive good, with the relative proportions being determined by the embodied carbon content of that good. We take the estimates of embodied carbon for goods in the CEX from Hassett et al. (2007), who calculate

---

[14] The CEX uses a rotating panel design: each calendar quarter, one fourth of the households in the sample are rotated out, so each household appears in the data for four quarters. To obtain a full year of data on each household that completed its time in the panel during 2008 therefore requires using data from both 2007 and 2008.

[15] Because income in the CEX is known to be poorly measured for low-income households, it is customary for incidence studies to drop some very low-income households to minimize the effects of that measurement problem. We follow the same approach as Grainger and Kolstad (2010), dropping households with reported income below \$7,500, but not altering the quintile cutoff levels of income and expenditure as a result of dropping those households.

Parry and Williams: The Costs of Meeting Distributional Objectives for Climate Policy

those estimates based on input-output tables from the US Bureau of Economic Analysis.

Stock ownership. We calculate stock ownership by aggregating for each quintile the value of estimated market value of all stocks, bonds, mutual funds, and other such securities reported by each household in the CEX. According to this calculation, the top-income quintile owned 59.8 percent of stockholdings, while the lowest income quintile owned 2.6 percent. Stock ownership shares are assumed to be the same in 2020 as in 2007-08.[16]

Household tax rates and transfers. Marginal and average rates of federal and state income taxes (accounting for the earned income tax credit and child tax credits), and sales taxes, were obtained by running each household from the CEX data for 1997-1999 through the NBER's TAXSIM model, using the tax laws for those years, and then aggregating for each quintile.[17] We add to these employer and employee payroll taxes (based on statutory rates) to obtain the overall income tax rates (assumed to apply in 2020) shown in Table 1. Marginal tax rates vary from 0.17 for the bottom income quintile to 0.41 for the top income quintile and average (weighted by taxable income) 0.40.

Taxable income elasticities. During the last decade, a substantial number of papers have estimated the elasticity of taxable income with respect to tax rates, both for the economy as a whole, and for high-income taxpayers. Although initial estimates of this elasticity were quite large (e.g., Feldstein 1999), more recent estimates are considerably smaller, in part reflecting better data and improved methods of controlling for non-tax factors affecting changes in taxable income. Based on a careful review of evidence for the United States and other countries, Saez et al. (2009) put the taxable income elasticity (for the economy as a whole) at 0.12 to 0.40, with more estimates closer to the top end of this range than the bottom.[18]

---

[16] Ideally, we would use information on the pattern of stock ownership across energy-intensive firms (rather than all firms) to gauge the distribution of rent income under grandfathered permits. However, this is difficult to obtain given that most stocks are owned indirectly through large investment firms.

[17] These years pre-date the federal income tax cuts of 2001 and 2003, and thus will approximate the taxes that will apply if those tax cuts are allowed to expire (which seems increasingly likely over time, given budgetary pressures). We feel this provides a better approximation to the tax laws in 2020 than would using the 2008 tax laws. To the extent that this over- or underestimates marginal tax rates in 2020, the magnitudes of the revenue-recycling and tax-interaction effects will also be over- or underestimated.

[18] One caveat here is that there is a slight mismatch between our representation of the taxable income elasticity and empirical estimates of that elasticity. In our model, this elasticity reflects tax-induced reductions in labor supply and shifting to tax-preferred consumption, whereas

The B.E. Journal of Economic Analysis & Policy, Vol. 10 [2010], Iss. 2 (Symposium), Art. 9

   Although we assume that all taxpayers have the same labor supply elasticity, higher income taxpayers tend to have more scope for exploiting tax preferences, so they tend to have higher taxable income elasticities. We assume taxable income elasticities with respect to tax rates vary between 0.2 for the low income quintile and 0.35 for the top income quintile, as shown in Table 1, where (weighting by household shares in taxable income) the average elasticity is 0.31.

   An increase in the general price level (in response to carbon policy) has a comparable effect on labor supply to a reduction in the real household wage—that is, households substitute leisure for work, though this is partly offset by an income effect in the opposite direction. The effect on tax-preferred spending should be small since the price of ordinary spending does not change relative to the price of tax-favored spending.[19] For the same reason, the change in taxable income in response to higher lump-sum payments should be similar to the change in labor earnings. We therefore use standard values for the uncompensated labor supply elasticity, and income elasticity of labor supply, namely 0.15 and -0.2 (e.g., Blundell and MaCurdy, 1999), as proxies for $\varepsilon_i^{\bar{p}}$ and ε_i' for all households.

Emissions reductions and prices. These are based on a policy simulation of representative climate bills reported in Krupnick et al. (2010) using a variant of the Energy Information Administration's National Energy Modeling System (NEMS). NEMS is a dynamic, economic-engineering model of the economy, with considerable detail on a wide spectrum of existing and emerging technologies across the energy system. Its projections are widely used by other energy modeling groups.

   Without climate policy, this model projects CO_2 emissions to be about 6 billion metric tons in 2020. The electricity and transport sectors account for about 40 and 33 percent of these emissions respectively, and other sources (i.e., non electricity emissions from the industrial, commercial, and residential sectors) account for the remaining 27 percent. With climate policy, energy-related CO_2 emissions are reduced by about 9 percent below business-as-usual levels in 2020, or by 0.54 billion tons, and the allowance price is $33 per ton of CO_2 in current dollars (Krupnick et al. 2010).[20] Thus, the Harberger triangle, WC^{HT}, is about $9

---

empirical studies summarize a broader range of responses including shifting to tax-sheltered saving. This mismatch (which we believe is not too important for our purposes) is due to our characterizing the income tax system as a tax on labor alone, rather than a tax on both labor and capital income.

[19] That is, although tax-preferred spending will fall in response to higher energy prices the resulting efficiency gain (due to offsetting the tax subsidy) is modest relative to the efficiency loss from the reduction in labor supply. This is because the "market" for tax-preferred spending is small (about 10-15 percent) relative to the labor market.

[20] This policy run involves a more aggressive emission reduction target, but allows domestic and international offsets (under an intermediate assumption about the availability of such offsets).

billion and policy rents, τZ, are about $180 billion. The carbon policy increases energy prices in our model by 7.9 percent or the general price level by 0.5 percent.

4. Results

A. Cost Comparison

Figure 1 shows the overall cost of imposing a carbon policy with an allowance price of $33 per ton of CO_2, under each of the four options for the use of allowance rents: a proportional income tax cut, lump sum rebates to households, free allocation of permits to firms, and a distribution-neutralizing tax cut. For this last option, the cost varies substantially based on whether the tax cut neutralizes distributional effects across income quintiles or across consumption quintiles, so these are presented separately. Note that Figure 1 provides just an estimate of policy costs: it does not incorporate any estimates of the climate benefits from lower CO_2 emissions.

For each case, the figure also shows the four components of welfare cost: the Harberger triangle term (the partial-equilibrium cost, ignoring any interactions with the tax system); the revenue-recycling effect (the gain from using allowance rents to finance tax rate cuts); the tax-interaction effect (the general-equilibrium loss resulting when higher energy prices discourage labor supply); and the income effect on labor supply (a further general-equilibrium loss as the income effect from the distribution of policy rents further discourages labor supply). The Harberger triangle is the same across all the policy simulations ($8.9 billion a year in 2020) as is the tax-interaction effect ($25.0 billion). However, the revenue-recycling effect, in particular, differs across policies, as does the income effect on labor supply.

When allowance rents are used to finance a proportional income tax cut, the overall cost of the policy is -$6.4 billion/year. This case yields a "strong" double dividend (Goulder, 1995): even ignoring the benefits of reduced carbon emissions, the cost of the policy is still negative. As discussed in Parry and Bento (2000), when certain categories of spending are tax-favored, the gains from the revenue-recycling effect are magnified (relative to a case without such tax preferences), but the losses from the tax-interaction effect are approximately unaffected. As a result, the revenue-recycling effect (a gain of $41.3 billion) exceeds the tax-interaction effect, and by enough to more than offset the Harberger triangle. This qualitative result is different than in earlier literature that

---

Here we assume that the same emissions price projected by this run is imposed, but is applied only to CO_2 with no offsets. In either case, the reduction in domestic, energy-related CO_2 should be essentially the same.

The B.E. Journal of Economic Analysis & Policy, Vol. 10 [2010], Iss. 2 (Symposium), Art. 9

focused only on the labor market distortion caused by the tax system (e.g., Goulder, 1995). In the latter models, the revenue-recycling effect typically falls short of the tax-interaction effect and the overall costs of auctioned cap-and-trade systems (or carbon taxes) is positive, and exceeds the Harberger triangle.[21]

Figure 1. Decomposition of Efficiency Costs under Alternative Policies

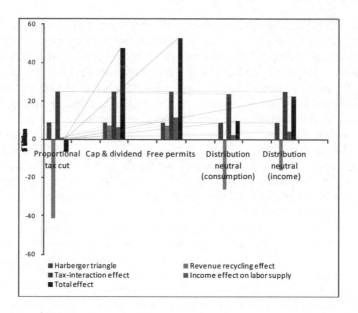

The welfare cost of the cap-and-trade policy is much higher under either the cap-and-dividend or free permit allocation cases: under each of these options the cost is roughly $50 billion a year in 2020. By giving away the permit rents, these two options give up the large gain from the revenue-recycling effect ($41.3 billion). In addition they produce a smaller, but still significant, loss of $6.4 billion (under cap-and-dividend) or $11.6 billion per year (under grandfathered permits) because the resulting lump-sum income to households reduces labor supply. This effect is stronger for the grandfathered permit case as higher income households — who face higher marginal income tax rates — receive a disproportionately larger share of the policy rents in this case. The overall welfare

---

[21] Similarly, if tax preferences were entirely justified by market failures then the revenue-recycling effect would be smaller. In this case, the overall cost of our proportional tax cut case would be positive (roughly $15 billion).

Parry and Williams: The Costs of Meeting Distributional Objectives for Climate Policy

cost of the cap-and-dividend and grandfathered permit polices is over five times the Harberger triangle, underscoring the bias in cost analyses that omit interactions with the broader tax system.[22] Dividing total costs by the CO_2 reduction (0.54 billion tons) the average cost per ton reduced is around $90 under cap-and-trade, compared with *minus* $12 in the proportional tax reduction case.

Inflation-indexing of taxes and transfers in the above policy cases requires revenue outlays of $30 billion/year in 2020, or one-sixth of the policy rents. The implied increase in marginal tax rates to cover this revenue loss adds $7.4 billion/year to overall welfare costs. In the proportional income tax case this cost leads to a smaller overall gain from the revenue-recycling effect, while in the grandfathered permit and cap-and-dividend cases (which give away the entire $180 billion of policy rents) it is reflected in a negative revenue-recycling effect.[23]

The overall cost of the distribution-neutral options fall between the cost under the proportional tax cut and the cost under the cap-and-dividend policy — welfare costs are $9.9 and $22.6 billion in the consumption-based and annual income-based cases for measuring inequality, respectively. The distribution-neutralizing tax change lowers tax rates rather than providing a lump-sum transfer, so it generates a gain from the revenue-recycling effect, which causes the overall cost to be substantially lower than the overall cost under the cap-and-dividend and grandfathered permits. However, because the tax cuts are larger for lower-income households (to offset the regressive incidence of higher energy prices), the marginal tax rate reductions and corresponding efficiency gains from revenue recycling are smaller than in the proportional tax cut case. This effect is

---

[22] These results implicitly assume that none of the permit rents are taxed under either policy option. This is very likely to be true in the cap-and-dividend case. However, in the free permit allocation case, the permit rents could well be taxed, at least in part. To the extent that these rents accrue to stocks held in taxable accounts (as opposed to retirement or other tax-sheltered accounts), they will be subject to tax. Under either policy option, if the rents are taxed, then the overall cost will be a linear combination of the cost under that policy option (which assumes that none of the rents are taxed) and the cost under the proportional tax cut (which is equivalent to a case in which all of the permit rents are taxed at a rate of 100 percent).

[23] Note that because the tax system is indexed such that tax rates depend on real income, the overall results – both for efficiency and distribution – do not depend on the particular normalization used for prices (i.e., the choice of numeraire). However, the decomposition of those results into "inflation indexing" and other components does depend on the price normalization. We use a price normalization that holds the pre-tax wage constant, so a rise in the price of consumer goods relative to labor shows up as an increase in the overall price level, which produces the "inflation indexing" results shown here. If instead we were to use a normalization that holds the price of consumer goods constant, that same relative price change would show up as a reduction in the pre-tax wage. In this case, there would be no revenue cost for inflation indexing, but instead there would be a reduction in income tax revenue resulting from the lower wages. Thus, the overall effect would be the same, but the decomposition of that effect between "inflation indexing" and other effects on revenue would differ.

even more pronounced when inequality is measured using annual income rather than consumption. In the former case, higher energy prices are more regressive, requiring an even greater concentration of tax cuts among lower income households. Thus, the revenue-recycling gains are $15.8 billion in this case, compared with $25.7 billion when inequality is measured using the consumption approach.

B. Distributional Effects

Figures 2 and 3 summarize the distributional burdens of the different policies across income quintiles, based on income and consumption quintiles, respectively. Burdens are expressed as a percent of household income. In addition, the net burden is decomposed into the burden of the higher energy prices caused by the cap-and-trade program; the gain to households resulting from inflation-indexing of the tax and transfer system; any gain from reductions in marginal income tax rates; and possible gains from permit rents or lump-sum dividends.[24]

All policies impose the same pattern of burdens across households due to higher energy prices. This burden component falls with income—for example, the bottom quintile bears a burden of 6.0 percent when inequality is measured on an income basis, falling steadily to 1.2 percent for the top quintile. Lower income groups spend a greater fraction of their income on energy-intensive consumption. One reason for this is that energy-intensive goods make up a slightly larger share of spending for lower-income households. Another is that lower-income households spend a larger fraction of their income due to lower or even negative saving rates (on average lower income households spend more than their income). Dividing households into consumption quintiles rather than income quintiles greatly reduces this second effect, and hence the distributional burden is considerably less regressive when measured based on consumption quintiles. In this case, the bottom quintile bears a burden of 3.3 percent while the top quintile bears a burden of 1.5 percent.

---

[24] The results discussed below are broadly consistent with those in Burtraw et al. (2009) who look at the incidence of federal cap-and-trade policies under alternative revenue uses using a model with considerable disaggregation by households and region.

Parry and Williams: The Costs of Meeting Distributional Objectives for Climate Policy

Figure 2. Decomposition of Net Burden by Income Quintile under Alternative Policies

(a) Proportional tax cut

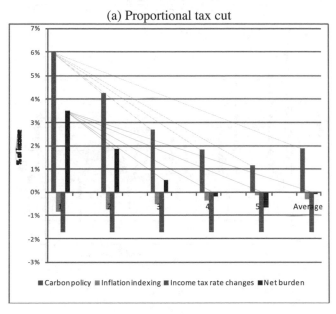

(b) Cap-and-Dividend

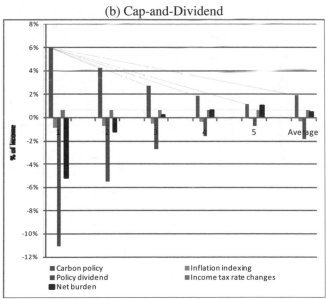

The B.E. Journal of Economic Analysis & Policy, Vol. 10 [2010], Iss. 2 (Symposium), Art. 9

Figure 2, continued

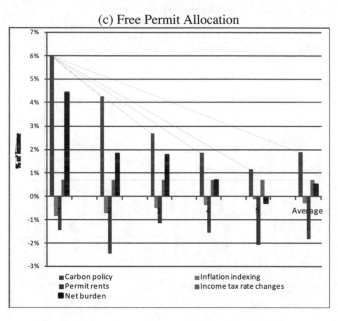

(c) Free Permit Allocation

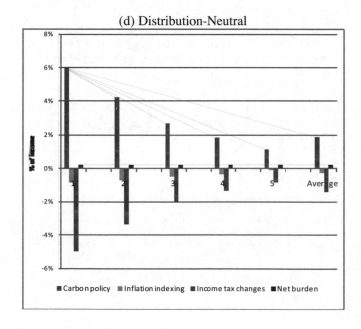

(d) Distribution-Neutral

Inflation-indexing of the tax and transfer system provides an automatic offset for part of the burden of the policy and again the pattern of this gain across households is the same across policy scenarios. As the prices of energy-intensive goods rise, inflation-indexing causes an increase in transfers (or decrease in taxes paid, for indexing of tax brackets and other elements of the tax system). Averaged across all households, the benefit from inflation indexing is about 0.3 percent of income, or about one-seventh of the average burden from higher energy prices (1.9 percent). The portion of the burden of energy prices offset by inflation indexing varies relatively little across income quintiles—for example, it offsets roughly one-seventh of the burden for the bottom quintile, about one-fifth for the middle, and one-eleventh for the top quintile.[25]

For the proportional income tax case (Figures 2a and 3a), by definition the gains from the tax cut are the same proportion of income for each quintile, roughly 1.7 percent of income for each group. The overall net burden under this policy as a percentage of income is highest for the bottom quintile and falls steadily as one moves up the income distribution, given that revenue-recycling does nothing to offset the regressive effect of higher energy prices. For income-based quintiles the net burden varies between 3.5 and minus 0.6 percent across the lowest and highest quintiles, while for consumption-based quintiles it varies from 1.2 to minus 0.3 percent.

For the cap-and-dividend policy (Figures 2b and 3b) the distribution of the policy dividend is highly progressive, given that all households receive the same absolute cash rebate. As a result, even though the overall cost of this policy is much higher than in the proportional income tax cut case, the bottom three quintiles experience smaller net burdens under cap-and-dividend. In fact, the bottom two quintiles actually have a negative net burden (under either income- or consumption-based quintiles). This illustrates the stark tradeoff between efficiency and distribution: lower-income groups are much better off under cap-and-dividend than they are under a proportional income tax cut, but the overall welfare cost of the policy is dramatically higher.

---

[25] For reasons analogous to those discussed in footnote 22, above, the net burden on each quintile is independent of the choice of price normalization. But the decomposition of that net burden into "inflation indexing" and other components does depend on the normalization.

The B.E. Journal of Economic Analysis & Policy, Vol. 10 [2010], Iss. 2 (Symposium), Art. 9

Figure 3. Decomposition of Net Burden by Consumption Quintile under Alternative Policies

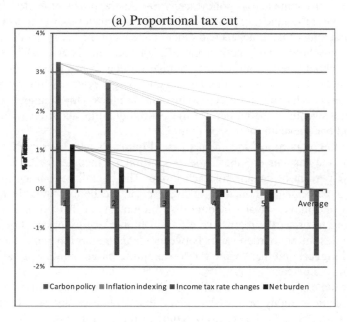

(a) Proportional tax cut

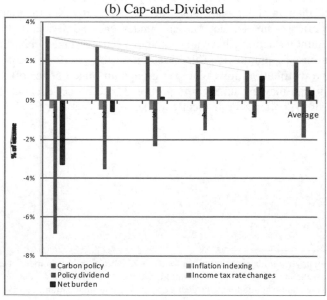

(b) Cap-and-Dividend

Parry and Williams: The Costs of Meeting Distributional Objectives for Climate Policy

Figure 3, continued

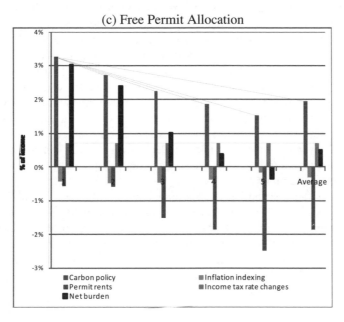

(c) Free Permit Allocation

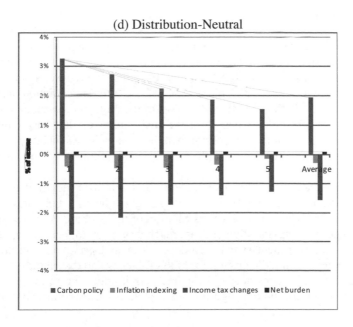

(d) Distribution-Neutral

The B.E. Journal of Economic Analysis & Policy, Vol. 10 [2010], Iss. 2 (Symposium), Art. 9

The policy with free allowance allocation to firms (Figures 2c and 3c) does nothing to offset the regressive effect of higher energy prices—in fact the distribution of rent income is itself regressive, for the most part, given that for better off households capital is typically a larger share of their income. Comparing this case to either the proportional income tax cut or the cap-and-dividend case reveals no tradeoff between efficiency and distribution. Freely allocating permits to firms results in a similarly regressive distribution of net burden to the proportional income tax case, but a much higher overall cost because it fails to exploit the revenue-recycling effect. In fact, every income quintile is better off under the proportional income tax cut than under free permits. Conversely, the cap-and-dividend case results in a much more progressive distribution of net burdens than free permits, but has a similar (even slightly lower) overall cost.

Under the distribution-neutralizing change (Figures 2d and 3d) the tax change is designed to be sufficiently progressive to exactly offset the regressive burden of higher energy prices, thus making the distribution of the net burden of the policy proportional to income. This case falls between the proportional tax cut and cap-and-dividend cases, in terms of both efficiency and distribution: the overall cost is lower than under cap-and-dividend but higher than under the proportional tax cut, and the distribution of burden is less progressive than under cap-and-dividend but less regressive than under the proportional tax cut.

5. Conclusions

The allocation and use of rents created under cap-and-trade programs (or revenues under an emissions tax) can hugely affect the overall efficiency and distributional effects of carbon policy. Within the range of options we consider for the use of revenue, in a case where the direct cost of the carbon restriction itself is $9 billion/year in 2020, the overall welfare cost of the policy ranges from negative $6 billion/year to $53 billion/year. And the distribution of that cost ranges from highly progressive (with the bottom two income quintiles bearing a negative burden) to highly regressive (with the top two quintiles bearing a negative burden).

In general, there is a clear tradeoff between efficiency and distribution. Using policy revenues to fund cuts in marginal income tax rates is highly efficient, but leads to a regressive distribution of the net burden. At the other end of the range, the cap-and-dividend approach has a far higher overall cost, but leads to a highly progressive distribution. And a distribution-neutralizing tax change represents a middle ground for both efficiency and distributional effects.

This tradeoff does not hold for free permit allocation to firms in affected industries, which has both a high overall cost and regressive distribution of that cost. But this can be viewed as an efficiency-distribution tradeoff along a different dimension, addressing distributional concerns across firms in different industries rather than across households in different income groups. Thus, while this type of allocation is both inefficient and regressive, it may have more political traction.

More generally, our discussion underscores that the case for cap-and-trade without revenue recycling is more fragile than generally realized (at least for medium term levels of emissions controls envisioned in recent climate bills). Our calculations imply the average cost of reducing domestic, energy-related CO_2 when policy rents are not used to cut distortionary taxes is around $90 per ton— far above most estimates of the benefits per ton of CO_2 reductions (e.g., Aldy et al., 2010, US IAWG, 2010).[26] In contrast, the average cost is *negative* $12 per ton when efficiency gains from the revenue-recycling effect are fully exploited (without even including any climate benefits). In our compromise cases where the carbon policy is neither regressive nor progressive the average cost per ton reduced is $18 to $42.

Given these tradeoffs, and the potentially huge consequences for both efficiency and distribution, more attention should be paid to the use of revenue when analyzing carbon policy. And future research on creative policy designs that balance efficiency, distribution, and political feasibility could be highly valuable.

Appendix: Analytical Derivations

Deriving equation (10)

We follow the usual two-step procedure for obtaining the marginal welfare effects of policy changes. First we solve the household's optimization problem. Then we obtain the welfare effects of a marginal change in the emissions price by totally differentiating the household's indirect utility, accounting for the household's behavior, and changes in prices, taxes, and lump-sum income. Finally, we integrate over marginal welfare effects to obtain the effects of non-marginal policy changes.

Household Optimization. From (1)-(3), the optimization problem for household group i is given by:

---

[26] Nonetheless, there are many arguments for moving ahead with a cap-and-trade program, even if it fails a narrowly defined cost-benefit test in the early years. For example, the benefit to cost ratio could easily become favorable over time as the policy is tightened and a greater share of policy rents are used in efficiency-enhancing ways; putting a price on carbon could have important long run benefits in terms of encouraging clean technology development; and action by the United States on climate policy could help to spur similar programs in other countries.

The B.E. Journal of Economic Analysis & Policy, Vol. 10 [2010], Iss. 2 (Symposium), Art. 9

(A1) $V_i(p_E, p_F, p_C, t_i, G_i, \pi_i) =$

$$\underset{X_{iE}, X_{iF}, X_{iC}, L_i}{\text{Max}} \quad u_i(X_{iE}, X_{iF}, X_{iC}, L_i, G^{PUB})$$

$$+\lambda_i \left[(1-t_i)\Gamma_i + G_i + \pi_i - \sum_j p_j X_{ij} \right]$$

where $V_i(.)$ denotes the indirect utility function and λ_i is the marginal utility of income. Using the definition of taxable income in (3a), this optimization yields the first order conditions, demand, and labor supply functions:

(A2) $\dfrac{\partial u_i}{\partial X_{iE}} = \lambda_i p_E, \quad \dfrac{\partial u_i}{\partial X_{iF}} = \lambda_i p_F (1-t_i), \quad \dfrac{\partial u_i}{\partial X_{iC}} = \lambda_i p_C, \quad \dfrac{\partial u_i}{\partial L_i} = \lambda_i (1-t_i) w_i,$

$$X_{ij} = X_{ij}(p_E, p_F, p_C, t_i), \qquad L_i = L_i(p_E, p_F, p_C, t_i)$$

To manipulate the analytical derivations below, we obtain the following additional expressions by totally differentiating the expression in (A1) with respect to arguments of the indirect utility function, and using the conditions in (A2). This gives:

(A3) $\dfrac{\partial V_i}{\partial p_j} = -\lambda_i X_{ij}, \quad \dfrac{\partial V_i}{\partial t_i} = -\lambda_i \Gamma_i, \quad \dfrac{\partial V_i}{\partial G_i} = \dfrac{\partial V_i}{\partial \pi_i} = \lambda_i$

Welfare Effects of Marginal Policy Changes

Aggregate welfare is given by the sum of individual utilities, $\sum_i V_i$. Totally differentiating this with respect to a change in the emissions price, and expressing in monetary units, gives:

(A4) $\displaystyle\sum_i \frac{1}{\lambda_i}\frac{dV_i}{d\tau} = \sum_i \frac{1}{\lambda_i}\left[\sum_j \frac{\partial V_i}{\partial p_j}\frac{dp_j}{d\tau} + \frac{\partial V_i}{\partial t_i}\frac{\partial t_i}{\partial \tau} + \frac{\partial V_i}{\partial G_i}\frac{dG_i}{d\tau} + \frac{\partial V_i}{\partial \pi_i}\frac{d\pi_i}{d\tau} \right]$

Next, we discuss expressions for some of the individual terms in (A4).

First consider $dp_j / d\tau$. Differentiating (5) with respect to τ gives:

(A5) $\quad \dfrac{dp_H}{d\tau} = (\tau - c')\dfrac{dz}{d\tau} + z$

However the first term cancels, assuming that firms equate marginal abatement costs c' to the emissions tax. Differentiating (4) with respect to τ, and using (A5), gives:

(A6) $\quad \dfrac{dp_j}{d\tau} = \rho_j z$

Using (A6), (A3) and (6), the first term in (A4) simplifies as follows:

(A7) $\quad \sum_i \dfrac{1}{\lambda_i} \sum_j \dfrac{\partial V_i}{\partial p_j}\dfrac{dp_j}{d\tau} = -Z$

Next, take the second and third terms in (A4), and substitute using (A3) to give:

(A8) $\quad \sum_i \left\{ \dfrac{\partial V_i}{\partial t_i}\dfrac{\partial t_i}{\partial \tau} + \dfrac{\partial V_i}{\partial G_i}\dfrac{dG_i}{d\tau} \right\}\dfrac{1}{\lambda_i} = \sum_i \left\{ \dfrac{dG_i}{d\tau} - \Gamma_i \dfrac{dt_i}{d\tau} \right\}$

Now we totally differentiate the government budget constraint in (8) with respect to τ, holding G^{PUB} constant but allowing t_i and G_i to vary, where $dt_i / d\tau = dt / d\tau$. This gives, after expressing changes in Z as a total differential:

(A9) $\quad \sum_i \dfrac{dG_i}{d\tau} = (1 - \phi)\left(Z + \tau\dfrac{dZ}{d\tau} + \sum_i t_i \dfrac{\partial \Gamma_i}{\partial \tau} \right) + \sum_i \left(\Gamma_i + t_i \dfrac{\partial \Gamma_i}{\partial t_i} \right)\dfrac{dt}{d\tau} + \sum_i t_i \dfrac{\partial \Gamma_i}{\partial G_i}\dfrac{dG_i}{d\tau}$

From the definition of the marginal excess burden in (10b):

(A10) $\quad \sum_i \left(\Gamma_i + t_i \dfrac{\partial \Gamma_i}{\partial t_i} \right) = \dfrac{\sum_i \Gamma_i}{1 + MEB}$

Substituting (A10) in (A9), multiplying through by $1+MEB$ and subtracting $\sum_i \Gamma_i dt / d\tau$ gives:

The B.E. Journal of Economic Analysis & Policy, Vol. 10 [2010], Iss. 2 (Symposium), Art. 9

(A11) $\sum_i \left(\frac{dG_i}{d\tau} - \Gamma_i \frac{dt}{d\tau} \right) = (1 + MEB)\left(\left(Z + \tau \frac{dZ}{d\tau} \right)(1-\phi) + \sum_i t_i \frac{\partial \Gamma_i}{\partial \tau} \right)$

$$+ (1 + MEB)\sum_i t_i \frac{\partial \Gamma_i}{\partial G_i} \frac{dG_i}{d\tau} - MEB \sum_i \frac{dG_i}{d\tau}$$

Finally, from differentiating (7) with respect to τ, and using (A3), the last term in (A4) can be expressed

(A12) $\sum_i \frac{1}{\lambda_i} \frac{\partial V_i}{\partial \pi_i} \frac{d\pi_i}{d\tau} = \phi\left(Z + \tau \frac{dZ}{d\tau} \right)$

where we have used $\sum_i \theta_i = 1$.

Substituting (A7), (A8), (A11) and (A12) in (A4), and expressing utility losses as a positive number (i.e., welfare cost) gives:

(A13) $-\sum_i \frac{1}{\lambda_i} \frac{dV_i}{d\tau} = -\tau \frac{dZ}{d\tau} - MEB \cdot \left[(1-\phi)\left(Z + \tau \frac{dZ}{d\tau} \right) - \sum_i \frac{dG_i}{d\tau} \right]$

$$-(1 + MEB) \cdot \sum_i t_i \frac{\partial \Gamma_i}{\partial \tau} - (1 + MEB) \cdot \sum_i t_i \frac{\partial \Gamma_i}{\partial G_i} \frac{dG_i}{d\tau}$$

Welfare Effects of Non-Marginal Policy Changes

Integrating the first term in (A13) between 0 and τ, assuming $dZ / d\tau$ is constant, gives:

(A14) $-\frac{dZ}{d\tau} \frac{\tau^2}{2}$

Again, given $dZ / d\tau$ is constant, $-(dZ / d\tau)\tau = Z_0 - Z$. Hence we obtain WC^{HT} in (10a).

We take the *MEB* as constant over the relevant range, which is reasonable given that proportional changes in income tax rates are relatively small. Integrating marginal emissions tax revenue $(1-\phi)(Z + \tau \cdot dZ / d\tau)$ over an emissions tax rising from 0 to τ simply gives revenue raised by the tax, $(1-\phi)\tau Z$. And integrating the marginal change in the transfer payment for household group i over the tax increase simply gives the total change in transfer payment $G_i - G_i^0 = \Delta G_i$. Hence we obtain WC^{RR} in (10a).

Parry and Williams: The Costs of Meeting Distributional Objectives for Climate Policy

From the third component of (A13), $\partial\Gamma_i / \partial\tau = (\partial\Gamma_i / \partial\bar{p})(\partial\bar{p} / \partial\tau)$. Taking a small change in the emissions price this expression becomes $(\partial\Gamma_i / \partial\bar{p})\Delta\bar{p}$. Hence we obtain the first expression for WC^{TI} in (10a).

Substituting expressions for $\eta_i^{\bar{p}}$ and α_i^T from (10b), gives:

(A15) $\quad WC^{TI} = (1 + MEB)\sum_i \alpha_i^T t_i \eta_i^{\bar{p}} \Delta\bar{p} \sum_i I_i$

where the general price level is normalized to unity. The burden of the emissions price $\tau(Z + \Delta Z / 2)$ is fully passed forward into higher product prices, therefore $\Delta\bar{p} = \tau(Z + \Delta Z / 2) / \sum_{ij} X_{ij}$. Making this substitution and $\sum_{ij} X_{ij} = \sum_i I_i$ in (A15) gives the second expression for WC^{TI} in (10a).

Finally, again if we approximate by taking t_i and $\partial\Gamma_i / \partial G_i$ as constant over the relevant range, then integrating over the last term in (A13) gives

(A16) $\quad -(1 + MEB)\cdot\sum_i t_i \frac{\partial\Gamma_i}{\partial G_i}\Delta G_i$

Substituting out $\partial\Gamma_i / \partial G_i = \partial\Gamma_i / \partial I_i$ using the taxable income elasticity with respect to income in (10b), gives the last expression WC^{INC} in (10a).

Deriving equation (11)

Equation (11) comes from the effect on utility from small changes in each of the arguments of $V_i(.)$ in equation (A1), after substituting (A3), and dividing by $\lambda_i I_i$.

References

Aldy, Joseph, Alan J. Krupnick, Richard G. Newell, Ian W.H. Parry and William A. Pizer, 2010. "Designing Climate Mitigation Policy." *Journal of Economic Literature*, forthcoming.

Blundell, Richard and Thomas MaCurdy, 1999. "Labor Supply: A Review of Alternative Approaches." in *Handbook of Labor Economics*, Volume III, O. Ashenfelter and D. Card (eds). New York: North-Holland.

The B.E. Journal of Economic Analysis & Policy, Vol. 10 [2010], Iss. 2 (Symposium), Art. 9

Bovenberg, A. Lans, and Lawrence H. Goulder, 2001. "Neutralizing the Adverse Industry Impacts of CO2 Abatement Policies: What Does It Cost?" In *Behavioral and Distributional Effects of Environmental Policy*, C. Carraro and G. Metcalf (eds). Chicago: University of Chicago Press, 45-85.

Burtraw, Dallas, Richard Sweeney, and Margaret A. Walls, 2009. "The Incidence of U.S. Climate Policy: Alternative Uses of Revenues from a Cap-and-Trade Auction." *National Tax Journal* 62: 497-518.

Dinan, Terry M. and Diane L. Rogers, 2002. "Distributional Effects of Carbon Allowance Trading: How Government Decisions Determine Winners and Losers." *National Tax Journal* 55: 199-222.

Fullerton, Don and Gilbert Metcalf, 2001. "Environmental Controls, Scarcity Rents, and Pre-Existing Distortions." *Journal of Public Economics* 80: 249-67.

Goulder, Lawrence H., 1995. "Environmental Taxation and the 'Double Dividend': A Reader's Guide." *International Tax and Public Finance* 2: 157-183.

Goulder, Lawrence H., Ian W. H. Parry, Roberton C. Williams III, and Dallas Burtraw, 1999. "The Cost-Effectiveness of Alternative Instruments for Environmental Protection in a Second-Best Setting," *Journal of Public Economics*, 1999, 72: 329-360.

Grainger, Corbett A. and Charles D. Kolstad, 2010. "Who Pays a Price on Carbon?" *Environmental and Resource Economics* 46: 359-376.

Hassett, Kevin A., Aparna Mathur, and Gilbert E. Metcalf, 2007. "The Incidence of a U.S. Carbon Tax: A Lifetime and Regional Analysis." NBER Working Paper 13554.

Kaplow, Louis, 2004. "On the (Ir)Relevance of Distribution and Labor Supply Distortion to Government Policy." *Journal of Economic Perspectives* 18: 159-175.

Krupnick, Alan J., Ian W.H. Parry, Margaret Walls, Tony Knowles and Kristin Hayes, 2010. *Toward a New National Energy Policy: Assessing the Options*. Resources for the Future and National Energy Policy Institute, Washington, DC.

Parry and Williams: The Costs of Meeting Distributional Objectives for Climate Policy

Parry, Ian W.H., and Antonio M. Bento, 2000, "Tax Deductions, Environmental Policy, and the "Double Dividend" Hypothesis," *Journal of Environmental Economics and Management* 39: 67-96.

Parry, Ian W.H., 2004. "Are Emissions Permits Regressive?" *Journal of Environmental Economics and Management* 47: 364-387.

Parry, Ian W.H., 2005. "Fiscal Interactions and the Costs of Pollution Control from Electricity." *RAND Journal of Economics* 36: 849-869.

Pizer, William A., 2003. "Combining Price and Quantity Controls to Mitigate Global Climate Change." *Journal of Public Economics* 85: 409-434.

Saez, Emanuel, Joel Slemrod, and Seth Giertz, 2009. "The Elasticity of Taxable Income with Respect to Marginal Tax Rates: A Critical Review." NBER Working Paper no. 15012.

Schöb, Ronnie, 2006. "The Double Dividend Hypothesis of Environmental Taxes: A Survey." In H. Folmer and T. Tietenberg (eds.), *The International Yearbook of Environmental and Resource Economics 2005/2006*. Cheltenham, United Kingdom: Edward Elgar.

Smith, Anne E., Martin E. Ross, and W. David Montgomery, 2002. "Implications of Trading Implementation Design for Equity-Efficiency Tradeoffs in Carbon Permit Allocations." Working paper. Washington, D.C.: Charles River Associates.

Sijm, Jos, Karsten Neuhoff and Yihsu Chen, 2006. "CO_2 Cost Pass-Through and Windfall Profits in the Power Sector." *Climate Policy* 6: 49-72.

US IAWG, 2010. *Technical Support Document: Social Cost of Carbon for Regulatory Impact Analysis Under Executive Order 12866*. Interagency Working Group on Social Cost of Carbon, United States Government, Washington, DC.

Williams, Roberton C., 2009a. "An Estimate of the Second-Best Optimal Gasoline Tax, Considering Both Efficiency and Equity." Working paper, University of Maryland.

Williams, Roberton C., 2009b. "Distribution, Distortionary Taxation, and the Evaluation of Public Goods." Working paper, University of Maryland.

[10]

The B.E. Journal of Economic Analysis & Policy

Symposium

| *Volume* 10, *Issue* 2 | 2010 | *Article* 10 |

DISTRIBUTIONAL ASPECTS OF ENERGY AND CLIMATE POLICY

Comment on "What are the Costs of Meeting Distributional Objectives for Climate Policy?"

William Randolph*

*Congressional Budget Office, william.randolph@cbo.gov

Randolph: Comment on Parry and Williams

The excellent article by Parry and Williams (2010) develops an elegantly simplified static general equilibrium model and applies it to examine the effects on economic efficiency and the distribution of household burdens when a cap-and-trade system of emissions regulation is combined with each of four alternative ways to distribute, or 'recycle,' the rents or revenues created under the system. Households in the model economy consume three types of goods plus leisure and government services. Each household can belong to one of five quintiles in the distribution of either household income or household consumption. To approximate the system of transfers and taxes in the United States, each type of household receives a lump-sum transfer from the government and is subject to its own single marginal tax rate that applies to additional wage income plus profits minus expenditures on the tax-favored good. Goods are produced under perfect competition with a technology that exhibits constant returns to scale. Producers thus pass the price of emissions permits and the costs of emissions abatements forward to consumers through higher output prices. Using those basic economic ingredients plus a few additional important details, the authors are able to answer the efficiency and distributional questions in a transparent and intuitively sensible manner. The authors' use of a general equilibrium model to analyze the combined efficiency and distributional effects of both the emissions regulation and alternative methods of revenue recycling represents an important advance over previous similar analyses that use only a partial equilibrium approach, such as Burtraw et al. (2009) and Dinan and Rogers (2002).

The authors calibrate the model to represent the U.S. economy and then apply it to estimate the efficiency and distributional effects, in 2020, of four alternative rent or revenue recycling options: (1) an equal percentage point reduction in all marginal tax rates on wage income, (2) equal lump-sum payments to all households, (3) free allocation of emissions permits to the regulated entities, or (4) reductions in marginal tax rates on wage income such that the net burdens (with emissions restrictions and tax reductions) are distributionally neutral. Parameters of the cap-and-trade system are chosen to represent policies under consideration in 2010.

For the overall effects on efficiency, the authors find that the choice of a revenue recycling method matters considerably more than the emissions regulation itself. The range over which the overall effect on efficiency differs across the alternative revenue recycling policies is almost twice as wide as the size of the efficiency cost measured before accounting for the method of recycling. An emissions policy that yields an allowance price of \$33 (2008 dollars) per ton of CO_2 is estimated to impose an efficiency cost of about \$34

The B.E. Journal of Economic Analysis & Policy, Vol. 10 [2010], Iss. 2 (Symposium), Art. 10

billion before accounting for the effects of revenue recycling.[1] After accounting for revenue recycling, however, the overall effect on efficiency differs over a range of $59 billion: from a $6 billion benefit when revenues from a permit auction are used to finance an equal reduction in all marginal tax rates to a $53 billion cost when permits are given away for free to the regulated entities.

In summary of the main findings, the authors emphasize the large apparent tradeoff between efficiency and the distribution of burdens associated with the choice of a revenue recycling method. Such a tradeoff often exists in the choices made about government policies for taxes, spending, and transfers. The general equilibrium model allows the burdens of the cap-and-trade policy to be combined with burdens of the rest of the tax system and the distribution of income or consumption so that the full tradeoff between efficiency and distribution can be estimated within a single consistent framework. An important contribution of the analysis is to unify and measure such a complicated set of economic policy issues in a relatively simple and transparent manner.

A particularly striking result not sufficiently emphasized by the authors is that the distributionally neutral method of recycling revenues can be used to make the total efficiency cost almost as small as the direct cost, as measured by the Harberger triangle of a partial equilibrium analysis. Explicit taxes on wage income are reduced in a way that almost completely offsets the efficiency loss and offsets the regressive distribution of burdens associated with the implicit tax on wage income imposed by the cap-and-trade policy. I find that outcome intuitive because the distributionally neutral changes in explicit taxes are structured to provide a mirror image to the pattern of implicit taxes, virtually eliminating both the regressivity and the tax-interaction effects. Those insights, for me, are among the most useful insights provided by the article.

As the authors mention in the text, the analysis ignores capital income taxation. Consideration of capital income taxes would be necessary for analyzing a full range of policy options for revenue recycling, and would most likely increase the apparent tradeoff between efficiency and distribution. Even if capital income taxes were included, however, revenue-recycling options that reduce tax rates on wage income would still be most closely related to the tax interaction effects of the emissions policy. Because adequate treatment of the capital income tax issues would also add substantial complexity, the authors made a reasonable choice to leave those issues out of their analysis.

---

[1] That $34 billion is the sum of a direct efficiency cost of $9 billion – the Harberger triangle – and an efficiency cost of about $25 billion that results from the effects on labor supply of increased prices of goods – the tax interaction effect.

Randolph: Comment on Parry and Williams

Finally, for other non-specialists in climate policy like me, it is necessary to point out the obvious fact that the authors don't try to value any environmental benefits of the emissions policy. They only try to explain and measure the costs. The valuation of benefits raises a vast array of difficult issues that have been examined elsewhere. Economists who specialize in climate policy understand that standard omission from analysis of policy costs so well that they rarely bother to mention it, like the group of longtime traveling friends who now only tell their jokes by number.

References

Burtraw, Dallas, Richard Sweeney, and Margaret A. Walls (2009), "The Incidence of U.S. Climate Policy: Alternative Uses of Revenues from a Cap-and-Trade Auction," *National Tax Journal* 62: 497-518.

Dinan, Terry M. and Diane L. Rogers (2002), "Distributional Effects of Carbon Allowance Trading: How Government Decisions Determine Winners and Losers," *National Tax Journal* 55: 199-222.

Parry, Ian W. H. and Roberton C. Williams III (2010), "What are the Costs of Meeting Distributional Objectives for Climate Policy?", *The B.E. Journal of Economic Analysis & Policy*, Vol. 10, No. 2.

These comments are those of the author and should not be interpreted as those of the Congressional Budget Office.

[11]

The B.E. Journal of Economic Analysis & Policy

Symposium

Volume 10, Issue 2	2010	Article 1

DISTRIBUTIONAL ASPECTS OF ENERGY AND CLIMATE POLICY

Distributional Implications of Alternative U.S. Greenhouse Gas Control Measures

Sebastian Rausch*   Gilbert E. Metcalf[†]

John M. Reilly[‡]   Sergey Paltsev**

*Massachusetts Institute of Technology, rausch@mit.edu

[†]Tufts University, gilbert.metcalf@tufts.edu

[‡]Massachusetts Institute of Technology, jreilly@mit.edu

**Massachusetts Institute of Technology, paltsev@mit.edu

Recommended Citation

Sebastian Rausch, Gilbert E. Metcalf, John M. Reilly, and Sergey Paltsev (2010) "Distributional Implications of Alternative U.S. Greenhouse Gas Control Measures," *The B.E. Journal of Economic Analysis & Policy*: Vol. 10: Iss. 2 (Symposium), Article 1.

Available at: http://www.bepress.com/bejeap/vol10/iss2/art1

Distributional Implications of Alternative U.S. Greenhouse Gas Control Measures*

Sebastian Rausch, Gilbert E. Metcalf, John M. Reilly, and Sergey Paltsev

Abstract

We analyze the distributional and efficiency impacts of different allowance allocation schemes motivated by recently proposed U.S. climate legislation for a national cap and trade system using a new dynamic computable general equilibrium model of the U.S. economy. The USREP model tracks nine different income groups and twelve different geographic regions within the U.S. We find that the allocation schemes in all proposals are progressive over the lower half of the income distribution and proportional in the upper half of the income distribution. Scenarios based on the Cantwell-Collins allocation proposal are less progressive in early years and have lower welfare costs due to smaller redistribution to low income households and, consequently, lower income-induced increases in energy demand and less savings and investment. Scenarios based on the three other allocation schemes tend to overcompensate some adversely affected income groups and regions in early years, but this dissipates over time as the allowance allocation effect becomes weaker. Finally, we find that carbon pricing by itself (ignoring the return of carbon revenues through allowance allocations) is proportional to modestly progressive. This striking result follows from the dominance of the sources over uses side impacts of the policy and stands in sharp contrast to previous work that has focused only on the uses side. The main reason is that lower income households derive a large fraction of income from government transfers, and we hold the transfers constant in real terms, reflecting the fact that transfers are generally indexed to inflation. As a result, this source of income is unaffected by carbon pricing while wage and capital income is affected.

KEYWORDS: U.S. Greenhouse Gas Policy, distributional effects, regional effects, income class, allowance allocation, computable general equilibrium models

---

*This paper was written for the Energy Policy Symposium on Distributional Aspects of Energy and Climate Policy held in Washington, D.C. on Jan. 22-23, 2010, organized by the University of Chicago Energy Initiative, Resources for the Future, and the University of Illinois. Without implication, we would like to thank Shanta Devarajan, Denny Ellerman, Don Fullerton, one anonymous referee, and participants at the Energy Policy Symposium on Distributional Aspects of Energy and Climate Policy held in Washington, D.C., and the CEEPR Spring 2010 Workshop for helpful comments. We thank Dan Feenberg for providing data from the NBER TAXSIM simulator on marginal income tax rates. We thank Tony Smith-Grieco for excellent research assistance. We acknowledge support of MIT Joint Program on the Science and Policy of Global Change through a combination of government, industry, and foundation funding, the MIT Energy Initiative, and additional support for this work from a coalition of industrial sponsors.

1. INTRODUCTION

U.S. Senate proposals for cap and trade legislation and the House-passed Waxman Markey Bill focus on similar overall cuts in greenhouse gases. The biggest difference among them is how allowances, and the revenue from their auction, would be distributed. Different uses of revenue or different allowance allocations would not in the first instance affect the direct cost of achieving emissions reductions but they can have important implications for how costs are borne by different regions and among households of different income levels. Different uses of revenue may have indirect effects on the overall welfare cost of a policy to the extent revenue is used to offset other distortionary taxes. In addition the allowance allocation has efficiency impacts to the extent that it creates further distortions or prevents pass through of the full CO_2 price in some products, or is used in some way that does not create value for U.S. citizens. Rausch *et al.* (2009) investigated some generic allocation schemes with a multi-region, multi-household static general equilibrium model of the U.S., the U.S. Regional Energy Policy (USREP) model. Here we extend the USREP model to a recursive dynamic formulation and design allocation schemes intended to approximate more closely specific cap and trade proposals.

In extending the USREP model to a recursive dynamic formulation we borrow the dynamic structure of the MIT Emissions Prediction and Policy Analysis (EPPA) model (Paltsev *et al.* (2005)). With this extension we are able more closely to represent features of revenue use and allowance allocation in specific legislative proposals and contrast their distributional implications. As with previous analyses of greenhouse gas legislation conducted with the EPPA model such as that in Paltsev *et al.* (2009) we attempt to capture key features of the cap and trade provisions in the proposals but are not able to address many other provisions of the bills that deal with energy efficiency standards and the like. The added value here is that we can consider distributional effects of proposed legislation. We contrast the allowance allocation schemes of the House legislation (Waxman-Markey) with those of the Senate proposals of Kerry and Boxer and of Cantwell and Collins. As a result of negotiations in the Senate the Kerry-Boxer bill has stalled and been replaced by a discussion draft by Senators Kerry and Lieberman. The bill contains a variety of new features but is similar to Waxman-Markey in its allocation of allowance value. To isolate the effects of different allocation schemes, we formulate a cap and trade policy designed to limit cumulative emissions over the control period in all scenarios to 203 billion metric tons (bmt). The cap and trade provisions of the proposals we consider would lead to somewhat different cumulative emissions because of differences in the timing of reductions, sectoral coverage, and whether outside credits were allowed.

The B.E. Journal of Economic Analysis & Policy, Vol. 10 [2010], Iss. 2 (Symposium), Art. 1

Waxman-Markey and Kerry-Boxer are part auction, part free allocation with a complex allowance and revenue allocation designed to achieve many different purposes. In contrast, Cantwell and Collins proposal auctions all allowances and distributes most of the revenue with a very straightforward lump sum allocation to individuals. Extending our analysis to distributional issues requires further interpretation, especially for those proposals with complex allocation schemes, of how allocation of allowances and auction revenue would actually occur if current proposals were implemented.

Our analysis shows a number of results. First, scenarios based on the Waxman-Markey and Kerry-Boxer (or Kerry-Lieberman) allowance allocation schemes are more progressive (i.e., a larger welfare loss is imposed on higher income households) in early years than scenarios based on the Cantwell-Collins proposal. We emphasize, however, that the overall distributional impact of these proposals depend on *all* the features of these legislative proposals and not just the cap and trade programs. Nonetheless the allowance allocation schemes are important determinants of the overall distributional impact of these bills. Second, scenarios based on the Cantwell-Collins allocation proposal have lower welfare costs due to lower redistribution to low income households and consequent lower income-induced increases in energy demand. Third, we find that the Waxman-Markey and Kerry-Boxer (or Kerry-Lieberman) allocation schemes appear to overcompensate some adversely affected income groups and regions early on, though this dissipates over time as the allocation scheme evolves to something closer to lump sum distribution. Fourth, the allocation schemes in all proposals are progressive over the lower half of the income distribution and essentially proportional in the upper half of the income distribution. Finally we find that carbon pricing by itself, ignoring the return of carbon revenues through allowance allocations, is proportional to modestly progressive. We trace our result to the dominance of the sources side over the uses side impacts of the policy. It stands in sharp contrast to previous work that has focused only on the uses side, and has hence found energy taxation to be regressive. It is worth pointing out that our model framework provides only an analysis of welfare *costs* of climate policy and does not attempt to incorporate any *benefits* from averting climate change. Any welfare changes reported in this paper therefore refer to changes in costs.

The paper is organized as follows: Section 2 briefly describes the recursive dynamic version of the USREP model. Section 3 provides some background on incidence theory. Section 4 discusses the legislative proposals we evaluate, mapping the allowance and revenue allocation in the Bills to specific distributional schemes in the model. Section 5 defines policy scenarios based on the proposed greenhouse gas control measures. Section 6 investigates the distributional implications across regions and income classes of allocation scenarios reflecting our interpretation of proposed policies, and Section 7 reports

Rausch et al.: Implications of Alternative U.S. GHG Control Measures

the results of a counterfactual analysis that allows us to trace the source of distribution effects we observe. Section 8 concludes.

2. A RECURSIVE-DYNAMIC U.S. REGIONAL ENERGY POLICY MODEL

USREP is a computable general equilibrium model of the U.S. economy designed to analyze energy and greenhouse gas policies.[1] It has the capability to assess impacts on regions, sectors and industries, and different household income classes. As in any classical Arrow-Debreu general equilibrium model, our framework combines the behavioral assumption of rational economic agents with the analysis of equilibrium conditions, and represents price-dependent market interactions as well as the origination and spending of income based on microeconomic theory. Profit-maximizing firms produce goods and services using intermediate inputs from other sectors and primary factors of production from households. Utility-maximizing households receive income from government transfers and from the supply of factors of production to firms (labor, capital, land, and resources). Income thus earned is spent on goods and services or is saved. The government collects tax revenue which is spent on consumption and household transfers. USREP is a recursive-dynamic model, and hence savings and investment decisions are based on current period variables.[2]

The USREP model is built on state-level economic data from the IMPLAN dataset (Minnesota IMPLAN Group, 2008) covering all transactions among businesses, households, and government agents for the base year 2006. The detailed representation of existing taxes captures effects of tax-base erosion, and comprises sector- and region-specific *ad valorem* output taxes, payroll taxes and capital income taxes. IMPLAN data has been augmented by incorporating regional tax data from the NBER tax simulator to represent marginal personal income tax rates by region and income class. Energy data from the Energy Information Administration's State Energy Data System (SEDS) are merged with the economic data to provide physical flows of energy for greenhouse gas

---

[1] As in any standard computable general equilibrium model, our framework adopts a full-employment assumption and further assumes that money is neutral, i.e. production and consumption decisions are solely determined by relative prices.

[2] Experience from a forward-looking version of the EPPA model (Babiker *et al.* (2008)) suggests that energy sector and CO_2 price behavior are similar to those derived from a recursive-dynamic model. Consumption shifting as an additional avenue of adjustment to the policy may, however, lower overall policy costs. On the other hand, inter-temporal optimization with perfect foresight poorly represents the real economy where agents face high levels of uncertainty that likely lead to higher costs than if they knew the future with certainty. We leave for future work the careful comparison of how alternative approaches to expectations formation may influence model results.

The B.E. Journal of Economic Analysis & Policy, Vol. 10 [2010], Iss. 2 (Symposium), Art. 1

accounting. Non-CO_2 greenhouse gases are based on the EPA inventory data, and are included as in the EPPA model with endogenous costing of the abatement (Hyman *et al.*, 2003).

The basic structure and data used in the USREP model are described in some detail in Rausch *et al.* (2009) with the dynamic structure borrowed from EPPA (Paltsev *et al.*, 2005). We focus discussion here on elements of the model that differ from that described in these two previous papers and on the data sources and calibration needed to regionalize the model. The underlying state level data base provides flexibility in the regional detail of the model. Here we use the regional structure shown in **Figure 1**. This structure separately identifies larger states, allows representation of separate electricity interconnects, and captures some of the diversity among states in use and production of energy.

Figure 1. Regional Aggregation in the USREP Model.

Table 1 provides an overview of the sectoral breakdown and the primary factors of production. Consistent with the assumption of perfect competition on product and factor markets, production and consumption processes exhibit constant-returns-to-scale and are modeled by nested constant-elasticity-of-substitution (CES) functions. A detailed description of the nesting structure for each production sector and household consumption is provided in Rausch *et al.* (2009).

Rausch et al.: Implications of Alternative U.S. GHG Control Measures

Table 1. USREP Model Details: Regional and Sectoral Breakdown and Primary Input Factors.

Region[a]	Sectors	Primary Input Factors
Alaska (AK)	**Non-Energy**	Capital
California (CA)	Agriculture (AGR)	Labor
Florida (FL)	Services (SRV)	Land
New York (NY)	Energy-Intensive (EIS)	Crude Oil
New England (NENGL)	Other Industries (OTH)	Shale Oil
South East (SEAST)	Transportation (TRN)	Natural Gas
North East (NEAST)	**Energy**	Coal
South Central (SCENT)	Coal (COL)	Nuclear
Texas (TX)	Convent. Crude Oil (CRU)	Hydro
North Central (NCENT)	Refined Oil (OIL)	Wind
Mountain (MOUNT)	Natural Gas (GAS)	
Pacific (PACIF)	Electric: Fossil (ELE)	
	Electric: Nuclear (NUC)	
	Electric: Hydro (HYD)	
	Advanced Technologies	

[a]Model regions are aggregations of the following U.S. states: NENGL = Maine, New Hampshire, Vermont, Massachusetts, Connecticut, Rhode Island; SEAST = Virginia, Kentucky, North Carolina, Tennessee, South Carolina, Georgia, Alabama, Mississippi; NEAST = West Virginia, Delaware, Maryland, Wisconsin, Illinois, Michigan, Indiana, Ohio, Pennsylvania, New Jersey, District of Columbia; SCENT = Oklahoma, Arkansas, Louisiana; NCENT = Missouri, North Dakota, South Dakota, Nebraska, Kansas, Minnesota, Iowa; MOUNT = Montana, Idaho, Wyoming, Nevada, Utah, Colorado, Arizona, New Mexico; PACIF = Oregon, Washington, Hawaii.

There are nine representative households in each region differentiated by income levels as shown in **Table 2**. Households across income classes and regions differ in terms of income sources as well as expenditures. State-specific projections through 2030 are from the U.S. Census Bureau (2009a).[3] Labor supply is determined by the household choice between leisure and labor. We calibrate compensated and uncompensated labor supply elasticities following the approach described in Ballard (2000), and assume for all income groups that the uncompensated (compensated) labor supply elasticity is 0.1 (0.3). Labor is fully mobile across industries in a given region but is immobile across U.S. regions.

---

[3] The USREP model incorporates demographic data on the population and number of households in each region and income class for the base year 2006 based on U.S. Census Data (2009b). We apply state-specific population growth rates uniformly to all income groups.

The B.E. Journal of Economic Analysis & Policy, Vol. 10 [2010], Iss. 2 (Symposium), Art. 1

Savings enters directly into the utility function which generates the demand for savings and makes the consumption-investment decision endogenous. We follow an approach by Bovenberg, Goulder and Gurney (2005) distinguishing between capital that is used in production of market goods and services and capital used in households (e.g. the housing stock). We assume income from the former is

Table 2. Annual Income Classes Used in the USREP Model and Cumulative Population.

Income class	Description	Cumulative Population for whole U.S. (in %)[a]
hhl	Less than $10,000 per year	7.3
hh10	$10,000 to $15,000 per year	11.7
hh15	$15,000 to $25,000 per year	21.2
hh25	$25,000 to $30,000 per year	31.0
hh30	$30,000 to $50,000 per year	45.3
hh50	$50,000 to $75,000 per year	65.2
hh75	$75,000 to $100,000 per year	78.7
hh100	$100,000 to $150,000 per year	91.5
hh150	$150,000 plus per year	100.0

[a]Based on data from U.S. Census Bureau (2009a).

subject to taxation while the imputed income from housing capital is not, and so households can shift investment between market and housing capital in response to changing capital taxation. Lacking specific data on capital ownership, households are assumed to own a pool of U.S. capital—that is they do not disproportionately own capital assets within the region in which they reside.

We adopt the vintage capital structure of the EPPA model. Malleable capital is mobile across U.S. regions and industries, while vintaged capital is region and industry specific. As a result there is a common rate of return on malleable capital across the U.S. The accumulation of both malleable and non-malleable capital is calculated as investment net of depreciation according to the standard perpetual inventory assumption. Given base year data about investment demand by sector and by region, we specify for each region an investment sector that produces an aggregate investment good equal to the sum of endogenous savings by different household types. Foreign capital flows are fixed as in the EPPA model. We assume an integrated U.S. market for fossil fuel resources and that the regional ownership of resources is distributed in proportion to capital income. Rausch *et al.* (2009) explored the implications of assuming instead that resource ownership was regional. Such an assumption amplifies regional differences in the impacts of climate legislation, resulting in greater costs for

Rausch et al.: Implications of Alternative U.S. GHG Control Measures

regions with significant energy production but we believe that assumption overestimates regional differences because equity ownership in large energy companies is broadly owned.

Labor-augmenting technical change is a key driver of economic growth as in EPPA. Regional labor productivity growth rates were calibrated to match AEO2009 GDP growth through 2030. Beyond 2030, population and labor productivity growth rates are extrapolated by fitting a logistic function that assumes convergence in growth rates in 2100. The 2100 targets for annual labor productivity growth and for annual population growth are two and zero percent, respectively.

Energy supply is regionalized for USREP by incorporating data on regional fossil fuel reserves from the U.S. Geological Service and the Department of Energy[4]. The resource depletion model and elasticities of substitution between resource and non-resource inputs in fossil fuel production are identical to those in EPPA. As in EPPA, we specify a range of advanced energy supply technologies.[5]

The markups, share parameters and elasticity parameters for the advanced energy supply technologies are those from Paltsev *et al.* (2009) and the same cost mark-ups apply in all regions except for renewables. For renewables the cost shares are taken from Paltsev *et al.* (2009) but regional mark-ups and elasticity parameters are derived from regional supply curves. Regional wind supply curves for each technology have been estimated based on high-resolution wind data from NREL (2009) and a levelized cost model described in Morris (2009) that was also the basis for cost estimates in Paltsev *et al.* (2009). The TrueWinds model (NREL, 2009) provides data on the capacity factors for wind turbines if they were located at sites across the U.S., allowing construction of a regional wind supply curve that depends on the quality of wind resources in each region. We derive regional supply curves for biomass from data from Oakridge National Laboratories (2009) that describes quantity and price pairs for biomass supply for each state.

Non-price induced improvements in energy efficiency are represented by an Autonomous Energy Efficiency Improvement (AEEI) parameter as in EPPA, and represent technological progress that reduces at no cost the energy needed in consumption and production activities, thus resulting in reduced energy use per unit of activity and general productivity improvement over time. Reference case energy use is calibrated to the updated AEO2009 reference case (Energy Information Administration (2009)). The baseline thus includes both the impacts of the American Recovery and Reinvestment Act (ARRA) and the Energy Independence and Security Act (EISA).

---

[4] Source for crude oil and natural gas reserves: Department of Energy (2009). Source for shale oil reserves: John R. Dyni (2006). Source for coal resources: USGS (2009).
[5] A list of these technologies can be found in Paltsev *et al.* (2009).

The B.E. Journal of Economic Analysis & Policy, Vol. 10 [2010], Iss. 2 (Symposium), Art. 1

Sectoral output produced in each region is converted through a constant-elasticity-of-transformation function into goods destined for the regional, national, and international market. All goods are tradable. Depending on the type of commodity, we distinguish three different representations of intra-national regional trade. First, bilateral flows for all non-energy goods are represented as "Armington" goods (Armington (1969)), where like goods from other regions are imperfectly substitutable for domestically produced goods. Second, domestically traded energy goods, except for electricity, are assumed to be homogeneous products, i.e. there is a national pool that demands domestic exports and supplies domestic imports. This assumption reflects the high degree of integration of intra-U.S. markets for natural gas, crude and refined oil, and coal. Third, we differentiate six regional electricity pools that are designed to provide an approximation of the existing structure of independent system operators (ISO) and the three major NERC interconnections in the U.S. More specifically, we distinguish the Western, Texas ERCOT and the Eastern NERC interconnections and in addition identify AK, NENGL, and NY as separate regional pools.[6] [7] Within each regional pool, we assume that traded electricity is a homogenous good, where no electricity is traded between regional pools.

Analogously to the export side, we adopt the Armington (1969) assumption of product heterogeneity for imports. A CES function characterizes the trade-off between imported, from national and international sources, and locally produced varieties of the same goods. Foreign closure of the model is determined through a national balance-of-payments (BOP) constraint.

3. BACKGROUND ON DISTRIBUTIONAL ANALYSIS

Carbon pricing through a cap-and-trade system has very similar impacts to broad based energy taxes – not surprising since over eighty percent of greenhouse gas emissions are associated with the combustion of fossil fuels (U.S. Environmental Protection Agency (2009)). The literature on distributional implications across income groups of energy taxes is a long and extensive one and some general conclusions have been reached that help inform the distributional analysis of

---

[6] We identify NY and NENGL as separate pools since electricity flows with contiguous ISOs represent only a small fraction of total electricity generation in those regions. For example, based on own calculation from data provided by ISOs, net electricity trade between ISO New England and ISO New York account for less than 1% of total electricity produced in ISO New England. Interface flows between the New York and neighboring ISOs amount to about 6% of total electricity generation in ISO New York.

[7] The regional electricity pools are thus defined as follows: NENGL, NY, TX, AK each represent a separate pool. The Western NERC interconnection comprises CA, MOUNT, and PACIF. The Eastern NERC interconnection comprises NEAST, SEAST, and FL.

carbon pricing. First, analyses that rank households by their annual income find that excise taxes in general tend to be regressive (e.g. Pechman (1985) looking at excise taxes in general and Metcalf (1999) looking specifically at a cluster of environmental taxes).

The difficulty with this ranking procedure is that many households in the lowest income groups are not poor in any traditional sense that should raise welfare concerns. This group includes households that are facing transitory negative income shocks or who are making human capital investments that will lead to higher incomes later in life (e.g. graduate students). It also includes many retired households which may have little current income but are able to draw on extensive savings.

That current income may not be a good measure of household well being has long been known and has led to a number of efforts to measure lifetime income. This leads to the second major finding in the literature. Consumption taxes – including taxes on energy – look considerably less regressive when lifetime income measures are used than when annual income measures are used. Studies include Davies, St Hilaire and Whalley (1984), Poterba (1989, (1991), Bull, Hassett and Metcalf (1994), Lyon and Schwab (1995) and many others.[8]

The lifetime income approach is an important caveat to distributional findings from annual incidence analyses but it relies on strong assumptions about household consumption decisions. In particular it assumes that households base current consumption decisions knowing their full stream of earnings over their lifetime. While it is reasonable to assume that households have some sense of future income, it may be implausible to assume they have complete knowledge or that they necessarily base spending decisions on income that may be received far in the future.[9] It may be that the truth lies somewhere between annual and lifetime income analyses. Moreover, if one were to use a lifetime income approach, one would like to track consumption over the lifecycle to capture any lifecycle changes in the consumption of carbon intensive products and compare lifetime carbon pricing burdens rather than a single-year snapshot. This paper takes a current income approach to sorting households.

Turning to climate policy in particular a number of papers have attempted to measure the distributional impacts of carbon pricing across household income groups. Dinan and Rogers (2002) build on Metcalf (1999) to consider how the

---

[8] Most of these studies look at a snapshot of taxes in one year relative to some proxy for lifetime income – often current consumption based on the permanent income hypothesis of Friedman (1957). An exception is Fullerton and Rogers (1993) who model the lifetime pattern of tax payments as well as income.

[9] On the other hand casual observation of graduate students in professional schools (business, law, medicine) make clear that many households are taking future income into account in their current consumption decisions.

The B.E. Journal of Economic Analysis & Policy, Vol. 10 [2010], Iss. 2 (Symposium), Art. 1

distribution of allowances from a cap and trade program affects the distributional outcome. Both these papers emphasize that focusing on the distributional burden of carbon pricing (either a tax or auctioned permits) without regard to the use of the revenue raised (or potentially raised) from carbon pricing provides an incomplete distributional analysis. How the proceeds from carbon pricing are distributed have important impacts on the ultimate distributional outcome.

The point that use of carbon revenues matters for distribution is the basis for the distributional and revenue neutral proposal in Metcalf (2007) for a carbon tax swap. It is also the focus of the analysis in Burtraw, Sweeney and Walls (2009). This latter paper considers five different uses of revenue from a cap and trade auction focusing on income distribution as well as regional distribution. A similar focus on income and regional distribution is in Hassett, Mathur and Metcalf (2009). This last paper does not consider the use of revenue but does compare both annual and lifetime income measures as well as a regional analysis using annual income. Grainger and Kolstad (2009) do a similar analysis as that of Hassett, Mathur and Metcalf (2009) and note that the use of household equivalence scales can exacerbate the regressivity of carbon pricing. Finally Burtraw, Walls and Blonz (2009) consider the distributional impacts in an expenditure side analysis where they focus on the allocation of permits to local distribution companies (LDCs). Rausch *et al.* (2009) also investigate the welfare costs of allocations to LDCs and find that allocations that lead to real or perceived reductions in electricity prices by consumers have large efficiency costs.

With the exception of the last paper, all of the papers above assume that the burden of carbon pricing is shifted forward to consumers in the form of higher energy prices and higher prices of energy-intensive consumption goods and services. That carbon pricing is passed forward to consumers follows from the analysis of a number of computable general equilibrium models. Bovenberg and Goulder (2001), for example, find that coal prices rise by over 90 percent of a $25 per ton carbon tax in the short and long run (Table 2.4).[10] This incidence result underlies their finding that only a small percentage of permits need be freely allocated to energy intensive industries to compensate shareholders for any windfall losses from a cap and trade program. See also Bovenberg, Goulder and Gurney (2005) for more on this issue.

Metcalf *et al.* (2008) consider uses side effects as a result of higher consumer prices and sources side effects as a result of lower factor returns, over different time periods for a carbon tax policy begun in 2012 and slowly ramped

---

[10] They assume world pricing for oil and natural gas so that the gross of tax prices for these two fossil fuels rise by the full amount of the tax.

up through 2050.[11] The tax on carbon emissions from coal are largely passed forward to consumers in all years of the policy in roughly the same magnitude found by Bovenberg and Goulder (2001). Roughly ten percent of the burden of carbon pricing on crude oil is shifted back to oil producers initially with the share rising to roughly one-fourth by 2050 as consumers are able to find substitutes for oil in the longer run. Interestingly the consumer burden of the carbon tax on natural gas exceeds the tax. This reflects the sharp rise in demand for natural gas as an initial response to carbon pricing is to substitute gas for coal in electricity generation. By 2050 the producer price is falling for reasonably stringent carbon policies.[12]

Fullerton and Heutel (2007) construct an analytic general equilibrium model to identify the various key parameters and relationships that determine the ultimate burden of a tax on a pollutant.[13] While the model is not sufficiently detailed to provide a realistic assessment of climate change impacts on the U.S. economy it illustrates critical parameters and relationships that drive burden results.

The general equilibrium models discussed above all assume a representative agent in the U.S. thereby limiting their usefulness to considering distributional questions. Metcalf, *et al.* (2008) apply results from a representative agent model to data on U.S. households that allows them to draw conclusions about distributional impacts of policies but the household heterogeneity is not built into the model.[14]

Several computable general equilibrium (CGE) models have been constructed to investigate regional implications of climate and energy in the U.S. For example, the ADAGE model, documented in Ross (2008), has a U.S. regional module which is usually aggregated to five or six regions. The MRN-NEEM model described in Tuladhar *et al.* (2009) has nine U.S. regions. Both these models use a single representative household in each region.

---

[11] Sources side effects refer to burden impacts arising from changes in relative factor prices, while uses side effects refer to burden impacts arising from change in relative product prices. This terminology goes back to Musgrave (1959).

[12] Distributional results depend importantly on the stringency of policy. How stringent the policy is affects whether carbon free technologies are adopted in the EPPA model and therefore what the relative demand for fossil fuels is. In the text above we are reporting carbon tax results for a policy that limits emissions to 287 billion metric tons over the control period.

[13] The paper also provides a thorough summary of the literature on the incidence impacts of environmental taxes.

[14] A recent paper by Bento *et al.* (2009) marks an advance in the literature by allowing for household heterogeneity over income and location. That paper considers the impact of increased U.S. gasoline taxes taking into account new and used car purchases along with scrappage and changes in driving behavior.

The B.E. Journal of Economic Analysis & Policy, Vol. 10 [2010], Iss. 2 (Symposium), Art. 1

Rausch *et al.* (2009) does an explicit CGE analysis of carbon pricing in a single-period CGE model. That analysis considers a variety of possible allocations of the revenue and/or allowances from cap-and-trade system and finds that the use of revenues affects the overall progressivity of the policy substantially. It also finds that a significant portion of the carbon price is passed back to factors of production — most notably owners of natural resources and capital. This contributes to a greater progressivity of carbon pricing than found in literature that assumes full forward shifting.

4. U.S. CAP AND TRADE PROPOSALS: ALLOWANCE ALLOCATION

Below we carry out distributional analyses of cap and trade policies based on alternative proposals for greenhouse gas control legislation currently under consideration in the U.S. These are the house-passed American Clean Energy and Security Act (H.R. 2454) sponsored by Reps. Waxman and Markey, the Clean Energy Jobs and American Power Act (S. 1733) a Senate bill similar to H.R. 2454 and sponsored by Senators Kerry and Boxer, and now replaced by the American Power Act (APA) draft bill by Kerry and Lieberman, and the Carbon Limits for America's Renewal (CLEAR) Act, a competing Senate Bill sponsored by Senators Cantwell and Collins. All proposals seek an overall reduction of GHG emissions in the U.S. to 83% below 2005 levels by 2050 with intervening targets. Cap and trade components of the bills cover most of the economy's emissions but not necessarily all of them, with other measures directed toward uncapped sectors. For example, estimates are that Waxman-Markey covers between 85% and 90% of emissions with a cap and trade system. Waxman-Markey has a slightly looser target for sectors covered by the cap and trade in 2020 than does Kerry-Boxer, issuing allowances at a level 17% below 2005 emissions in 2020, whereas the economy-wide goal is a 20% reduction by that date. Kerry-Lieberman would sell as many allowances as needed to refineries at a fixed price but would adjust over time to meet quantity targets. In our simulations of the effects of these bills, we assume the national goals are met, and we achieve them with a cap and trade system that covers all U.S. emissions except for land use CO_2 sources (or sinks). All of these proposals including banking and limited borrowing provisions and hence the time profile of reductions described in the bills are better thought of as the time profile of allowance allocation, with actual emissions levels in each year determined by how allowances are banked or borrowed (to the extent borrowing is allowed). In our simulations we find that the allocations result in net banking with no borrowing. Of course, in actuality borrowing may occur to the extent that unexpected costs make it attractive to bring permits forward in time.

While the stated national targets are identical across the bills, the Cantwell and Collins proposal has no provision for the use of offsets from outside the capped

12

Rausch et al.: Implications of Alternative U.S. GHG Control Measures

sectors to be used in lieu of the cap. Reductions similar in nature to the offsets allowed in the other bills are to be funded from a portion of the auction revenues that are subject to future appropriations. The other two proposals allow up to two billion tons per year of outside credits from a combination of domestic and foreign sources. In our simulations the domestic credits would need to come from a combination of reduced land use emissions and increased land use sinks. Foreign credits would come from qualified reductions abroad. As shown in

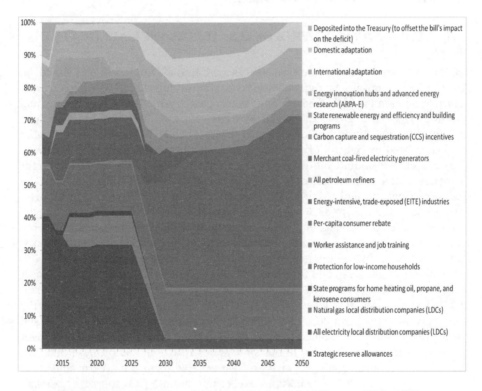

Figure 2. The Allocation of Allowance Value in the Waxman-Markey Bill.

Paltsev *et al.* (2009) if these credits are available at reasonable costs they would significantly reduce the CO_2 price and expected welfare cost of the legislation. Emissions from the capped sectors then are reduced much less than the target levels in the bill because available allowances are supplemented with external credits. Our main interest in this paper is the consequences of alternative distribution of allowances, and so we simulate the Cantwell-Collins allocation scheme allowing for the same level of outside credits as the other two bills. Any

The B.E. Journal of Economic Analysis & Policy, Vol. 10 [2010], Iss. 2 (Symposium), Art. 1

differences are the result of the allowance distribution mechanisms rather than the level of the cap.

The proposals are not always clear as to whether allowances are auctioned by some central Federal Agency and the revenue distributed or the allowances are distributed to entities who then can sell them. For example, designations to States could involve either a portion of allowance revenue or direct allocation of allowances leaving it up to the State to sell them into the allowance market. For our modeling purposes it does not matter whether it is revenue or the allowances that are distributed. We thus focus in our analysis on the allocation of "allowance value" in the different proposals to allow for distribution of allowances or the revenue from an auction.

Figure 2 shows the allowance allocation scheme as it is proposed in the Waxman-Markey bill. We do not show graphically the Kerry-Lieberman, Kerry-Boxer and Cantwell-Collins allowance allocation schemes here. The Cantwell-Collins bill calls for 75% of allowance revenue to be returned in a lump sum manner and 25% retained to meet several objectives but without specifying percentages for each. In terms of Figure 2, that bill would be simply two bars dividing allowance value among these two purposes. The allocation schemes in Kerry-Boxer and Kerry-Lieberman are similar to Waxman-Markey. The main difference is in terms of allowances set aside to offset the impact of the bill on the deficit. Waxman-Markey allocates at most 10% of the allowances for this purpose, in part directly and in part by directing how revenues obtained through early auction would be used, whereas Kerry-Boxer allocates a percentage that grows to 25%.[15] The allocation of revenue for deficit impacts in Kerry-Lieberman is much closer to Waxman-Markey. The increasing share devoted to this purpose proportional reduces the allocation to all other purposes. For example, Kerry-Boxer is able to allocate less than 50% of allowance value directly to households through either the low income energy assistance or the consumer rebate fund— whereas Waxman-Markey is able to allocate about 65% to households by 2050 through these two programs.

Both Kerry-Boxer and Waxman-Markey have a small strategic reserve of allowances and both allocate a substantial portion of allowances to local electricity and natural gas distribution companies in early years on the basis that these regulated entities will turn allowance value over to ratepayers, thus offsetting some of the impact of higher energy prices. This turns these LDCs into the mechanism for distribution as opposed to a government auction agency as in Cantwell-Collins. The other bills transition to a system closer to Cantwell-Collins over time, replacing the LDC distribution with a consumer rebate fund. Both

---

[15] This depends in part on whether future vintage allowances are sold early in the 2014-2020 period. If so, the share of allowances allocated to deficit reduction rises to roughly 12 percent of total allowances (current allowances and future allowances brought forward).

Rausch et al.: Implications of Alternative U.S. GHG Control Measures

retain a separate allocation to focus specifically on low income energy consumers. Both also then distribute allowances to different industries that are expected to be particularly affected by the legislation, but these allocations phase out by 2030. Use of allowances as an extra incentive for carbon capture and sequestration is also identified in both. A next set of allowances are allocated to fund various domestic energy efficiency programs. The next grouping of allocations is for international mitigation and adaptation and for domestic adaptation programs. Waxman-Markey contains a large set of allowances in later years designated for prior year use. This use possibly reallocates allowances through time, allowing the possibility of Federal borrowing if allowance prices rise too much. Of more relevance here is that the bill prescribes about one-half of this allowance value to go to the Treasury to offset impacts on the deficit and the other half as a consumer rebate. These amounts are shown in Figure 2 combined with the other provisions that direct revenue to the Treasury and to the consumer rebate. That value is allocated in the year in which the allowances would be originally issued, i.e. assuming the Federal government does not borrow them or if it does, the income is not rebated immediately. The Kerry-Boxer bill does not have this provision.

We do not represent the many different programs to which these allowances or allowance value would go, and the exact recipients will depend on program decisions yet to be made. However, we approximate the impact on regions and households of different income levels by distributing the allowance value based on data we have within the model, and that approximates what we believe to be the intent of the different distributions or how they would tend to work in practice. The distributional instruments we have at our disposal in the USREP Model and the correspondence to allocations called out in the bills are given in **Table 3**. For example, we allocate to households the proposed distribution of allowances to LDCs based on emissions and respective electricity and natural gas consumption. To determine the regional distribution, we allocate 50% of LDCs allowances based on historic sectoral emissions for the electricity and natural gas electrictiy sector, respectively. The other half is allocated to regions based on household electricity and natural gas consumption.[16] Within a region, allowances to LDCs are allocated based on respective fuel consumption. Allocations designated for low income households are distributed to households with incomes of less than $30,000 per year.

Distributions to industries other than LDCs go to households based on their capital earnings on the basis that this value will be reflected in the equity value of firms, and so households that own capital, for example, through stock

---

[16] Rausch, et al. (2009) consider the efficiency implications of a misperception by households that this lump-sum transfer lowers the marginal price of electricity and natural gas.

Table 3. Correspondence between Proposals Allowance Value Allocations and Distribution Instruments in USREP.

ALLOWANCE RECIPIENTS	MODEL INSTRUMENT
Mitigating Price Impacts on Consumers	
All electricity local distribution companies (LDCs)	Lump-sum transfer to consumers. Allocated to regions based on GHG emissions (50%) and based on value of electricity consumption (50%). Within a region, allocated to households based on the value of electricity
Additional allowances for small electricity LDCs	Lump-sum transfer to consumers. Allocated to regions based on GHG emissions (50%) and based on value of gas consumption (50%). Within a region, allocated to households based on the value of gas consumption.
Natural gas LDCs	Lump-sum transfer to consumers based on value of gas consumption
State programs for home heating oil, propane, and kerosene consumers	Lump-sum transfer to consumers based on value of oil consumption (excluding oil consumed for transportation purposes)
Assistance for Households and Workers	
Protection for low-income households	Lump-sum transfer to households with annual income less than $30k.
Worker assistance and job training	Distributed to regions based on value of energy production (coal, crude oil and refined oil). Within a region, distributed across households base on wage income.
Per-capita consumer rebate	Lump-sum transfer based on per-capita
Nuclear working training[1]	Distributed to regions based on value of nuclear electricity generation. Within a region, distributed across households based on wage income.
Allocations to Vulnerable Industries[2]	Lump-sum transfer based on capital income.
Technology Funding[3]	Distributed to regions based on energy use (industrial and private). Within a region, distributed based on household energy consumption.
International Funding[4]	Transferred abroad.
Domestic Adaptation	Distributed to government.
Other Uses	
Deposited into the Treasury (to offset the bill's impact on the deficit)	Distributed to government.
Grants to state and local agencies for transportation planning and transit[1]	Distributed to government.
Compensation for "early action" emission reductions prior to cap's inception	Distributed to households on a per capita basis
Allowances already auctioned in prior years	46% distributed to households on a per-capita basis, 54% distributed to government.[5]
Strategic reserve allowances	Distributed to households on a per capita basis.

Note to Table 3:[1]This allowance category only applies to the Kerry-Boxer bill.[2]Allocations to vulnerable industries include: Energy-intensive trade-exposed (EITE) industries, all petroleum refiners, additional allowances for small refiners, merchant coal-fired electricity generators, generators under long-term contracts without cost recovery, cogeneration facilities in industrial parks. [3]Technology Funding includes: Carbon capture and sequestration (CCS) incentives, state renewable energy and efficiency programs, state building retrofit programs, incentives for renewable energy and agricultural emissions reductions, clean vehicle technology incentives, energy innovation hubs, energy efficiency and renewable energy worker training fund, advanced energy research, supplemental reductions from agriculture, abandoned mine land, and reneable energy.[4]International Funding includes: International avoided deforestation, international clean technology deployment, and international adaptation.[5]We allocate allowances that are already auctioned in prior years to the goverment and to households according to the respective average share over the period from 2012-2050.

ownership, will be the beneficiaries.[17] Allowances distributed for energy efficiency and such are distributed by region based on regional energy consumption and then within a region by energy consumption by household on the basis that regions and households that consume more energy have more opportunities to take advantage of these programs. Allowances designated for worker assistance are distributed to regions based on oil and coal production on the basis that these industries are most likely to be affected by unemployment as the country shifts away from fossil fuels. We distribute funds devoted to CCS along with other energy R&D funds.

Given this mapping of the allocation provisions in the various legislative proposals we construct **Figure 3** that is similar to Figure 2 but showing instead the allocation of allowance value mapped to the instruments we use in USREP. The distribution instruments for all of these uses, except Foreign and Government, direct revenue to households but the particular instrument determines how the allowance value is allocated among households in different regions and in different income classes. As modeled, allowance value allocated abroad has no value for U.S. households. In the proposed legislation, most of the allowance value distribution is a pure transfer but some of these program expenditures are intended to incentivize energy savings and the like. Our allocation approach treats all of these program expenditures as pure transfers.[18] To the extent these programs overcome barriers that are not addressed by the CO_2 price, additional efficiency gains would reduce the welfare costs we estimate. To the extent these programs create double-incentives for particular activities, then they are redirecting abatement to activities that are not the most cost effective and that would increase the welfare cost we estimate. The assumption that they are pure transfers is therefore a neutral assumption. Furthermore, note that transfers of allowance value to households are treated as being non-taxable, with the effect of increasing how much allowance value must be set aside relative to a scenario where such transfers are taxed.

---

[17] An output-based rebate (OBR) to energy-intensive and trade-exposed (EITE) industries may result in a greater pass-through of allowance value to end consumers relative to our allocation based on capital earnings. How this allowance should be divided up, however, depends on factors such as the degree of pass-through in downstream sectors, given that EITE products are often intermediate goods, e.g. steel. Our approach treats all vulnerable industries symmetrically, and we allocate allowances lump sum proportional to capital income. Our approach does not capture incentive effects that would arise if allowance value was used to subsidize output in these industries. Such output subsidies would incur additional efficiency costs that we do not capture.

[18] We assume that the 25% of allowances in the Cantwell-Collins bill that go to a dedicated trust to fund climate mitigation and adaptation, clean energy and efficiency, and transition assistance programs, are allocated according to residual shares for similar categories (Energy use, Foreign, Government) in the Waxman-Markey bill. We understand that additional legislation would be

The B.E. Journal of Economic Analysis & Policy, Vol. 10 [2010], Iss. 2 (Symposium), Art. 1

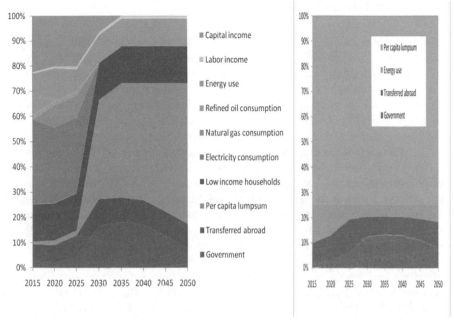

(a) Targeted Allowance Allocation Scheme   (b) Per Capita Dividend Scheme

Figure 3. The Allocation of Allowance Value according to Model Distribution Instruments.

Allowances allocated to government reduce the need for capital and labor taxes to be raised as much to meet the revenue neutrality assumption we impose[19], and so affect the distribution to households based on how increases in taxes affect different regions and income classes.

---

needed to appropriate this allowance revenue to the purposes identified in the legislation, and absent that the revenue would be returned to the Treasury.

[19] See Section 5 for a discussion of our treatment of revenue neutrality.

Rausch et al.: Implications of Alternative U.S. GHG Control Measures

5. SCENARIO DESIGN

We distinguish two sets of scenarios that differ with respect to the underlying allowance allocation scheme. Scenarios labeled *TAAS* represent a Targeted Allowance Allocation Scheme that is based on the Waxman-Markey or Kerry-Lieberman proposal. The *TAAS_DR* scenario sets aside a larger amount of allowances for the purpose of Deficit Reduction (DR) as in the allocation rule proposed by Kerry-Boxer. Scenarios labeled *PCDS* model a simple Per Capita Dividend Scheme as is described in the Cantwell-Collins proposal.

For each of the proposed allocation schemes, we design two scenarios that differ with respect to how the revenue neutrality requirement is met.[20] Our base case assumption is that sufficient allowance revenue is withheld by the government to cover the deficit impact and the remaining revenue is allocated at the percentages shown in Figure 3. An alternative case, denoted *TAX*, assumes that only the amount of allowance revenue specifically designated for deficit reduction in the bills is allocated to the government. We then raise capital and labor taxes uniformly across regions and income classes (in percentage points) to offset revenue losses from carbon pricing. This is separate from any allowance revenue targeted to deficit reduction. All scenarios assume the medium offset case from the analysis carried out in Appendix C of Paltsev *et al.* (2009) with identical assumptions about supply and costs of domestic and international offsets. We further assume that offsets have a cost to the economy, and implement this assumption by transferring abroad the value of allowances purchases internationally. Our assumption is that the average cost of these credits is $5 per effective ton of offsets of CO_2-e in 2015, rising at 4% per year thereafter.[21] Also note that since we create more allowance revenue for the government by increasing the allowances to account for credits coming from outside the system, we assume that the income transferred abroad to account for permit prices is taken from the allowance revenue. Finally, our assumptions about the supply of offsets imply a 203 bmt cumulative emissions target for 2012-205, which underlies all of the scenarios we consider here.

---

[20] We fix real government spending in the policy scenarios to match government spending under the reference scenario. Since government spending does not enter household utility functions, we did not want to confound welfare impacts from changes in the size of government with welfare impacts of climate policy. We discuss the implications of this assumption in section 6.1 below. Government spending in the reference scenario is assumed to grow in proportion with aggregate income.

[21] The Waxman-Markey bill specifies that 1.25 tons of foreign reductions are required to produce 1 ton of effective offsets. The $5/ton initial offset price means the actual payment per ton of foreign reduction is $4. For all proposals analyzed, we treat offsets costs symmetrically.

The B.E. Journal of Economic Analysis & Policy, Vol. 10 [2010], Iss. 2 (Symposium), Art. 1

Our analysis also takes banking and borrowing into consideration. In the Waxman-Markey bill, banking of allowances is unlimited and a two-year compliance period allows unlimited borrowing from one year ahead without penalty. Limited borrowing from two to five years ahead is also allowed, but with interest. In general, we find no need for aggregate borrowing, and so there is no need to implement an explicit restriction on it.

Our scenarios draw on features of the proposed pieces of legislation described above but in no way purport to model them in their entirety. Our focus is on the efficiency and distributional consequences of allowance allocation schemes and our scenarios model allowance trading along with their allocation over time. In that regard, we have had to interpret how we believe various allocations would work in practice when the exact allocation approach has not yet been fully described, and would only be completely determined by executive branch agencies responsible for these programs if the legislation were implemented. In addition, we do not model other components of the various pieces of legislation dealing with other policy measures such as renewable portfolio standards.

6. ANALYSIS OF SCENARIOS

Figure 4 shows the *Reference* and *Policy with offsets* emissions for the period 2012 to 2050. Projected *Reference* cumulative emissions over the 2012-2050 period are 298 bmt. In the *Policy with offsets case*, cumulative emissions are 203 bmt, a reduction of nearly one-third from Reference. The emissions path shown in Figure 4 for the *Policy with offsets case* is the result from the scenario *TAAS*. Cumulative emissions are identical under all six policy simulations, and the actual emissions paths are nearly identical. Slight differences in the emissions paths exist because of different overall welfare costs and distributional effects that can lead to a slightly different allocation of abatement over time, but these differences are so small that they would be imperceptible if plotted in Figure 4. Emissions in the *Reference* include estimates of the effects of existing energy policies under the Energy Independence and Security Act and the American Recovery and Reinvestment Act as they are projected to affect greenhouse gas emissions. Note, that while the allowance allocation for 2050 is set at 83% below 2005, our projected emissions in 2050 in the *Policy with offsets* case are only 35% below 2005 emissions because of the availability of offsets and banking. Before turning to distributional analyses by income group or region, we consider the aggregate U.S. welfare impacts of the various policies we model. **Figure 5** presents the change in welfare relative to the *Reference* scenario, measured in equivalent

Rausch et al.: Implications of Alternative U.S. GHG Control Measures

variation as a percentage of full income[22], for the various bills. One key result we see is that the *_TAX* scenarios lead to higher welfare costs than the scenarios where a fraction of the allowance revenue is withheld to satisfy revenue neutrality. Considering the *TAAS* scenario, for example, the welfare cost is 1.38 percent of full income by 2050 under the lump-sum scenario and 1.60 percent

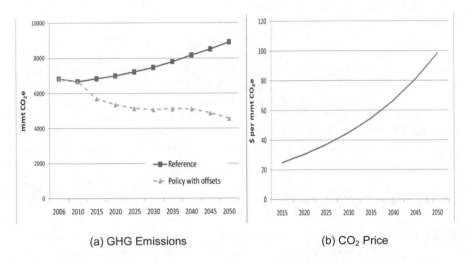

(a) GHG Emissions   (b) CO$_2$ Price

Figure 4. U.S. Greenhouse Gas Emissions and Carbon Price (Scenario *TAAS*).

under the tax scenario. Similar results hold for *TAAS_DR* and *PCDS*. This occurs because the *_TAX* scenarios create more deadweight loss from capital and labor taxation. Many economists have focused on a double-dividend effect where allowance revenue is used to lower capital and labor taxes, but here we have the reverse effect. Not enough of the revenue is retained to offset the deficit effects of the bill so that capital and labor taxes need to be increased, thereby increasing the cost the bill.[23]

Conditional on the treatment of revenue shortfalls, the three scenarios have very similar aggregate costs. *TAAS_DR_TAX* is somewhat less costly than *TAAS_TAX* because the former scenario reserves more of the allowance to offset the deficit and thus capital and labor taxes do not need to be increased as much.

---

[22] Full income is the value of consumption, leisure, and the consumption stream from residential capital.

[23] This follows from our particular assumption about how taxes are raised to maintain revenue neutrality. It is certainly possible that lump-sum taxes could be employed or some other configuration of tax increases that is less distortionary than the tax increases we model. Therefore one should not conclude that our result is general.

The B.E. Journal of Economic Analysis & Policy, Vol. 10 [2010], Iss. 2 (Symposium), Art. 1

The costs of *PCDS* and *PCDS_TAX* are slightly lower than the *TAAS* scenarios. The lower costs of the *PCDS* scenarios at first blush are surprising. These scenarios retain less of the allowance value to offset the deficit, and hence in the *_TAX* case it requires somewhat higher increases in capital and labor taxes to offset the deficit. The lower costs in *PCDS* scenarios arise from the distributional outcomes as they affect energy expenditures and savings. In particular, *TAAS* and *TAAS_DR*, through the low income energy assistance programs allocate more of

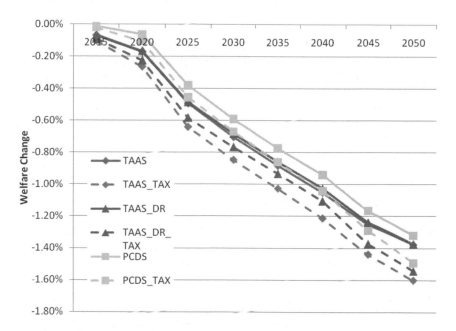

Figure 5. Welfare Change for Different GHG Control Proposals (U.S. Average).

the revenue value to poorer households. Lower income households spend a larger fraction of their income on energy and they save less. Thus, the abatement effect of pricing carbon is offset to greater extent by an income effect among poorer households in the *TAAS* and *TAAS_DR* than in the *PCDS* scenarios. In addition, there is less saving and therefore less investment in *TAAS* and *TAAS_DR* because less is saved for each additional dollar allocated to poorer households. Note that our aggregate welfare estimates are a simple sum of the welfare of each income class across all regions. An aggregate welfare function that weighted the welfare of lower income households higher, giving welfare benefit to more progressive outcomes would change these results, showing better results for *TAAS* and *TAAS_DR*. How much to value more progressive outcomes is a judgment. Here

Rausch et al.: Implications of Alternative U.S. GHG Control Measures

we leave it to the policy community to decide whether the more progressive outcome of *TAAS* and *TAAS_DR* is worth the extra welfare cost.

6.1. Distributional Impacts across Income Groups

Aggregate impacts obscure differential effects across households. Ideally we would construct a measure of the lifetime burden of carbon pricing and relate that to a measure of lifetime income. Our data do not allow us to do that. Our recursive-dynamic model has households of different income groups in each year but we have no data that allow us to track the transition of households from one income group to another. Instead we report burden impacts for different income groups at different points of time to show how the relative burden shifts over time.

Figure 6 shows the burden for a representative household in each income group for 2015, 2030, and 2050 for *TAAS* measured as equivalent variation Positive values indicate that a household benefits from the carbon policy. Households in the two lowest income groups, *hh1* and *hh10*, benefit in all periods as the return of permit revenue through various mechanisms more than offsets the higher cost of goods and services due to carbon pricing and any effects on their wages and capital income. Households *hh15* and *hh25* initially benefit but eventually bear net costs, *hh15* only in the final period. The effect of allocating an increasing amount of allowances on a per-capita basis is particularly strong for the lowest income group relative to higher income households since a dollar of additional revenue makes up a larger fraction of full income for these households.[24] The five highest income households bear net costs throughout the period though the burden through 2030 is less than 1 percent of income for all income groups. Over time, the burden of the policy grows for wealthier households with the burden ranging from 1 to roughly 1.5 percent by 2050.

In all years the cap and trade policy combined with the *TAAS* allocation scenarios is sharply progressive over the first five income groups though the burden for each income group, except that of the lowest, grows over time as the policy begins to impose larger reductions in emissions. The difference in burdens

---

[24] Pechman (1985) realized that income data for the low income groups suffered from substantial income mismeasurement. Since then, the approach adopted by him and many others is to omit the lowest income group from distributional analyses. Given the interest of the policy community for impacts on low income households, we decided to report results for households with annual income less than $10k, but we want to point out that in light of likely measurements problems we do not have the same degree of confidence in results as we do for other income groups.

over the lowest five income groups grows over time as does the spread between the burden for the lowest income group relative to the highest income group. The

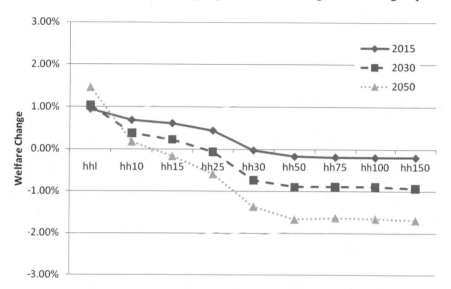

Figure 6. Welfare Change by Income Group, U.S. Average (Scenario *TAAS*).

policy is essentially neutral over the top income groups in all periods, as we will show below over time sources side effects become more important in shaping the distributional outcomes than do uses side effects.

Table 4 reports the annual cost in dollar terms for different households, in different years. On average the per-household costs are relatively modest in the early years of the program. While the costs appear large by 2050, it is important to keep in mind that incomes are growing so that these costs are still modest relative to household income. The average over time is the net present value (NPV) average. Note that Waxman-Markey allows considerable borrowing of allowances from the future by the Federal government if necessary to moderate CO_2 prices in the early years.

If these were auctioned in earlier years then the allowance revenue would accrue to the government earlier and in principle it could be used earlier. We have assumed the revenue is only available when the allowances were originally scheduled to be auctioned. If borrowing occurred and the revenue was used as specified in the bill—to reduce deficit impacts and as a lump sum rebate to consumers - that could blunt some of the progressivity in earlier years.

Costs and distributional impacts for *TAAS_DR* are very similar to *TAAS* and so we do not report them here. Rather we turn to the *PCDS* in **Figure 7.** Like

Rausch et al.: Implications of Alternative U.S. GHG Control Measures

TAAS and *TAAS_DR*, *PCDS* has modest to negative burdens (positive gains in Figure 7) initially with burdens rising over time. In comparison to the former bills the burden spreads across income groups in any given year are smaller. Lower income households benefit in the early years but not as much as in *TAAS*

Table 4. Annual Cost per Household by Income Group (Scenario *TAAS*).

	hhl	hh10	hh15	hh25	hh30	hh50	hh75	hh100	hh150	Avg.
2015	-614	-472	-467	-426	36	261	328	344	401	87
2020	-563	-412	-418	-349	230	386	532	501	532	221
2025	-450	-248	-207	-64	631	920	1109	1098	1237	646
2030	-603	-240	-168	63	950	1420	1589	1650	2031	939
2035	-763	-304	-190	110	1170	1842	2050	2192	2758	1195
2040	-851	-307	-156	203	1397	2222	2456	2636	3304	1451
2045	-827	-216	-13	411	1658	2661	2916	3141	3918	1774
2050	-778	-109	129	594	1853	2974	3246	3482	4278	2008
NPV Average[a]	-291	-150	-119	-33	331	538	614	642	780	347

Note: Table reports annual dollar costs per household by income group in various years. All dollar amounts are in 2006 dollars. [a] Net Present Value (NPV) average of welfare costs discounted to 2010 at 4% per annum.

and *TAAS_DR*. This is reflected in the flatter distributional curves for different years. By 2050 the *PCDS* scenario and the *TAAS* scenario have more similar distributional effects because by that time the allocation formula in *TAAS_DR* has become similar to that of the *PCDS*, with 65 percent of revenue distributed on per capita basis. The remaining difference is the continued allocation to low income consumers.

Distributional outcomes are altered when the full value of allowances is allocated as specified in the bills and revenue losses in the federal budget are instead made up by raising personal income tax rates. In general, the distributional burden across household groups is more progressive in the *_TAX* cases. Consider the burden snapshots for three different years as shown in **Figure 8** for *TAAS_TAX*. Lower-income households fare better under this approach with benefits to the lowest income group rising from 1 to about 1.5 percent of full

The B.E. Journal of Economic Analysis & Policy, Vol. 10 [2010], Iss. 2 (Symposium), Art. 1

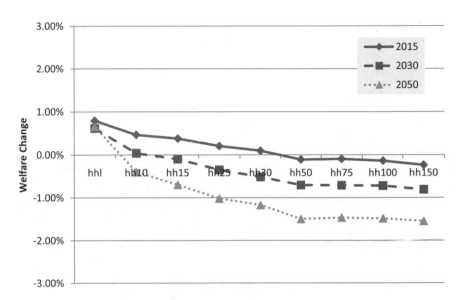

Figure 7. Welfare Change by Income Group, U.S. Average (Scenario *PCDS*).

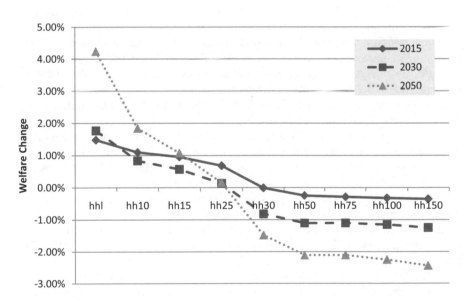

Figure 8. Welfare Change by Income Group, U.S. Average (Scenario *TAAS_TAX*).

Rausch et al.: Implications of Alternative U.S. GHG Control Measures

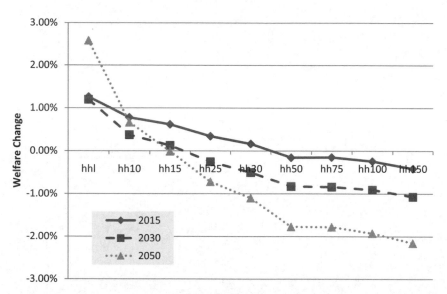

Figure 9. Welfare Change by Income Group, U.S. Average (Scenario *PCDS_TAX*).

income in 2015, while the highest income groups are only slightly affected. Lower income groups continue to do better – and in some cases are better off – when tax rates are raised to recoup lost tax revenues than when allowance value is withheld. In general they remain better off through 2050 because of the tax changes. By raising taxes to offset the deficit, more revenue remains available to be distributed, and the increase in transfers to lower-income groups more than offsets increases in taxation to these households. A similar result holds for the *PCDS* allocation proposal (see **Figure 9**).

The different treatments of revenue neutrality illustrates a classic equity-efficiency trade-off, where the withholding of allowances to preserve revenue neutrality yields higher efficiency but less progressive outcomes than if taxes are raised to maintain revenue neutrality in the government budget.

The impact of climate policy on government tax revenues is significant and helps explain why the different approaches to maintaining revenue neutrality matter. The initial loss of tax revenue due to higher costs for firms and reduced economic activity is 31.3 percent of the value of allowances in 2015. The percentage begins rising in 2040 and by 2050, the loss in tax revenue rises to one-half. The high tax revenue loss is in part an artifact of the assumption in the model that fixes the path of government spending to match that of the reference (no policy) scenario. We refer to this as absolute revenue neutrality. Lower GDP growth increases the size of government relative to GDP and magnifies the loss in

The B.E. Journal of Economic Analysis & Policy, Vol. 10 [2010], Iss. 2 (Symposium), Art. 1

tax revenue relative to allowance value. We make this assumption because the government sector in USREP does not produce explicit public goods that have any welfare value. By keeping revenue neutral changes in government we do not release or consume more resources that otherwise would be available to private sector.

An alternative approach would be to fix the ratio of government spending to GDP in the policy scenarios. To assess the distributional implications of this would then require production of a public good and an estimate of how that public good created welfare for different income classes in different regions, so that when government spending was increased or decreased we would have an estimate of how that was affecting distribution compared with how distribution was affected by changes in resources available to the private sector. If the government were kept at the same size in relative rather than absolute terms, the revenue needed to offset impacts on the deficit would not increase and would generally be at about the percentage we see in 2015. The difference would then be additional allowance value that could be used for distributional or other purposes.

We note that the Congressional Budget Office scores bills on their impact on the deficit, using a standard procedure for all legislation that is accepted by Congress. The CBO methodology is described in Congressional Budget Office (2009). That approach will not be consistent with our approach that endogenously calculates the deficit, and the revenue needed to close the deficit. The two approaches do lead to reasonably close estimates of the allowance value that must be set aside in early years (25 percent for CBO and 30 percent in this analysis) before the results diverge due to the different modeling approach taken by CBO from the approach taken here.

With absolute revenue neutrality, the need to make up substantial revenue losses leads to fairly large increases in marginal personal income tax rates under the tax-based make-up. In 2015, tax rates under *TAAS_TAX*, *TAAS_DR_TAX*, and *PCDS_TAAS* increase by 0.52, 0.34, and 0.48 percentage points, respectively. Respective tax rate increases in 2050 are 1.50, 0.79, 1.48 percentage points. The *TAAS_DR_TAX* increases are much less than the other two scenarios because more of the revenue is explicitly allocated to deficit effects of the proposal. This just illustrates one way to make up revenue losses. Other approaches could be undertaken that could enhance efficiency or equity goals.[25]

Summing up, we find that the *TAAS* and *TAAS_DR* scenarios on the one hand and the *PCDS* scenarios on the other have quite different distributional impacts across households, especially in the early years of the program. In

---

[25] This is simply a variant on the green tax swap idea analyzed by Metcalf (1999) and others.

addition, policy decisions on how to close the budget deficit arising from decreased tax collections have both efficiency and distributional implications.

Using higher personal income taxes to close the deficit incurs an efficiency cost but increases the progressivity of the programs because more of the allowance revenue is available for distribution to households. We next turn to regional impacts.

6.2. Distributional Impacts across Regions

Policy makers have also expressed concern over the regional impacts of climate policy. In this section we explore how regional impacts change over time for the allocation scenarios we have designed. **Figure 10** shows that the greenhouse gas emission reductions differ substantially among regions. Results are shown for the *TAAS* scenario. These differences reflect different shares of emissions from different sectors (electricity, transportation, industry) and different electric generation technologies (nuclear, hydro, coal, natural gas). The energy and emissions intensive regions (MOUNT, SEAST, SCENT, NCENT) show the largest reductions. States in the Mountain, Southeast, Northeast and North Central regions all experience reductions in GHG emissions relative to the business as usual scenario in excess of 50 percent by 2050.

Figure 11 shows the welfare impact of the *TAAS* scenario for each region. Initially California, Texas, Florida and states in the South Central, Pacific, and New England regions gain from the policy while other states suffer losses. By 2050 all states are bearing costs, ranging from about one-half of one percent (New England) to about one and three-quarters percent.

Welfare impacts for Alaska are not shown in Figure 11 to better visualize relative welfare impacts for other regions. Under the *TAAS* scenario Alaska's welfare effects are as follows: 2015: -0.42%; 2020: -1.15%; 2025: -2.26%; 2030: -2.57%; 2040: -3.52%; 2050: -5.27%. The substantial welfare impacts for Alaska can be attributed to the fact that Alaska exhibits by far the highest energy intensity among all regions and is a large energy producing state with a small population (see Figure 14). In earlier years of the policy, welfare effects are relatively modest compared to, e.g., 2030 and 2050. Alaska actually receives by far the highest allowance revenue per household among all regions under the *TAAS* scheme since many of the allowances are allocated on the basis of either energy consumption or production, but this is far from sufficient to offset the large costs the economy bears. As we note below, over time the allowance allocation effect becomes less important in determining

With absolute revenue neutrality, the need to make up substantial revenue losses leads to fairly large increases in marginal personal income tax rates under

The B.E. Journal of Economic Analysis & Policy, Vol. 10 [2010], Iss. 2 (Symposium), Art. 1

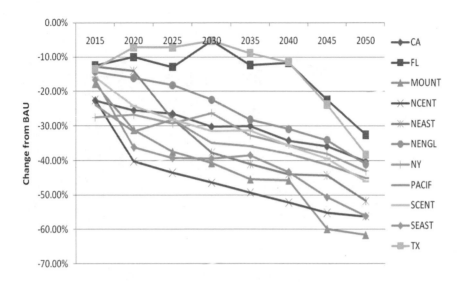

Figure 10. GHG Emissions Reductions by Region (Scenario *TAAS*).

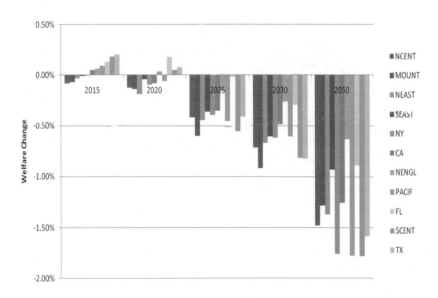

Figure 11. Welfare Change by Region (Scenario *TAAS*).

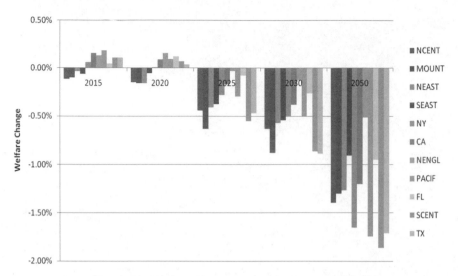

Figure 12. Welfare Change by Region (Scenario *PCDS*).

overall policy costs, and relative regional welfare differences are increasingly shaped by energy characteristics and income sources. This explains why welfare effects for Alaska become more negative over time both relative to earlier periods of the policy and in comparison to other regions. The Alaska case is an interesting one in that it is a small state in terms of population and GDP with relatively unique energy use and production attributes. Our other regions, by aggregating more states, tend to average out so that there is less disparity. The Alaska results are illustrative of within region effects that we do not capture because of our aggregation.

Regional impacts under *PCDS* are less balanced initially (**Figure 12**). The standard deviation of welfare impacts under *PCDS* is slightly larger (0.11) than under the *TAAS* scenario (0.09). Recall that *PCDS* deliberately takes a per-capita approach premised on the view that regional disparities do not matter, while *TAAS* includes a number of provisions (such as LDC allocations) that are explicitly intended to address regional disparities. While the regional dispersion of welfare impacts is slightly larger under *PCDS*, one interesting result of this analysis is that the much simpler per-capita based approach is almost as effective in achieving a balanced regional outcome as the targeted allocation scheme. By 2050, the impacts under *PCDS* are quite similar to those under *TAAS*. Differential regional impacts due to differences in allowance allocation schemes dissipate over time. Section 7.1 provides a discussion of this effect.

Impacts under *TAAS_DR* are very similar to those under *TAAS* and are not reported here. Figure 15 also shows that the relative impacts across regions are

The B.E. Journal of Economic Analysis & Policy, Vol. 10 [2010], Iss. 2 (Symposium), Art. 1

fairly stable over the policy period under the PCDS allocation. South Central, North Central and Northeast states bear a larger impact of the policy though the maximum difference across the period is less than two percentage points.[26]

We do not show here the _TAX scenarios because the results are broadly similar to the scenarios where a fraction of the allowance value is withheld to satisfy revenue neutrality. The main differences are that the overall welfare costs are larger for the U.S. as whole and thus regional losses tend to be somewhat larger. In terms of distribution, the _TAX cases tend to favor lower income regions (South and middle of the country) at the expense of higher income regions (mainly the east and west coasts) because higher income regions pay more taxes.

Summing up the regional results, all allocation scenarios lead to modest differential impacts across most regions. The *TAAS* and *TAAS_DR* proposals show greater gains to several regions in the initial years of the policy and higher costs to other regions than do the *PCDS* scenarios. One of the political economy realities of climate change is that the East and West Coast regions have pushed harder for climate legislation, while the middle of the country and much of south has resisted such legislation. With high energy intensity in these regions and the significant presence of fossil industry, one might expect greater economic impacts of GHG mitigation legislation in these regions. The Cantwell-Collins bill has not been subject to as much debate and negotiation as the other two bills, and has been able to retain a simple allocation formula. The much richer set of allocation mechanisms in Markey-Waxman and Kerry-Boxer are likely the result of negotiation among representatives of these regions. To the extent our analysis captures the regional distributional intent of these bills it suggests that the allocation formula are not completely effective in evening out regional effects. Some states like Texas and those in the South Central region that might have been expected to suffer higher costs have those costs blunted significantly and actually come out ahead in early years. Other regions such as the Mountain and North Central states remain the biggest losers in early years. Over time the allocation mechanisms evolve, and regional impacts are driven more directly by other factors.

---

[26] Welfare impacts for Alaska under the *PCDS* scenario are as follows: 2015: -0.60%; 2020: -1.10%; 2025: -2.31%; 2030: -3.25%; 2040: -4.61%; 2050: -5.95%. Note that under the *PCDS* allocation scheme Alaska receives less allowance revenue as compared to the *TAAS* case. This lowers savings and investment, and hence brings about even larger welfare losses in later periods of the policy as for the *TAAS* scenario.

7. FURTHER ANALYSIS OF THE BURDEN RESULTS

In a CGE model it is difficult to attribute differences in results by region and income class to specific causes because the possible sources of differences are many and they interact in complex ways. This section provides an analysis of the results to provide greater insight into why we see differences in effects.

7.1. The Importance of the Allowance Allocation Effect over Time

In order to isolate the impact of the allowance allocation on welfare, we run a scenario assuming the allowance value in a given period is not recycled while allowances in antecedent periods are allocated according to the scheme described in the TAAS scenario.

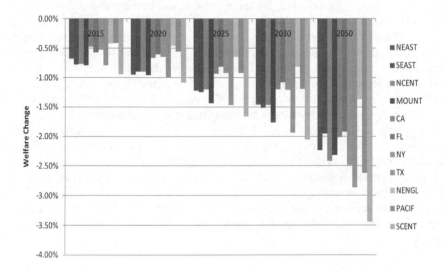

Figure 13. Regional Welfare Impacts without Allowance Allocation (Scenario *TAAS*).

Note that welfare costs will be higher in this case because the unrecycled revenue increases government expenditure which as described earlier does not, as modeled, enter household utility functions. The intent here is to use this exercise to isolate the effects of higher energy costs caused by pricing from those distributional impacts that result from the allowance allocation. **Figure 13** shows

The B.E. Journal of Economic Analysis & Policy, Vol. 10 [2010], Iss. 2 (Symposium), Art. 1

regional welfare impacts under this "no-recycling" case. As expected, welfare costs for each region and each period are higher as compared with the corresponding scenario that assumes revenue allocation (compare with Figure 11). For 2015, the distribution of regional costs is due to differences in regional abatement costs. In later years, the results are driven by abatement costs for that year, and the economic growth effects from previous years through the impact on Gross Regional Product and savings and investment. We see from Figure 13 that the pattern of regional welfare costs corresponds closely to differences in regional energy intensity (energy consumption per dollar of GDP). **Figure 14** shows an index of energy intensity by region over time (normalized to the current period U.S. value). The patterns of regional welfare impacts and relative energy intensities largely coincide, and are stable over time.

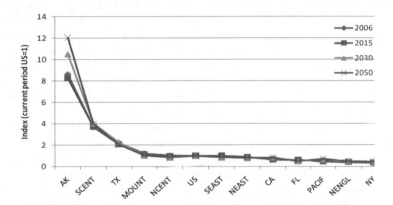

Figure 14. Regional Energy Intensity over Time (Scenario *TAAS*).

Comparing Figure 13 with Figure 11 now provides a way of disentangling the effect of the current period allowance allocation on welfare. The key result is that the allocation effect becomes less important over time, and that regional welfare impacts are eventually driven more by differences in the energy intensity. One reason for this result is that over time there is less allowance value to be distributed relative to the rising CO_2 price as the carbon policy becomes tighter. The number of allowances decreases over time and, in addition to that, the erosion of the tax base is steadily increasing which means that more of the allowance value has to be retained to maintain revenue-neutrality. This effect explains why initially in periods 2015-2025 the allocation of allowances has a strong effect on regional welfare impacts of the policy. As noted, regional effects of *TAAS* bear little relationship to factors like energy intensity and energy production that should factor into the cost of the policy. Some of the regions that display relative high energy intensity are actually overcompensated in 2015 and 2020 (viz. the South Central and North Central region, and Texas). The results suggest that any implemented allocation scheme will prove to be less effective over time in muting the regional variation in welfare impacts.

7.2. Sources vs. Uses Side Impacts of Carbon Pricing

A well-established observation is that carbon pricing incorporates a regressive element because lower income households spend a higher proportion of their income on energy. Most estimates of the distributional impact of carbon and energy pricing focus on this "cost-push analysis" element of carbon pricing by using an Input-Output framework to trace price increases through a make-and-use matrix to evaluate the policy cost on different households based on expenditure shares (e.g., Dinan and Rogers (2002), Parry (2004), Burtraw *et al.* (2009) and Hassett *et al.* (2009)). Such an approach neglects behavioral responses to relative price changes and does not take into account sources side effects. Rausch *et al.* (2009) found that even in a static model the sources side effects were important in determining the distributional effects of carbon pricing. Here we repeat their counterfactual analysis in our recursive dynamic simulation.

Figure 15 provides welfare impacts across income groups for three scenarios designed to disentangle the contribution of sources and uses side effects on welfare *across* the income distribution. The logic of our counterfactual analysis is as follows. If households in different income groups are characterized by identical income shares i.e., have equal ratios of capital, labor, and transfer income, then a change in relative factor prices affects all households equally. This counterfactual analysis isolates the distributional impacts of the uses of income effects of a policy. If households are assumed to have identical expenditure shares for all goods and services, a change in relative product prices produces an equal impact on consumers in different income classes. In that case, we isolate the distributional impacts by effects on sources of income of a policy. Any differential burden impacts of a policy across households from the counterfactual case that eliminates differences among households in how they spend their income are then determined by sources of income effects. Results that eliminate differences in income sources, allows us to focus on how uses side factors shape the relative burden of carbon pricing.

The two counterfactual cases do not eliminate these drivers of incidence but by eliminating household heterogeneity they suppress *differential* impacts across the income distribution. Harberger (1962) uses a similar analysis to identify the incidence of a corporate income tax. Note that as we measure the *real* burden, i.e., the change in equivalent variation, our incidence calculation is independent from the choice of numéraire.

Panel *a* shows results for 2015, panel *b* for 2030 and panel *c* for 2050. In each panel results for three cases are shown. The line labeled "carbon pricing burden" shows the welfare effect that combines income and expenditure heterogeneity.

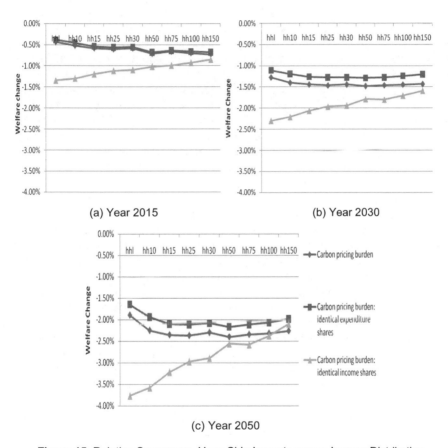

(a) Year 2015   (b) Year 2030

(c) Year 2050

Figure 15. Relative Sources vs. Uses Side Impacts across Income Distribution.

This is the welfare effect, without any recycling, given observed income sources and expenditures shares as they vary among households. The line labeled "identical income shares" eliminates heterogeneity of income sources to isolate the uses side effect of the policy. The line labeled "identical expenditure shares"

Rausch et al.: Implications of Alternative U.S. GHG Control Measures

eliminates expenditure heterogeneity to isolate the sources side effect. A downward slope indicates a progressive result and an upward slope a regressive result.

To eliminate the muddying effect of allowance allocation we assume that the carbon revenue is not recycled to households.[27] Non-recycled revenue increases government spending on goods and services which, by assumption, is not utility enhancing. As a result, the costs to households are much larger because the allowance revenue is not available to them. Still, however, we see the striking result that carbon pricing is modestly progressive initially and, for income groups above the two lowest becomes essentially neutral by 2030. For the counterfactual analysis we hold real government transfers to households constant at the no-policy level.

The uses side impacts are sharply regressive in all years in accord with previous analyses that focus on expenditure side burdens only. Sources side impacts, on the other hand, are modestly progressive in 2015 and essentially proportional in the other years. In all years, combined effects in the line "carbon pricing burden" track closely the line "identical expenditure shares". This suggests that relative welfare impacts across the income distribution are largely driven by sources side effects.

Table 5 reports sources of income by income class for the base year, and helps to explain why sources side effects are modestly progressive especially at low income levels. The relative income burden of carbon pricing depends on the change in relative factors prices and on differences in the ratio for the sources of income for households. We find that the capital rental rate increases over time relative to the price for labor. As the capital-labor ratio slightly increases in income, just looking at the relative income burden from changes in capital and labor income would imply that the sources side is slightly regressive. This finding is in line with Fullerton and Heutel (2010) who find that the capital and labor income for the lowest income households falls proportionally more than average. What makes the source-side incidence modestly progressive to proportional is the fact that low income households derive a large fraction of income from transfers relative to high income households, and we hold transfers constant relative to the no policy baseline. Transfer income thus insulates households from changes in capital and labor income. This effect is strongest for the two lowest income households where transfers account for about 80 and 60 percent of income as shown in Table 5.[28]

---

[27] We also looked at a scenario in which we assume that additional government revenue is spent according to private sector consumption. We find that this has second-order effects only.

[28] The sensitivity of distributional impacts of policies to the treatment of government transfers has been found in other work. Browning and Johnson (1979), for example, find that holding transfers fixed in real terms sharply increases the progressivity of the U.S. tax system.

The B.E. Journal of Economic Analysis & Policy, Vol. 10 [2010], Iss. 2 (Symposium), Art. 1

Table 5. Source of Income by Annual Income Class in USREP Model.

	Fraction of Income from Labor	Fraction of Income from Capital	Fraction of Income from Transfers	K/L ratio	Transfer / (Cap.+Labor) ratio
hhl	12.8%	6.5%	80.8%	0.5	4.2
hh10	28.6%	9.8%	61.6%	0.3	1.6
hh15	43.0%	18.2%	38.8%	0.4	0.6
hh25	48.3%	22.3%	29.5%	0.5	0.4
hh30	55.3%	24.7%	20.0%	0.4	0.3
hh50	60.4%	35.4%	4.2%	0.6	0.0
hh75	62.0%	37.5%	0.5%	0.6	0.0
hh100	59.4%	42.3%	-1.7%	0.7	0.0
hh150	57.6%	45.7%	-3.3%	0.8	0.0

Note: Based on IMPLAN data (Minnesota IMPLAN Group, 2008). Household transfers include social security, state welfare payments, unemployment compensation, veterans' benefits, food stamps, supplemental security income, direct relief, earned income credit. Note that transfers are net of household transfer payments to the rest-of-world (including cash transfers as well as goods to the rest-of-world).

Figure 15 also suggests that especially in a dynamic setting, the sources side effect is more important in determining the welfare impact than is the uses side effect for a *given* income class. The intuition for this result seems fairly obvious — over time the impacts of an ongoing mitigation policy cumulate through effects on overall economic growth and are reflected in general wage rates and capital returns. The annual abatement costs become an ever smaller share of the economic burden of the policy, and so are less important in determining the overall impacts. Furthermore, because the fraction of income derived from transfers increases over time, we find that the progressivity of the sources-side effect also slightly increases for the five lowest income groups.

Overall, this analysis demonstrates that it can be misleading to base the distributional analysis on uses side factors only. The virtue of our general equilibrium framework is the ability to capture effects both from the uses of income and through the sources of income.

8. SUMMARY

There has been much attention on the overall cost and efficiency of current legislative proposals for addressing climate change in the U.S. In this paper we focus on the distributional effects of the policies taking account of both the higher energy costs that carbon pricing implies and the distribution of allowance value described in the bills. Secondarily we are also interested in any efficiency effects of the allowance allocation approaches in the different bills. To focus on the effect of allowance allocation, we used approximations of the allowance allocation features of current proposals, but represented here as a comparable, comprehensive cap on all emissions in the U.S. with the same level of external credits allowed across all allocation scenarios. We, therefore, did not represent other features of the bills many of which may have strong efficiency and distributional consequences. While we try to adhere to the text of the various pieces of legislation as closely as possible when allocating allowance value, we note that we had to rely on our own interpretation of legislative intent in places where allocation mechanisms were not completely defined in the bills. While the scenarios are motivated by the various proposed pieces of legislation, none of the scenarios should be interpreted as an analysis of the complete legislation.

Focusing on efficiency first, we find that retaining more of the revenue to offset the deficit impacts of the legislation, as does the Kerry Boxer bill, improves the efficiency of mitigation policy because labor and capital taxes need to be raised less to maintain revenue neutrality. Economic efficiency is improved if all deficit impacts are offset with revenue retained from the allowance auction. The trade-off is that it would leave less revenue to affect desired distributional outcomes.

We also find that the scenarios designed to approximate the Cantwell-Collins allocation proposal to be less costly than those we used to approximate the other bills. We trace this result to the fact that the Cantwell-Collins allocation proposal distributes less of the allowance value to poor households. In the other allocation schemes, more money for poorer households produces a greater income effect on energy demand, and as a result abatement is more costly. Poorer households also save less, and so more allowance value going to poor households leads to less savings and investment. Economists have widely acknowledged that there is an equity-efficiency tradeoff between schemes with lump-sum distribution and those that would cut labor and capital taxes, reducing the distortions they create. Here we find a more subtle equity-efficiency tradeoff, where even under lump sum distribution of revenue there is an efficiency gain to distributing value to wealthier households because less is spent on energy and more of the allowance value ends up as savings and investment.

Our analysis of distribution by income class and region show that the Waxman-Markey and Kerry-Boxer (or Kerry-Lieberman) allocation schemes address the distributional impacts of the policy by redistributing more of the allowance value to poorer households and to central and southern regions of the U.S. in the early years of the policy, shifting allowance value away from wealthier households and the coasts. In fact the bills redistribute to such a degree that they tend to result in net economic benefits for the poorest households and for some

The B.E. Journal of Economic Analysis & Policy, Vol. 10 [2010], Iss. 2 (Symposium), Art. 1

regions of the country such as the South Central states, Texas, and Florida that would generally be expected to bear the highest costs. The very simple per capita allocation scheme of Cantwell-Collins tends to be more distributionally neutral by income class but produces slightly less balanced outcome by region. Over time the distribution schemes matter less. In part this is because over time all these bills convert to a consumer rebate and so are more like the Cantwell-Collins allocation approach. However, over time more of the annual cost of the policy is the result of economic growth effects—reductions in past Gross Regional Product, savings, and investment. The annual abatement costs become a smaller share of the total costs, and the available revenue to alter distributional effects shrinks relative to this increasing cost.[29]

An important finding of this paper is that sources side effects of carbon mitigation proposals dominate the uses side effect in terms of determining distribution outcomes. In the near term, the distributional consequences of the carbon pricing can be significantly affected by the distribution of allowance value. Over the longer term, however, the overall growth effects are more important determinants of distribution and the revenue available from the allowance auction may not be sufficient to have much effect in changing distributional outcomes. This point is reinforced by the finding that carbon pricing by itself, i.e., when carbon revenues are not recycled back to households, is neutral to modestly progressive. This follows from the dominance of sources over uses side impacts of the policy and stands in sharp contrast to previous work that has focused only on the uses side. We find sources side effects to be modestly progressive to proportional because low income households derive a relatively large fraction of their income from transfers which insulates them from changes in capital and labor income.

We emphasize that our scenarios focused solely on the distributional implications due to carbon pricing and the allocation of allowance revenue, and that we did not attempt to model each bill in its entirety. More precise representation of the many programs described in these bills could give different outcomes and there is inevitable uncertainty in economic forecasts of this type. We also must admit significant limitations in our ability to forecast relative effects on regions over the longer term. Climate policy will dramatically change energy technologies, and regions that aggressively develop these industries and attract investment could fare better even if they currently are heavily fossil energy dependent. However, such regions must overcome the initially higher costs of their fossil energy dependence.

---

[29] As noted above, the share of allowances that must be held back for revenue neutrality in the out years falls if government spending as a share of GDP is held fixed. A priori it is not obvious which assumption on government spending is more realistic.

9. REFERENCES

Babiker, M., G. Metcalf and J. Reilly, 2003: Tax Distortions and Global Climate Policy. *Journal of Environmental Economics and Management*, *46*, pp. 269-87.

Babiker, M., A. Gurgel, S. Paltsev and J. Reilly, 2008: A Forward Looking Version of the MIT Emissions Prediction and Policy Analysis (EPPA) Model. MIT Joint Program on the Science and Policy of Global Change, *Report 161*, Cambridge, MA, available at: http://globalchange.mit.edu/files/document/MITJPSPGC_Rpt161.pdf

Ballard, C., 2000: How many hours are in a simulated day? The effect of time endowment on the results of tax-policy simulation models. Working Paper, Michigan State University.

Bento, A., L. Goulder, M. Jacobsen and R. von Haefen, 2009: Distributional and Efficiency Impacts of Increased Us Gasoline Taxes. *American Economic Review*, *99*(3), pp. 667-99.

Bovenberg, A. and L. Goulder, 2001: Neutralizing the Adverse Industry Impacts of CO_2 Abatement Policies: What Does It Cost? C. Carraro and G. E. Metcalf, *Distributional and Behavioral Effects of Environmental Policy*. Chicago: University of Chicago Press, pp. 45-85.

Bovenberg, A., L. Goulder and D. Gurney, 2005: Efficiency Costs of Meeting Industry-Distributional Constraints under Environmental Permits and Taxes. *RAND Journal of Economics*, *36*(4), pp. 951-71.

Browning, Edgar K. and William R. Johnson, 1979. *The Distribution of the Tax Burden*, Washington, DC: American Enterprise Institute.

Bull, N., K. Hassett and G. Metcalf, 1994: Who Pays Broad-Based Energy Taxes? Computing Lifetime and Regional Incidence. *Energy Journal*, *15*(3), pp. 145-64.

Burtraw, D., R. Sweeney and M. Walls, 2009: The Incidence of U.S. Climate Policy: Alternative Uses of Revenue from a Cap and Trade Auction. Washington, DC: Resources for the Future.

Burtraw, D., M. Walls and J. Blonz, 2009: Distributional Impacts of Carbon Pricing Policies in the Electricity Sector. Washington, DC: Resources For the Future.

Congressional Budget Office, 2009: The Role of the 25 Percent Revenue Offset in Estimating the Budgetary Effects of Legislation. Washington, DC: Congressional Budget Office.

Davies, J., F. St Hilaire and J. Whalley, 1984: Some Calculations of Lifetime Tax Incidence. *American Economic Review*, *74*(4), pp. 633-49.

The B.E. Journal of Economic Analysis & Policy, Vol. 10 [2010], Iss. 2 (Symposium), Art. 1

Department of Energy, 2009: U.S. Crude Oil, Natural Gas, and Natural Gas Liquids Reserves, 1977 through 2007. Annual Reports, DOE/EIA-0216.

Dinan, T. and D. Rogers, 2002: Distributional Effects of Carbon Allowance Trading: How Government Decisions Determine Winners and Losers. *National Tax Journal*, *55*(2), pp. 199-221.

Dyni, J., 2006: Geology and Resources of Some World Oil-Shale Deposits. USGS Scientific Investigations Report 2005-5294, p 42.

Energy Information Administration, 2009: Annual Energy Outlook 2009. Washington, D.C., U.S. Energy Information Administration DOE/EIA-0383.

Friedman, M., 1957: *A Theory of the Consumption Function*. Princeton, NJ: Princeton University Press.

Fullerton, D.and G. Heutel, 2007: The General Equilibrium Incidence of Environmental Taxes. *Journal of Public Economics*, *91*(3-4), pp. 571-91.

Fullerton, D.and G. Heutel, 2010: Analytical General Equilibrium Effects of Energy Policy on Output and Factor Prices. NBER Working Paper 15788, Cambridge, MA.

Fullerton, D. and D. Rogers, 1993: *Who Bears the Lifetime Tax Burden?* Washington, D.C., Brookings Institution.

Grainger, C. and C. Kolstad, 2009: Who Pays a Price on Carbon? Cambridge, MA: National Bureau of Economic Research Working Paper No. 15239.

Hassett, K., A. Mathur and G. Metcalf, 2009: The Incidence of a U.S. Carbon Tax: A Lifetime and Regional Analysis. *The Energy Journal*, *30*(2), pp. 157-79.

Harberger, A., 1962: The Incidence of the Corporation Income Tax. *Journal of Political Economy 70*, pp. 215-40.

Hyman, R., J. Reilly, M. Babiker, A. De Masin and H. Jacoby, 2003: Modeling Non-CO_2 Greenhouse Gas Abatement. *Environmental Modeling and Assessment*, *8*(3), pp. 175-86.

Lyon, A. and R. Schwab, 1995: Consumption Taxes in a Life-Cycle Framework: Are Sin Taxes Regressive? *Review of Economics and Statistics*, *77*(3), pp. 389-406.

Metcalf, G., 1999: A Distributional Analysis of Green Tax Reforms. *National Tax Journal*, *52*(4), pp. 655-81.

Metcalf, G., 2007: A Proposal for a U.S. Carbon Tax Swap: An Equitable Tax Reform to Address Global Climate Change. Washington, DC: The Hamilton Project, Brookings Institution.

Metcalf, G., S. Paltsev, J. Reilly, H. Jacoby and J. Holak, 2008: Analysis of a Carbon Tax to Reduce U.S. Greenhouse Gas Emissions. Cambridge, MA: MIT Joint Program on the Science and Policy of Global Change Report No. 160.

Minnesota IMPLAN Group, 2008: State-Level U.S. Data for 2006 Stillwater, MN: Minnesota IMPLAN Group

Morris, J., 2009: Combining a Renewable Portfolio Standard with a Cap-and-Trade Policy: A General Equilibrium Analysis. MS in Technology and Public Policy, MIT, Cambridge, MA.

Musgrave, R.A., 1959: The Theory of Public Finance. McGraw-Hill, New York.

NREL, 2009: Eastern and Western Wind Datasets. Available at: http://nrel.gov/wind/integrationdatasets/

Oakridge National Laboratories, 2009: Estimated Annual Cumulative Biomass Resources Available by State and Price. Available at: http://bioenergy.oml.gov/main.aspx#Biomass%20Resources

Paltsev, S., J. Reilly, H. Jacoby, R. Eckaus, J. McFarland, M. Sarofim, M. Asadoorian and M. Babiker, 2005: The MIT Emissions Prediction and Policy Analysis (EPPA) Model: Version 4 MIT Joint Program on the Science and Policy of Global Change, *Report 125*, Cambridge, MA, available at: http://globalchange.mit.edu/files/doument/MITJPSPGC_Rpt125.pdf

Paltsev, S., J. Reilly, H. Jacoby and J. Morris, 2009: The Cost of Climate Policy in the United States MIT Joint Program on the Science and Policy of Global Change, *Report 173*, Cambridge, MA, available at: http://globalchange.mit.edu/files/document/MITJPSPGC_Rpt173.pdf

Parry, Ian W. H. 2004. Are Emissions Permits Regressive *Journal of Environmental Economics and Management* 47:364-387.

Pechman, J., 1985: *Who Paid the Taxes: 1966-85?* Washington D.C., Brookings.

Poterba, J., 1989: Lifetime Incidence and the Distributional Burden of Excise Taxes. *American Economic Review,* 79(2), pp. 325-30.

Poterba, J., 1991: Is the Gasoline Tax Regressive? *Tax Policy and the Economy,* 5, pp. 145-64.

Rausch, S., G. Metcalf, J. Reilly and S. Paltsev, 2011: Distributional Impacts of a U.S. Greenhouse Gas Policy: A General Equilibrium Analysis of Carbon Pricing. *U.S. Energy Tax Policy* ed. G. Metcalf, Cambridge University Press, MA.

Ross, M., 2008: Documentation of the Applied Dynamic Analysis of the Global Economy (ADAGE) Research Triangle Institute Working Paper 08-01.

Tuladhar, S., M. Yuan, P. Bernstein, W. Montgomery and A. Smith, 2009: A Top-Down Bottom-up Modeling Approach to Climate Change Policy Analysis *Energy Economics, in press.*

The B.E. Journal of Economic Analysis & Policy, Vol. 10 [2010], Iss. 2 (Symposium), Art. 1

U.S. Census Bureau, 2009a: American Community Survey 2006: Household Income in the Past 12 Months. Table B19001, accessed on 22 June 2009 via the American FactFinder website, http://factfinder.census.gov/

U.S. Census Bureau, 2009b: State Population Projections. Accessed on 11 January 2009, http://www.census.gov/population/www/projections/stproj.html

U.S. Environmental Protection Agency, 2009: Inventory of U.S. Greenhouse Gas Emissions and Sinks: 1990 – 2007. Washington, D.C., Environmental Protection Agency EPA 430 R-09-00-4.

U.S. Geological Survey, 2009: USGS USCOAL Coal Resources Database. Available at: http://energy.er.usgs.gov/coalres.htm and http://pubs.er.usgs.gov/djvu/B/bull_1412.djvu

[12]

The B.E. Journal of Economic Analysis & Policy

Symposium

Volume 10, Issue 2	2010	Article 2

DISTRIBUTIONAL ASPECTS OF ENERGY AND CLIMATE POLICY

Comment on 'Distributional Implications of Alternative U.S. Greenhouse Gas Control Measures'

Shanta Devarajan*

*The World Bank, sdevarajan@worldbank.org

Devarajan: Comment on Rausch et al.

This paper is a *tour de force*. It simulates the effects of three different proposals to curb greenhouse gas emissions using a multisectoral general-equilibrium model of the United States. To capture distributional effects, the model has nine household groups (stratified by income) and eleven regions. Finally, the model is solved recursively over time until 2050.

Now, many people can build and solve complex models. The beauty of the Rausch et al. paper is that they painstakingly interpret their results in terms of the underlying assumptions and data of their model. In fact, they undertake further simulations to confirm that their interpretations are correct. The net result is that we have learned not just how their model behaves, but also some economics, and—to an extent—something about the efficiency and equity implications (and hence political consequences) of the three proposals.

I have three sets of comments—on the modeling choices made, on the interpretation of the results, and on the uses to which we can put the paper's findings.

Modeling choices

The authors claim their modeling choices are driven by a desire to get a "realistic" assessment of the different cap-and-trade proposals. In 2010, with the U.S. economy reeling from a major recession, brought on by a financial crisis, and the unemployment rate at 10 percent, it is hard to claim that a full-employment model with no financial sector is "realistic." Even if one claims that climate change issues are long-term whereas the recession is short-term, much of the discussion around carbon taxes is around their effects on innovation and growth, none of which is captured (except as exogenous parameters) in this recursive, dynamic model. My point is not that the authors should have chosen a different model, but that they should not describe the goal of their exercise as "realism." Rather, what these general-equilibrium models provide is a logical consistency with which to ask some of the efficiency and distributional questions that are being asked here. But for logical consistency, sometimes one has to abstract from reality, and that is what is being done here.

At the same time, the authors refrain from taking logical consistency too far by, say, specifying a fully dynamic model (where producers and consumers make intertemporally optimal decisions) because they don't think economic agents have full information about their lifetime earnings. But they *are* assuming these same agents have full information about the entire vector of relative prices within the period. It seems rather arbitrary to distinguish between the two sets of information in deciding on the model. If the problem is that a fully dynamic model would mean that the model is too big to solve, the authors may wish to consider adding on a microsimulation model (Bourguignon and Spadao [2006])

The B.E. Journal of Economic Analysis & Policy, Vol. 10 [2010], Iss. 2 (Symposium), Art. 2

that would reduce the dimensionality on the household side without losing much information. Furthermore, as I will point out later, this particular modeling choice may limit the uses of the model's results.

One minor question about the data: What was the criterion for dividing households according to the income classes as defined in Table 1? They don't seem to correspond to deciles.

Results

The basic results of the model are that carbon pricing imposes a deadweight loss on the economy of the order of 1.5 – 2 percent of income; the welfare cost is higher the more of the revenue from allocating carbon permits is transferred to households rather than used to reduce the fiscal deficit (and hence reduce existing capital and labor taxes); the transfer scheme leads to a progressive distribution of income initially but, as the revenue from carbon permits declines over time, the distribution becomes approximately neutral. This is true both for the distribution across households and across regions. In fact, the distribution is roughly U-shaped, which is reminiscent of the result in Devarajan, Fullerton and Musgrave [1980] which, in turn, was due to the fact that extreme poor and rich households had comparable capital-labor ratios of their sources of income (in those days, the "poor" included pensioners).

Within these basic results, there are a number of interesting sub-results. For instance, the welfare cost rises with greater transfers to poor households because these households spend a larger share of income on energy, which is the good that is being "taxed", and so the size of the distortion rises. To take another example, over time, carbon prices affect household welfare more through the income effect than through the expenditure effect. The authors show this by running parallel simulations with each household having the same expenditure composition and another with the same sources of income. The underlying reason (not mentioned in the paper) is that the burden of a price on carbon shifts from the final consumer to factors of production over time, as consumers find alternative sources of energy. One other reason might be that, even though the poor spend a higher fraction of their income on energy directly, the non-poor consume more energy-intensive goods, as was the case in the Philippines (Devarajan and Hossein).

Uses of the results

Given that the paper examines actual House and Senate proposals for curbing GHGs, how can these results be used to influence policy? At one level, simply disseminating the results, and pointing out the tradeoff between efficiency and

Devarajan: Comment on Rausch et al.

equity could help policymakers with their decisions. At another level, these results could be used to anticipate the supporters and opponents of the legislation, because the paper shows very clearly which income groups and which regions will be adversely and favorably affected. But here is the problem with the modeling choice of the recursive, dynamic model. If, as the paper shows, the distributional results will be progressive in the near term but neutral in the long term, then it is not clear that the "winners" will necessarily embrace the legislation (because they know their gains will be short-lived) and likewise the losers may not oppose the legislation. But the model assumes the agents are not forward-looking, so we don't know how they would behave if they knew that the value of the transfers will diminish over time. It could be that they will not increase their energy consumption in the short-term, in which case the added efficiency cost may not be as high. It could also mean that the distribution of welfare may be more neutral in the short-run, as everybody anticipates that the transfers will decline, and adjusts their behavior accordingly. Ironically, to make use of one of the most interesting, central results of the model, we may need to change the model.

References

Bourguignon, François and Amedeo Spadaro, "Microsimulation as a Tool for Evaluating Redistribution Policies", *Journal of Economic Inequality*, 4(1), 77-106 (2006).

Devarajan, S., Don Fullerton and Richard Musgrave, "Estimating the Distribution of Tax Burdens: A Comparison of Different Approaches," *Journal of Public Economics,* 13(2), 155-82 (1980).

Devarajan, S. and Shaikh Hossein, "The Combined Incidence of Taxes and Public Expenditures in the Philippines, *World Development,* 26(6), (June 1998): 963-77

[13]

The B.E. Journal of Economic Analysis & Policy

Symposium

| *Volume* 10, *Issue* 2 | 2010 | *Article* 17 |

DISTRIBUTIONAL ASPECTS OF ENERGY AND CLIMATE POLICY

The Distributional Impact of Climate Policy

Dale W. Jorgenson*   Richard Goettle[†]   Mun S. Ho[‡]

Daniel T. Slesnick**   Peter J. Wilcoxen[††]

*Harvard University, djorgenson@harvard.edu

[†]Northeastern University, r.goettle@neu.edu

[‡]Resources for the Future, ho@rff.org

**University of Texas, slesnick@eco.utexas.edu

[††]Syracuse University, wilcoxen@maxwell.syr.edu

Recommended Citation

Dale W. Jorgenson, Richard Goettle, Mun S. Ho, Daniel T. Slesnick, and Peter J. Wilcoxen (2010) "The Distributional Impact of Climate Policy," *The B.E. Journal of Economic Analysis & Policy*: Vol. 10: Iss. 2 (Symposium), Article 17.

Available at: http://www.bepress.com/bejeap/vol10/iss2/art17

The Distributional Impact of Climate Policy*

Dale W. Jorgenson, Richard Goettle, Mun S. Ho, Daniel T. Slesnick, and Peter J. Wilcoxen

Abstract

The purpose of this paper is to present a new methodology for evaluating the distributional impacts of climate policy. This methodology builds directly on the framework introduced by Jorgenson, Slesnick, and Wilcoxen (1992), but generalizes it by including leisure time, as well as goods and services, in the measure of household welfare. We provide detailed results for 244 different types of households distinguished by demographic characteristics. In addition, we evaluate the overall impact of a cap-and-trade system, as represented in Energy Modeling Forum 22. While there is a wide range of outcomes for different demographic groups, the impact on economic welfare is regressive and generally negative but relatively small.

KEYWORDS: distribution, climate policy, leisure, goods, demographic groups

---

*We are grateful to Jon Samuels, Johns Hopkins University, for help with national accounting data and Hui Jin, International Monetary Fund, for estimation of the production model. We are also grateful for valuable comments from participants in Energy Modeling Forum 22, Climate Change Scenarios, the 2009 AERE Workshop on Energy and the Environment, and the 2009 Woods Hole meeting of the Governing Board of the National Research Council. Financial support for this research was provided by the Climate Economics Division of the Environmental Protection Agency. The usual disclaimer applies.

1. Introduction

In this paper we present a new approach to the evaluation of economic policies based on equivalent variations in full wealth. Full wealth is the present value of full consumption over all time periods. Full consumption includes the value of leisure, as well as goods and services, at a given point of time.[1] Our approach builds on the framework introduced by Jorgenson, Slesnick, and Wilcoxen (1992) for evaluating the introduction of a carbon tax in the U.S. economy. Their framework employs the wealth associated with the alternative policies to provide a money metric for the change in household welfare. Our principal innovation is to include leisure time, as well as goods and services, in the measure of household welfare.

Our starting point for policy evaluation is a reference policy or *base case* projection of the U.S. economy. This projection is associated with no change in policy. We then consider an *alternative case* projection with a change in policy. Each case is represented by an intertemporal general equilibrium. The equivalent variation for a given household is the difference between the full wealth required for the utility level associated with the alternative case and that required for the base case. Both are evaluated at the prices of the base case. As an illustration, we evaluate alternative policies for imposing a cap-and-trade system to control emissions of greenhouse gases in the United States.

We have employed the Intertemporal General Equilibrium Model (IGEM) used by the Environmental Protection Agency (EPA).[2] In this paper we introduce an updated and revised version of IGEM that gives similar results to the EPA version. We describe the new version of IGEM in Section 2, focusing on the household model developed by Jorgenson and Slesnick (2008).[3] This model includes present and future full consumption, as well as consumption of goods and services and leisure time. The model is implemented econometrically, using data from the Consumer Expenditure Survey (CEX), conducted by the Bureau of Labor Statistics (BLS).[4] Jorgenson and Slesnick (2008) have employed expenditure data for 150,000 households and have augmented these data with estimates of leisure time for each household, also based on the CEX.

One recent example of a carbon permit system is the Waxman-Markey (WM) bill that passed the House in 2009 but failed in the Senate. We evaluated a generic cap-and-trade proposal that is broadly similar to that bill, but we do not

---

[1] The term "full consumption" is employed by Becker (1965).
[2] See http://www.epa.gov/climatechange/economics/modeling.html#intertemporal and http://www.igem.insightworks.com/
[3] More details on modeling of consumer behavior are given by Jorgenson (1997a).
[4] See http://www.bls.gov/cex/ The CEX is the standard source of household data on expenditures for goods and services in the United States.

try to capture all the details (which are now irrelevant in any case). For example, the WM bill covered about 85% or carbon emissions in the U.S., whereas we apply the carbon price to all carbon emissions. Also, the WM bill employed a complicated scheme for the allocation of permit value, whereas we just assume that all permits are sold by government, with the revenue returned in a lump-sum fashion to all households in proportion to their full wealth. This choice neutralizes any distributional impacts of the use of permit revenue, in order to focus on the distributional aspects of the impact of cap-and-trade on commodity prices and on factor prices.[5]

In Section 3 we summarize the approach to policy evaluation originated by Jorgenson, Slesnick, and Wilcoxen (1992). We generalize this approach to treat leisure, as well as goods and services, as components of full consumption.[6] A key idea is to consider each of 244 different types of households, distinguished by demographic characteristics, as the progenitor of a *dynasty* with an infinite lifetime. Each household faces a given level of full wealth and a time path of prices for leisure, goods, and services. The household chooses a time path for full consumption that generates the maximum level of utility. We calculate the equivalent variation in full wealth as a money metric of the change in utility associated with a change in policy.

In Section 4 we present a detailed evaluation of alternative cap-and-trade policies.[7] We consider the impacts of alternative limits on emissions on the future time path of the U.S. economy, beginning with effects on the gross domestic product (GDP) and aggregates such as consumption, investment, government expenditures, imports, and exports. We then consider the effects on different industries, focusing on the five industries that make up the energy sector. Finally, we consider the welfare impacts of the legislation on individual households. We present effects on households with different demographic characteristics, including family size, region of residence, and the race and the gender of household head, as well as different levels of resources.

The policies raise the price of goods relative to leisure, causing households to substitute toward leisure and reduce labor supply. Output falls and lower investment cumulates into larger reductions in future GDP. The coal industry is

---

[5] Others have investigated distributional effects of alternative uses of permit revenue. If permit revenues used to make the same lump sum payment to every individual irrespective of income or wealth, for example, Burtraw *et al* (2009) find that the overall impact of cap-and-trade can be progressive. For another example, some or all permits might be "grandfathered" or handed out to firms in proportion to some past year's emissions. Since stockholders who benefit are in relatively high income brackets, Parry (2004) finds that allocation to be regressive.

[6] More details on measuring economic welfare are provided by Jorgenson (1997b).

[7] Recent evaluations of policies to reduce emissions of greenhouse gases in the United States are presented by Burtraw, Sweeney, and Walls (2009), Greenstein, Parrott, and Sherman (2008), Hassett, Mathur, and Metcalf (2009), Metcalf *et al* (2008), and Shammin and Bullard (2009).

heavily impacted by imposing limits on greenhouse gas emissions. The price of coal rises substantially and coal output drops precipitously. The impacts on other energy-producing sectors are lower in magnitude, but still very substantial The rise in leisure partially offsets the reduction in goods consumption and the welfare impact is modest. Section 5 concludes the paper.

2. Intertemporal General Equilibrium Model

The Intertemporal General Equilibrium Model (IGEM) of the U.S. economy was introduced by Jorgenson and Wilcoxen (1990). Applications of the original version of the model to the analysis of energy, environmental, tax, and trade policies are presented in detail in Jorgenson (1998). The version of the model we have employed is similar to that developed for EPA and documented on the EPA Climate Economics website. The model consists of four sub-models for the household, production, government, and rest of the world sectors of the U.S. economy. We begin with a description of the household sub-model constructed by Jorgenson and Slesnick (2008).

Household sector

The Jorgenson-Slesnick model has three stages. In the first stage full wealth is allocated to full consumption at different points of time. In the second stage full consumption is allocated to leisure and three commodity groups – nondurables, capital services, and services. In the third stage the three commodity groups are allocated among 35 individual commodities, including the five types of energy. To capture differences in preferences between households we distinguish among demographic groups. The groups are cross-classified by size, region and characteristics of the head of household as shown in Table 1.

Table 1. Demographic groups identified in household model

Number of children	0,1,2,3 or more
Number of adults	1,2,3 or more
Region	Northeast, Midwest, South, West
Location	Urban, Rural
Gender of head	Male, Female
Race of head	White, nonwhite

Table 1 gives a total of 384 household types. However, the number of types with a positive number of households is only 244. In the first stage, households are assumed to maximize an additively separable intertemporal utility

The B.E. Journal of Economic Analysis & Policy, Vol. 10 [2010], Iss. 2 (Symposium), Art. 17

function subject to a full wealth budget constraint. A full description of this utility function is presented in Section 3 where we discuss the method used to compute the lifetime welfare effects of climate policy. In the second stage of the model household k maximizes a utility function defined on three commodity groups – nondurables, capital services, services – and leisure, $U(C_{ND,k}, C_{K,k}, C_{SV,k}, C_{Rk}; A_k)$, where C_{Rk} is leisure and A_k denotes the demographic characteristics of household k. The composition of the commodity groups is given below.

To characterize substitutability among leisure and the three commodity groups, we represent the maximum level of utility as a function of prices and full consumption by means of the translog indirect utility function $V(p_k, m_k; A_k)$, where:

$$(1) \qquad -\ln V_k = \alpha_0 + \alpha^H \ln \frac{p_k}{m_k} + \tfrac{1}{2} \ln \frac{p_k}{m_k}{}' B^H \ln \frac{p_k}{m_k} + \ln \frac{p_k}{m_k}{}' B_A A_k$$

where $p_k = (P_{ND}^C, P_K^C, P_{SV}^C, P_R^C)'$ is a vector of prices faced by household k, α^H is a vector of parameters, and B^H and B_A are matrices of parameters that describe price, total expenditure, and demographic effects and A_k is a vector of variables that describe the demographic characteristics of household k. The value of full expenditure is:

$$(2) \qquad m_k = P_{ND}^C C_{NDk} + P_K^C C_{Kk} + P_{SV}^C C_{SVk} + P_R^C C_{Rk}.$$

An hour of leisure has a different opportunity cost for each member of each household. We assume that the quantity of leisure R_k^m for person m is non-work hours multiplied by labor quality given by the after-tax wage rate, relative to the base wage $q_k^m = p_R^m / p_R^0$. We assume a time endowment of $\bar{H} = 14$ hours a day for each adult. The annual quantity of effective leisure for an individual is the time endowment less hours worked LS weighted by labor quality, $R_k^m = q_k^m (\bar{H}_k^m - LS_k^m)$.

The quantity of leisure for household k is the sum over all adult members,

$$C_{Rk} = \sum_m R_k^m,$$

Jorgenson et al.: The Distributional Impact of Climate Policy

and the value is the nominal wage multiplied by leisure hours:

$$(3) \quad P_R^C C_{Rk} = p_R^0 \sum_m R_k^m = \sum_m p_R^m (\bar{H}_k^m - LS_k^m).$$

The demand functions for commodities and leisure are derived from the indirect utility function (1) by applying Roy's Identity:

$$(4) \quad \mathbf{w}_k = \frac{1}{D(p_k)} (\alpha^H + B^H \ln p_k - \iota' B^H \ln m_k + B_A A_k),$$

where $D(p_k) = -1 + \iota' B^H \ln p_k$, $\mathbf{w}_k$ is the vector of shares of full consumption, and ι is a vector of ones.

We require that the indirect utility function obeys the restrictions implied the theory of individual consumer behavior and the requirements for exact aggregation discussed below:

$$(5) \quad B^H = B^{H'}; \quad \iota' B^H \iota = 0, \quad \iota' B_A = 0, \quad \iota' \alpha^H = -1,$$

where B^H are the share elasticities, $\iota' B^H$ represents the full expenditure effect, and the kth column of B_A determines how the demands of demographic group k differs from the base group. These restrictions and their incorporation into the estimation process are described in detail by Jorgenson and Slesnick (2008).

Let n_k be the number of households of type k. Then the vector of aggregate demand shares for the U.S. economy is obtained by aggregating over all types of households:

$$(6) \quad w = \frac{\sum_k n_k m_k \mathbf{w}_k}{\sum_k n_k m_k}$$

$$= \frac{1}{D(p)} \left[\alpha^H + B^H \ln p - \iota B^H \xi^d + B_A \xi^L \right]$$

where M is the national level of full expenditures and the distribution terms are:

$$(7) \quad \xi^d = \sum_k n_k m_k \ln m_k / M; \quad M = \sum_k n_k m_k$$

$$\xi^L = \sum_k n_k m_k A_k / M$$

By constructing a model of aggregate consumer demand through exact aggregation over individual demands, we are able to incorporate the restrictions

The B.E. Journal of Economic Analysis & Policy, Vol. 10 [2010], Iss. 2 (Symposium), Art. 17

implied by the theory of individual consumer behavior. In addition, we include demographic information through the distribution terms in (7). For the sample period we observe the values of these terms. For the period beyond the sample we extrapolate them, using projections of the U.S. population by sex and race. That is, we project the number of households of type k, n_{kt}, by linking the age and race of the head of household to the projected population.

The estimated price and income elasticities are reported in Table 2; a full set of results is given by Jorgenson and Slesnick (2008). The elasticities are calculated for the reference household type – two adults, two children, Northeast, urban, male head, white. They are computed at $100,000 of full consumption in 1989. The compensated own-price elasticities are negative for all goods and services, as well as for leisure. Capital services are price elastic, while nondurables, consumer services, and leisure are price inelastic. The uncompensated wage elasticity of household labor supply is negative but close to zero, a common finding in modeling labor supply, while the compensated wage elasticity is 0.7. The full consumption elasticity for leisure is greater than one, so that leisure is classified as a luxury. Nondurables and capital services are necessities with full consumption elasticities less than one, while services are a luxury.

Table 2. Price and Income Elasticities of Demand

	Uncompensated Price Elasticity	Compensated Price Elasticity	Expenditure Elasticity
Nondurables	-0.727	-0.651	0.673
Capital Services	-1.192	-1.084	0.902
Consumer Services	-0.561	-0.490	1.067
Leisure	0.014	-0.305	1.063
Labor Supply	-0.032	0.713	-2.486

Table 3 gives the fitted shares of the four commodity groups at different levels of full consumption for the reference household. The share allocated to nondurables falls rapidly as expenditures rise while the share allocated to services rises a little. Leisure value is hours multiplied by wage rates; the share rises substantially with rising wage rates of the higher income households.

In the third stage of the household model quantities of nondurables, capital services, and other services (N^{ND}, N^K and N^{CS}) are allocated among 35 individual commodities and capital services. We do not employ demographic information for this part of the model and use a nested tier structure of homothetic indirect utility functions. This structure is given in Table 4. For example in the energy node, total energy consumption is allocated among gasoline, fuel oil and coal, electricity, and gas. The model tracks changes in the composition of consumption due to non-price effects using latent variables in the same way as in the production model (9) given below.

Table 3. Full expenditures and household budget shares

Full	Nondurables	Capital	Services	Leisure
Expenditures ($)				
7500	0.208	0.151	0.055	0.586
25000	0.164	0.137	0.06	0.626
75000	0.123	0.124	0.065	0.693
150000	0.098	0.116	0.068	0.713
275000	0.075	0.108	0.071	0.718
350000	0.066	0.106	0.072	0.716

The share demand functions at each node m of the tier structure are:

$$(8) \qquad SN_t^m = \alpha^{Hm} + B^{Hm} \ln PN_t^{Hm} + f_t^{Hm} \quad ,$$

where PN^{Hm} is the vector of prices at node m and f^{Hm} is the vector of latent variables. When $B^{Hm} = 0$ the demand function reduces to linear logarithmic form.

A full set of estimates of unknown parameters of the household model for all 16 nodes is given by Goettle, *et al.* (2009). Most of the estimated share elasticities (β_{ii}^{Hm}) are between -0.1 and 0.1. About half are negative, that is, the price elasticity is greater than one. The latent variables f_t^{Hm} representing changes in preferences have noticeable trends in the sample period. For example, the term for electricity rises between the late 1960s and 1990 but since has flattened.

Table 4. Tier structure of the household model, 2005 (bil $)

		Energy 503	gasoline & oil	284		
			Fuel-coal	21	Coal	0.3
					fuel-oil	21
			electricity	133		
			gas	65		
	Nondurables 2715	Food 1270	food	720		
			meals	449		
			meals-emp	12		
			tobacco	88		
		Cons. Goods 942	Clothing-shoe	342	shoes	55
					clothing	287
			Hhld articles	181	toilet art.; cleaning	138
					furnishings	43
			drugs	265		
			Misc goods	154	toys	66
					stationcry	20
					imports	7
					reading materials	61
Full consumption 23423	Capital svc 1972	Housing 536	rental housing	334		
			owner maintenace	202		
	Cons. svc 4303	HH operation 281	water	64		
			communications	133		
			domestic service	20		
			other household	64		
		Transportation 324	own transportation	263		
			transportation svc	62		
		Medical 1491	medical services	1350		
			health insurance	141		
		Misc svcs 1670	personal svcs	116		
			Business Svcs	646	financial svcs	499
					other bus. svcs	147
			Recreation	458	recreation	358
					foreign travel	100
			educ & welfare	451		
	Leisure 14432					

Jorgenson et al.: The Distributional Impact of Climate Policy

Table 5. Industry output, energy use in 2005 and historical growth

		Output (bil $)	Energy share (incl. feedstocks) (% output)	Output growth 1960-05 (% p.a.)	TFP growth 1960-05 (% p.a.)
1	Agriculture	424	4.4	2.00	1.40
2	Metal Mining	25	9.8	0.67	-0.60
3	Coal Mining	26	12.5	2.21	1.17
4	Petroleum and Gas	260	7.6	0.40	-0.58
5	Nonmetallic Mining	24	12.3	1.56	0.27
6	Construction	1356	2.7	1.60	-0.61
7	Food Products	595	1.8	2.01	0.52
8	Tobacco Products	31	0.7	-0.83	-1.52
9	Textile Mill Products	60	3.2	1.17	1.56
10	Apparel and Textiles	36	1.4	-0.28	0.97
11	Lumber and Wood	130	2.9	2.03	0.15
12	Furniture and Fixtures	101	1.9	3.27	0.69
13	Paper Products	168	4.4	2.04	0.47
14	Printing and Publishing	230	1.1	1.83	-0.15
15	Chemical Products	521	4.9	2.81	0.55
16	Petroleum Refining	419	51.3	1.63	0.08
17	Rubber and Plastic	188	2.5	4.21	0.87
18	Leather Products	6	2.7	-2.36	0.33
19	Stone, Clay, and Glass	129	5.9	1.90	0.54
20	Primary Metals	251	5.1	0.84	0.32
21	Fabricated Metals	296	2.2	1.94	0.51
22	Industrial Machinery	424	1.3	5.92	2.65
23	Electronic & Electric Equip	331	1.4	6.50	3.81
24	Motor Vehicles	442	0.9	3.22	0.27
25	Other Transportation Equip	227	1.3	1.91	0.28
26	Instruments	207	1.0	4.32	1.10
27	Miscellaneous Manufacturing	61	1.8	2.18	0.88
28	Transport and Warehouse	668	13.1	3.01	0.99
29	Communications	528	0.8	5.65	1.16
30	Electric Utilities	373	14.2	2.94	0.30
31	Gas Utilities	77	55.0	-0.45	-0.86
32	Trade	2488	3.2	3.72	0.84
33	Finance, Insurance, Real Estate	2752	1.2	4.19	0.77
34	Services	4354	1.7	3.93	-0.27
35	Government Enterprises	328	7.8	2.43	0.19

Production sector

A total of 35 industries are identified in IGEM, including five related to energy production. Jin and Jorgenson (2009) have modeled substitution among inputs in these 35 producing industries by a nested series of translog price functions.[8] The top tier determines the price of output as a function of the prices of capital, labor, energy and non-energy inputs (K, L, E, M):

$$(9) \quad \ln PO_t = \alpha_0 + \sum_i \alpha_i \ln p_{it} + \tfrac{1}{2} \sum_{i,k} \beta_{ik} \ln p_{it} \ln p_{kt} + \sum_i \ln p_{it} f_{it}^p + f_t^p$$

$$p_i, \; p_k = \{P_K, P_L, P_E, P_M\}$$

Jin and Jorgenson (2009) represent technical change by latent variables in a Kalman filter. The latent variables $f_{Kt}^p, f_{Lt}^p, f_{Et}^p, f_{Mt}^p, f_t^p$ are generated by a vector autoregression. This allows non-price determinants of changes in technology to be extrapolated into the projection period. For example, rapid price declines due to productivity growth in electrical machinery are maintained in our base case and alternative case projections. The wide range of past behavior for individual industries is summarized in Table 5.

The remaining sectors of IGEM are the government and rest-of-the-world sectors. The solution of this forward-looking model generates a cost-of-capital equation that links asset prices between two consecutive periods with the marginal product of capital and the rate of return. This solution also incorporates an Euler equation characterizing intertemporal choice. Translog price functions similar to (9) are used to represent the commodity demands by the government sector, as well as exports and imports. Latent variables are used to capture the non-price effects of changes in technology and preferences.

In the rest-of-the-world sector of IGEM imported commodities are imperfect substitutes with the domestic varieties. The current account deficit is specified exogenously and this constraint is met by adjusting the world relative price. The tax system is represented by exogenous tax rates on capital, labor, output, imports and property. The government budget deficit is set exogenously, transfers and interest payments are set exogenously, and government purchases of goods are the endogenous variables that satisfy the budget constraint.

In IGEM emissions of carbon dioxide (CO_2) are generated from fossil fuel consumption. In addition, the model includes four sets of variables representing other greenhouse gas (GHG) emissions –emissions of non-CO_2 GHG gases, CO_2 emissions from sources other than fossil fuel use, GHG emissions from sources covered by a particular policy, and emissions not covered by the policy. The

---

[8] More detail about econometric modeling of producer behavior is provided by Jorgenson (2000).

externality coefficients for the environment are derived from the detailed historical data in the EPA's *2010 U.S. Greenhouse Gas Inventory Report.*

Base case projection

To facilitate comparisons of our estimates of policy impacts with estimates from other models we calibrate the growth of energy utilization in IGEM to the *Annual Energy Outlook 2010* (AEO 2010) from the Energy Information Administration of the U.S. Department of Energy. The consumption of energy is targeted by adjusting productivity growth in the domestic energy production sectors. Import prices are set equal to those in AEO 2010.

The most important exogenous variables in IGEM are those related to the population, government finances and international financial flows. The population projection is taken from the U.S. Bureau of the Census[9], the quality of aggregate labor is projected assuming a modest improvement in educational attainment, and the federal government deficits, expenditures and implied tax rates are taken from the Congressional Budget Office (2010). We project a gradually declining path for the U.S. current account deficit.

3. Welfare Measurement

Our methodology for measuring the welfare effects of policy changes was introduced by Jorgenson, Slesnick, and Wilcoxen (1992). As presented in section 2 above, the household sector is comprised of infinitely-lived households that we refer to as dynasties. Each household takes commodity prices, wage rates, and rates of return as given. All dynasties are assumed to face the same vector of prices p_t at time t and the same nominal rate of return r_t. The quantity of a commodity, including leisure, consumed by dynasty d in period t is C_{ndt}, and the full expenditure of dynasty d on consumption in period t is m_{dt} as given in (2) above.

We assume that each dynasty maximizes an additive intertemporal utility function of the form:

$$(10) \quad V_d = \sum_{t=0}^{\infty} \delta^t \ln V_{dt}$$

where $\delta = 1/(1+\rho)$ and ρ is the subjective rate of time preference. The intratemporal indirect utility function (4) is expressed in terms of *household equivalent members*:

---

[9] Census Bureau projections of the U.S. population released in 2008 are available at http://www.census.gov/population/www/projections/natproj.html.

The B.E. Journal of Economic Analysis & Policy, Vol. 10 [2010], Iss. 2 (Symposium), Art. 17

(11) $$\ln V_{dt} = \alpha^H {}'\ln p_t + \tfrac{1}{2}\ln p_t {}'B^H \ln p_t - D(p)\ln \frac{M_{dt}}{N_{dt}}$$

where $N_{dt} = \dfrac{1}{D(p_t)}\ln p_t B_A A_d$, and A_d is a vector of attributes of the dynasty allowing for differences in preferences among households.

The utility function V_d is maximized subject to the lifetime budget constraint:

(12) $$\sum_{t=0}^{\infty} \gamma_t M_{dt}(p_t, V_{dt}, A_d) = \Omega_d$$

where $\gamma_t = \prod_{s=0}^{t}\dfrac{1}{1+r_s}$, and Ω_d is the full wealth of the dynasty. In this representation $M_{dt}(p_t, V_{dt}, A_d)$ is the intratemporal full expenditure function and takes the form:

(13) $$\ln M_{dt}(p_t, V_{dt}, A_d) = \frac{1}{D(p_t)}[\alpha^H {}'\ln p_t + \tfrac{1}{2}\ln p_t {}'B^H \ln p_t - \ln V_{dt}] + \ln N_{dt}$$

The necessary conditions for a maximum of the intertemporal utility function, subject to the wealth constraint, are given by the discrete time Euler equation:

(14) $$\ln V_{dt} = \frac{D_t}{D_{t-1}}\ln V_{dt-1} + D_t \ln\left(\frac{D_{t-1}\gamma_t N_{dt}P_t}{\delta D_t \gamma_{t-1}N_{dt-1}P_{t-1}}\right)$$

where we have used D_t to denote $D(p_t)$. The aggregate price term is:

(15) $$P_t = \exp\left(\frac{\alpha^H {}'\ln p_t + \tfrac{1}{2}\ln p_t {}'B^H \ln p_t}{D_t}\right)$$

The Euler equation implies that the current level of utility of the dynasty can be represented as a function of the initial level of utility and the initial and future prices and discount factors:

(16) $$\ln V_{dt} = \frac{D_t}{D_0}\ln V_{d0} + D_t \ln\left(\frac{D_0 \gamma_t N_{dt}P_t}{\delta' D_t N_{d0}P_0}\right)$$

Equation (16) enables us to represent dynastic utility as a function of full wealth and initial and future prices and interest rates. We begin by rewriting the intertemporal budget constraint as:

Jorgenson et al.: The Distributional Impact of Climate Policy

(17)
$$\sum_{t=0}^{\infty} \gamma_t N_{dt} P_t V_{dt}^{-1/D_t} = \Omega_d$$

Substituting (16) into (17) and simplifying yields the following:

(18)
$$\ln V_{d0} = -D_0 \ln\left(\frac{\Omega_d}{N_{d0}R}\right)$$

where $R = \dfrac{P_0}{D_0} \sum_{t=0}^{\infty} \delta^t D_t$.

Equation (18) enables us to evaluate dynastic utility in terms of full wealth:

(19)
$$
\begin{aligned}
V_d &= \sum_{t=0}^{\infty} \delta^t \ln V_{dt} \\
&= \sum_{t=0}^{\infty} \delta^t \left[\frac{D_t}{D_0} \ln V_{d0} + D_t \ln\left(\frac{D_0 \gamma_t N_{dt} P_t}{\delta^t D_t N_{d0} P_0}\right) \right] \\
&= \sum_{t=0}^{\infty} \delta^t \left[-D_t \ln\frac{\Omega_d}{R} + D_t \ln\left(\frac{D_0 \gamma_t N_{dt} P_t}{\delta^t D_t P_0}\right) \right] \\
&= S \ln R - S \ln \Omega_d + \sum_{t=0}^{\infty} \delta^t D_t \ln\left(\frac{D_0 \gamma_t N_{dt} P_t}{\delta^t D_t P_0}\right)
\end{aligned}
$$

where $S = \displaystyle\sum_{t=0}^{\infty} \delta^t D_t$.

Solving for full wealth as a function of prices and utility yields the intertemporal expenditure function of the dynasty:

(20)
$$\ln \Omega_d(\{p_t\},\{\gamma_t\},V_d) = \frac{1}{S}\left[S \ln R + \sum_{t=0}^{\infty} \delta^t D_t \ln\left(\frac{D_0 \gamma_t N_{dt} P_t}{\delta^t D_t P_0}\right) - V_d \right],$$

where $\{p_t\}$ is the time profile of prices and $\{\gamma_t\}$ is the profile of discount factors.

We employ the expenditure function (20) in measuring the monetary equivalent of the effect on welfare of a change in policy. We let $\{p_t^0\}$ and $\{\gamma_t^0\}$ represent the time profiles of prices and discount factors for the base case and V_d^0 the resulting level of welfare. Denoting the welfare of the dynasty after the imposition of the new policy by V_d^1, the equivalent variation in full wealth is:

(21) $\Delta W_d = \Omega_d(\{p_t^0\},\{\gamma_t^0\},V_d^1) - \Omega_d(\{p_t^0\},\{\gamma_t^0\},V_d^0)$

The B.E. Journal of Economic Analysis & Policy, Vol. 10 [2010], Iss. 2 (Symposium), Art. 17

The equivalent variation in full wealth (21) is the wealth required to attain the welfare associated with the new policy at prices of the base case, less the wealth required to attain base case welfare at these prices. If the equivalent variation is positive, the policy produces a gain in welfare; otherwise, the policy change results in a welfare loss. Equivalent variations in full wealth enable us to rank the base case policy and any number of alternative policies in terms of a money metric of dynastic welfare.

In Section 4 below we report the equivalent variations of full wealth resulting from introduction of a cap-and-trade policy for all 244 types. We calculate the equivalent variations at mean wealth, half the mean wealth and twice mean wealth.

4. Evaluation of Climate Policy

We next consider the evaluation of cap-and-trade policies to control greenhouse gas emissions in the United States. The policies impose "caps" or emissions limits on greenhouse gases (GHG) of 167 giga-tonnes (metric) of carbon dioxide equivalent (GtCO2-e) and 203 (GtCO2-e) over the period 2012-2050. These yield reductions in emissions levels of 50% and 80% below the 6148 GtCO2-e observed in 1990. The caps refer to economy-wide emissions of six GHG's – carbon dioxide, methane, nitrous oxide, hydrofluorocarbons, perfluorocarbons, and sulfur hexafluoride. These policies are described in more detail by Goettle and Fawcett (2009) in their contribution to Energy Modeling Forum 22.

A key assumption in setting the policy simulation is the treatment of the government budget. We could keep nominal revenues and expenditures equal to the base case or keep real expenditures on goods equal to the base case with an endogenous tax (or transfer). Although there is no price inflation in IGEM there are large changes in relative prices, especially energy prices, due to the policy, that is, a dollar buys a different basket of goods in year *t* in the policy case compared to the base case. Our procedure is to keep real government expenditures under the cap-and-trade policy equal to those of the base case.

Macroeconomic effects

Table 6 summarizes the macroeconomic impacts of the cap-and-trade policy. Real GDP falls substantially. There is a reallocation of activities, so that consumption falls by less than GDP, while exports and imports fall substantially more. Investment falls by a bit more than the fall in GDP.

Jorgenson et al.: The Distributional Impact of Climate Policy

Table 6. Macroeconomic Impacts of GHG caps

Cumulative Emissions Target, 2012-2050 Billions of MT of CO_2-Equivalent (BMT)	203 BMT	167 BMT
Allowance Prices		
2012 in $(2005)/MTCO_2$ Equivalent	$18.97	$40.10
2050 in $(2005)/MTCO_2$ Equivalent	$121.13	$256.06
% change from base case average, 2010-2050		
Real GDP	-2.26	-4.05
Consumption	-1.37	-2.53
Investment	-3.51	-5.89
Government	-0.09	-0.17
Exports	-5.36	-9.53
Imports	-4.13	-6.69
GDP Prices	2.17	4.01
Consumption	1.66	3.05
Investment	1.15	2.04
Government	1.01	1.83
Exports	2.24	4.20
Imports	0.74	0.70
Household Full Consumption (Goods, Services and Leisure)	-0.13	-0.28
Capital Stock	-1.96	-3.33
Labor Demand (Labor Supply)	-1.30	-2.25
Leisure Demand	0.52	0.89
Exchange Rate ($/Foreign Currency)	1.20	1.60

The B.E. Journal of Economic Analysis & Policy, Vol. 10 [2010], Iss. 2 (Symposium), Art. 17

The higher goods prices lower the relative price of leisure, so that leisure demand rises. The price of labor input also rises relative to the price of capital input. These shifts in labor supply and demand result in a fall in hours of work and a rise in leisure consumption. Keep in mind that full consumption is an aggregate of consumption of goods and services and leisure. In the initial periods, when the carbon prices are low, the increase in leisure dominates and full consumption is higher in the policy case. As the carbon prices rise over time the reduction in goods consumption becomes more substantial and full consumption becomes lower than in the base case.

Table 7 gives the effects for industry prices and output. The policies raise coal prices the most and substantially reduce coal output. The next largest impact is on refined petroleum products and this is followed by gas utilities and electric utilities. Given the fall in GDP due to a smaller capital stock , output of all sectors is lower than in the base case; the smallest reductions are in the non-energy intensive industries such as Services and Finance.

The impacts of the cap-and-trade system on the outputs of the various energy commodities depend on carbon intensity and the degree of substitutability for other inputs. The share elasticities given by the β_{ik} coefficients in (9) reveal relatively easy substitution for coal, especially in comparison to the parameter estimates in previous versions of IGEM. The share elasticities are the main determinant of the costs of abating carbon emissions.

To summarize the impacts of the alternative cap-and-trade policies, we plot allowance prices and abatement levels under each emissions limit in Figures 1a and 1b. The curves rise at an increasing rate at higher abatement levels, so that it is steadily more difficult to reduce the emissions.

Household welfare effects

We next report the impacts of a cap-and-trade policy on household welfare given by the equivalent variation in full wealth. Recall that the equivalent variation is the full wealth required to attain the welfare associated with the new policy at prices in the base case, less the wealth required to attain the welfare of the base case at the same prices. This money metric enables us to rank the base case policy and any number of alternative policies in terms of the impacts on dynastic welfare. If the equivalent variation is positive, the policy produces a gain in welfare. We present equivalent variations for each of the 244 household types, cross-classified by the demographic categories, presented in Table 1.

Jorgenson et al.: The Distributional Impact of Climate Policy

Table 7. Industry Effects (% change from base case, average, 2010-2050)

Cumulative Emissions Target, 2012-2050 Billions of MT of CO_2-Equivalent (BMT)	203 BMT		167 BMT	
	Price	Output	Price	Output
1 Agriculture, forestry, fisheries	6.98	-8.34	14.30	-15.76
2 Metal mining	2.64	-4.89	4.79	-9.41
3 Coal mining	99.26	-64.21	206.86	-78.50
4 Crude oil and gas extraction	-4.37	-5.04	-7.73	-9.07
5 Non-metallic mineral mining	3.09	-5.00	5.76	-8.80
6 Construction	1.40	-2.49	2.62	-4.33
7 Food and kindred products	2.80	-3.21	5.41	-5.99
8 Tobacco manufactures	1.38	-2.15	2.58	-4.14
9 Textile mill products	2.68	-3.38	5.19	-6.18
10 Apparel and other textile products	1.41	-2.02	2.62	-3.55
11 Lumber and wood products	1.86	-3.52	3.50	-6.22
12 Furniture and fixtures	1.45	-2.55	2.68	-4.40
13 Paper and allied products	2.00	-3.20	3.68	-5.80
14 Printing and publishing	0.85	-1.17	1.51	-2.22
15 Chemicals and allied products	6.18	-8.84	12.62	-16.41
16 Petroleum refining	19.21	-17.58	39.42	-29.76
17 Rubber and plastic products	4.48	-6.33	8.97	-11.34
18 Leather and leather products	1.63	-2.29	2.92	-3.90
19 Stone, clay and glass products	3.65	-5.07	7.88	-9.59
20 Primary metals	4.16	-6.94	7.44	-12.35
21 Fabricated metal products	1.86	-4.31	3.30	-7.34
22 Non-electrical machinery	1.09	-3.59	1.88	-6.15
23 Electrical machinery	1.13	-2.78	1.98	-4.75
24 Motor vehicles	1.56	-4.18	2.72	-6.65
25 Other transportation equipment	0.88	-0.47	1.53	-1.04
26 Instruments	0.74	-2.18	1.31	-3.90
27 Miscellaneous manufacturing	1.45	-2.39	2.59	-4.00
28 Transportation and warehousing	2.67	-5.72	5.15	-10.36
29 Communications	0.78	-1.97	1.35	-3.54
30 Electric utilities (services)	8.39	-5.63	14.47	-9.26
31 Gas utilities (services)	18.12	-18.86	37.71	-31.82
32 Wholesale and retail trade	1.05	-2.28	1.91	-4.06
33 Finance, insurance and real estate	0.89	-2.20	1.58	-4.00
34 Personal and business services	0.86	-2.02	1.57	-3.69
35 Government enterprises	1.66	-3.01	3.08	-5.63

The B.E. Journal of Economic Analysis & Policy, Vol. 10 [2010], Iss. 2 (Symposium), Art. 17

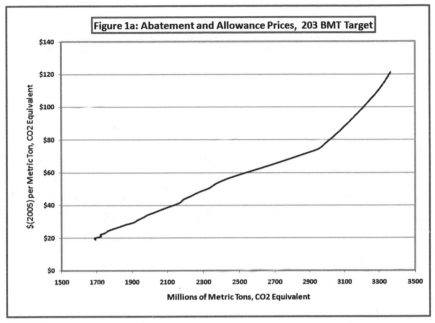

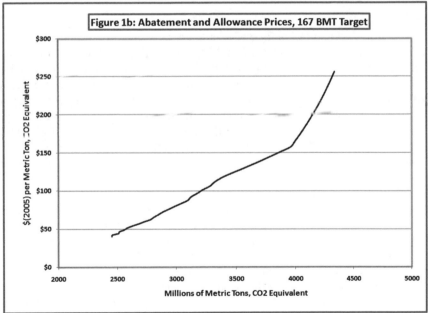

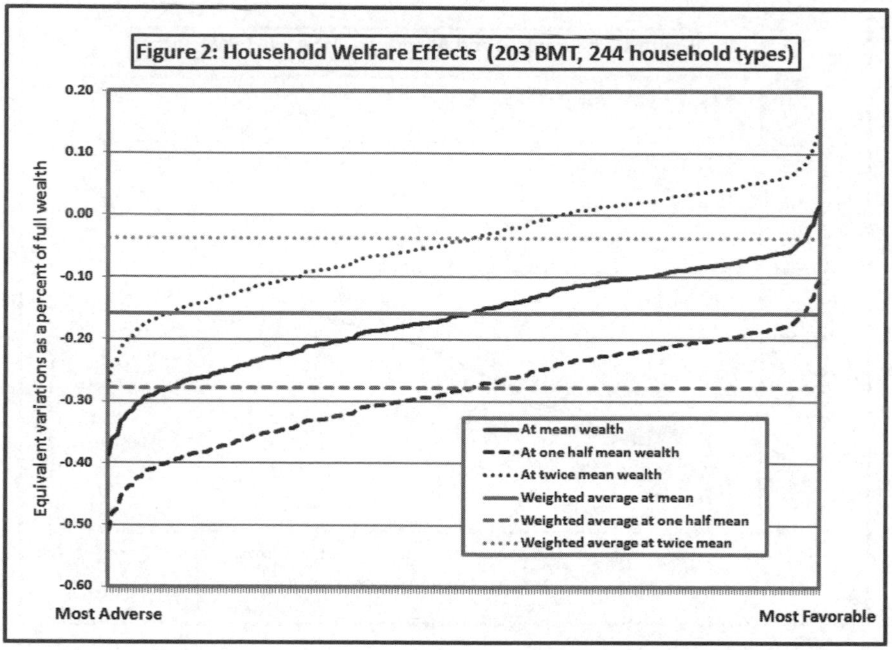

Jorgenson et al.: The Distributional Impact of Climate Policy

Figure 2: Household Welfare Effects (203 BMT, 244 household types)

Figure 2 shows the welfare effects of adopting the less stringent target -- a 50% reduction in emissions relative to 1990 levels. The 244 household types are arranged from the most adversely to the most favorably affected. The principal curve is the solid line labeled "At mean wealth" which shows the welfare impact on households with the mean wealth among those with the same demographic characteristics. For mean wealth, almost all households experience a welfare loss, while few households gain slightly. The most negatively affected households consist of one child with one adult living in the rural South and headed by non-white females. The most positively affected are large urban households in the West: three or more children with three or more adults headed by a nonwhite male.

To illustrate how the policy affects households with different levels of full consumption, the effects in Figure 2 also are shown for half and twice mean wealth. The population-weighted average welfare effects are -0.28, -0.16 and -0.04 percent of lifetime full wealth at the one half mean, mean, and twice mean levels, respectively, as represented by the horizontal lines in Figure 2. It should be emphasized that these are effects on individual households, not measures of social welfare. Figure 2 shows that the effects of the policy change are regressive; the equivalent variations become less negative (or more positive) as

The B.E. Journal of Economic Analysis & Policy, Vol. 10 [2010], Iss. 2 (Symposium), Art. 17

full wealth increases. However, it should be noted that in all cases the welfare effects are small, substantially less than one percent of wealth.[10]

Figure 3 decomposes the welfare effect by isolating the impact of price changes alone. The solid line in Figure 3 repeats the curve from Figure 2 for mean wealth. The dashed line below it shows the welfare effects due solely to output price changes, holding household full expenditure at its base case value. In the absence of changes in expenditure, all households would experience a net welfare loss. However, lump sum tax rebates required to hold real government spending at its base case level cause expenditure to rise.

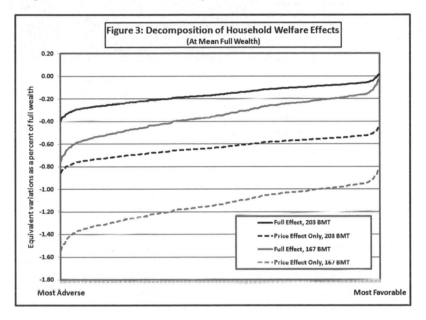

Figure 4 orders the welfare losses first by the number of adults, second by the number of children from most to least, and finally from the most adversely to most favorably affected. The first 75 bars correspond to 3+ adult households, the second 91 bars to 2-adult households, and the last 78 bars to 1-adult households. Clearly, households containing three or more adults are generally better off than those with two adults, which in turn are better off than single-adult households.

---

[10] Our household types are distinguished by region, but production is not. With one electricity market, all consumers face a higher price of electricity that is not distinguished by region. We capture the overall fraction of electricity that is produced using coal or other inputs, but not production differences by region that might lead to differential prices of electricity by region.

Among the single adult households, within each grouping based on the number of children, rural households headed by females fare worst.

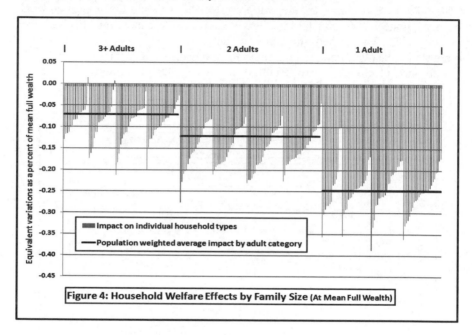

Figure 4: Household Welfare Effects by Family Size (At Mean Full Wealth)

Figure 5 shows the welfare losses by region of residence (the first 56 bars are for the households in the Northeast, ordered by the size of the welfare change). In the sample underlying these results, 18.9% of the population resides in the Northeast, with 23.0%, 36.5% and 21.6% residing in the Midwest, South and West, respectively. Most of the households with large welfare losses are located in the South or Midwest and the largest losses occur in the South. The households with gains are in the West and, on average, this region fares the best.

Figure 6 sorts the welfare changes first by gender and then by race of head. Households headed by non-white females comprise 7.4% of the sample population. Households headed by white females comprise 22.5% of the sample. Households headed by non-white and white males, comprise 10.3% and 59.8% of the sample, respectively. The household types with largest welfare losses are headed by females. Finally, Figure 7 orders the welfare losses by type of residence, urban and rural. The households with the largest losses are concentrated in rural areas.

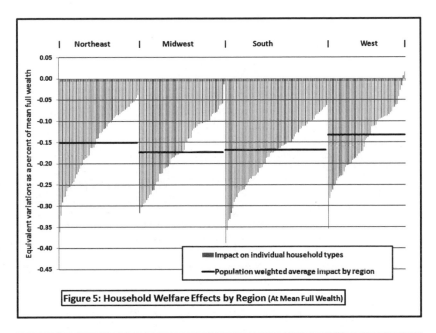

Figure 5: Household Welfare Effects by Region (At Mean Full Wealth)

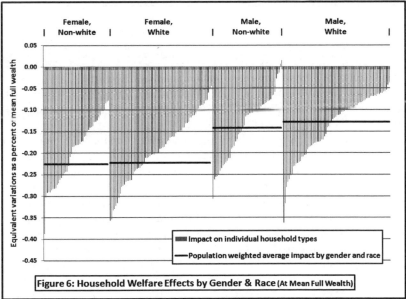

Figure 6: Household Welfare Effects by Gender & Race (At Mean Full Wealth)

Jorgenson et al.: The Distributional Impact of Climate Policy

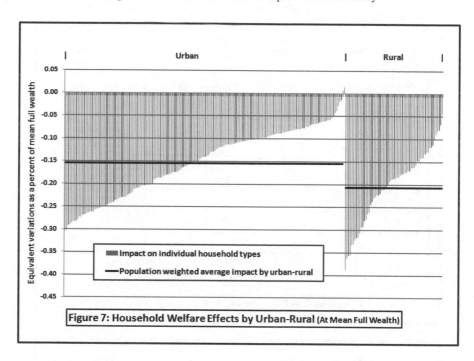

Figure 7: Household Welfare Effects by Urban-Rural (At Mean Full Wealth)

5. Conclusion

In this paper we have successfully incorporated labor-leisure choices, as well as choices among goods and services, into the evaluation of climate policy. This has required the construction of a new version of the Intertemporal General Equilibrium Model (IGEM), employed for evaluation of climate policy by the Environmental Protection Agency. This incorporates a new model of household behavior developed by Jorgenson and Slesnick (2008) that includes labor supply.

Like the models of household behavior used in previous versions of IGEM, the Jorgenson-Slesnick model encompasses all the restrictions implied by the theory of consumer behavior. The new model also satisfies the conditions required for exact aggregation, so that we construct a model of aggregate consumer behavior by aggregating over individual households. We then recover money measures of the impact on household welfare of changes in climate policy.

We provide results for 244 different types of households distinguished by demographic characteristics. We confirm the findings of previous studies of climate policy, including the study of a carbon tax by Jorgenson, Slesnick, and Wilcoxen (1992), that the impact of climate policy would be regressive and negative, but is a relatively small effect. Overall, our findings imply that

The B.E. Journal of Economic Analysis & Policy, Vol. 10 [2010], Iss. 2 (Symposium), Art. 17

incorporating labor-leisure choice into the evaluation of alternative climate policies is a very worthwhile addition to policy analysis. This can be done while preserving the well-established framework for policy evaluation introduced by Jorgenson, Slesnick, and Wilcoxen (1992). This earlier framework, however, must be augmented by a model of household behavior with the features introduced by Jorgenson and Slesnick (2008).

References

Becker, Gary S. (1965), "A Theory of the Allocation of Time," *Economic Journal,* Vol. 75, No. 299, September, pp. 493-517.

Bureau of the Census (2008), *2008 National Population Projections*, Washington, DC: U.S. Department of Commerce, August. http://www.census.gov/population/www/projections/natproj.html.

Bureau of Labor Statistics (2009), *Consumer Expenditure Survey,* Washington, DC, U.S. Department of Labor, October. http://www.bls.gov/cex/

Burtraw, Dallas, Richard Sweeney, Margaret Walls (2009), "The Incidence of U.S. Climate Policy: Alternative Uses of Revenues from a Cap-and-Trade Auction," *National Tax Journal*, Vol. 63, No. 3, September, pp. 497-518.

Congressional Budget Office (2010), *The Long-Term Budget Outlook,* Washington DC: U.S. Government Printing Office, June. http://www.cbo.gov/doc.cfm?index=11579

Energy Information Administration (2010), *Annual Energy Outlook 2010,* Washington, DC: U.S. Department of Energy, March. http://www.eia.doe.gov/oiaf/archive/aeo10/index.html

Environmental Protection Agency (2009), "Intertemporal General Equilibrium Model," Washington, DC: U.S. EPA, Office of Atmospheric Programs. http://www.epa.gov/climatechange/economics/modeling.html#intertemporal

_____ (2010), "2010 U.S. Greenhouse Gas Inventory Report," Washington, DC: U.S. Environmental Protection Agency, Office of Atmospheric Programs, April. http://www.epa.gov/climatechange/emissions/usinventoryreport.html

Goettle, Richard J., and Allen A. Fawcett (2009), "The Structural Effects of Cap and Trade Climate Policy," *Energy Economics,* Vol. 31, Supplement 2, December, pp. S244-S253.

Jorgenson et al.: The Distributional Impact of Climate Policy

Goettle, Richard J., Mun S. Ho, Dale W. Jorgenson, Daniel T. Slesnick and Peter J. Wilcoxen (2009), *Analyzing Environmental Policies with IGEM, an Intertemporal General Equilibrium Model of U.S. Growth and the Environment.* Washington, DC: U.S. EPA, Office of Atmospheric Programs, November. http://www.igem.insightworks.com

Greenstein, Robert., Sharon Parrott, and Arloc Sherman (2008), "Designing Climate-Change Legislation That Shields Low-Income Households from Increased Poverty and Hardship," Washington, DC: Center on Budget and Policy Priorities, May.
http://www.cbpp.org/cms/index.cfm?fa=view&id=770

Hassett, Kevin A., Apama Marthur, and Gilbert E. Metcalf (2009), "The Incidence of a U.S. Carbon Tax: A Lifetime and Regional Analysis," *The Energy Journal*, Vol. 30, No. 2, pp. 157-179.

Jin, Hui, and Dale W. Jorgenson (2010), "Econometric Modeling of Technical Change," *Journal of Econometrics,* Vol. 152, No. 2, August, pp. 205-219.

Jorgenson, Dale W. (1997a), *Aggregate Consumer Behavior,* Cambridge, MA: The MIT Press.

_____ (1997b), *Measuring Social Welfare,* Cambridge, MA: The MIT Press.

_____ (1998), *Energy, the Environment, and Economic Growth*, Cambridge, MA: The MIT Press.

_____ (2000), *Econometric Modeling of Producer Behavior,* Cambridge, MA: The MIT Press.

Jorgenson, Dale W, and Daniel T. Slesnick (2008), "Consumption and Labor Supply," *Journal of Econometrics*, Vol. 147, No. 2, pp. 326-335.

Jorgenson, Dale W., Daniel T. Slesnick, and Peter J. Wilcoxen (1992), "Carbon Taxes and Economic Welfare," *Brookings Papers on Economic Activity: Microeconomics 1992*, pp. 393-431.

Jorgenson, Dale W., and Peter J. Wilcoxen (1990), "Environmental Regulation and U.S. Economic Growth," *Rand Journal of Economics,* Vol. 21, No. 2, pp. 314-340.

Metcalf, Gilbert E., Sergey Paltsev, John M. Reilly, Henry D. Jacoby, and Jennifer F. Holak (2008), "Analysis of U.S. Greenhouse Gas Tax Proposals," *Environment and Development Economics*, forthcoming.

The B.E. Journal of Economic Analysis & Policy, Vol. 10 [2010], Iss. 2 (Symposium), Art. 17

Parry, Ian W.H. (2004), "Are Emissions Permits Regressive?" *Journal of Environmental Economics and Management*, Vol. 47, No. 2, pp. 364-387.

Shammin, Md Rumi, and Clark W. Bullard (2009), "Impact of Cap-and-Trade Policies for Reducing Greenhouse Gas Emissions on U.S. Households," *Ecological Economics,* Vol. 68, Nos. 8-9, pp. 2432-2438.

[14]

The B.E. Journal of Economic Analysis & Policy

Symposium

Volume 10, Issue 2	2010	Article 18

DISTRIBUTIONAL ASPECTS OF ENERGY AND CLIMATE POLICY

Comment on "The Distributional Impact of Climate Policy"

Thomas Hertel*

*Purdue University, hertel@purdue.edu

Hertel: Comment on "The Distributional Impact of Climate Policy"

As a starting point for discussion of the paper by Jorgenson *et al.* (2010), I think it is important to recap some of the highlights of this work. Here are a few points which I believe make this paper a valuable contribution to the literature on the distributional impacts of climate policy.

Strengths of the IGEM Framework: Firstly, the model is fully intertemporal. This stands in sharp contrast to most of the other models used to evaluate the costs of climate policy – and those presented at this conference in particular – which are either static, or recursive-dynamic. This permits a rigorous, intertemporal measure of welfare in terms of full household wealth. Having such a welfare measure is important in the context of climate policy, which is inherently a dynamic, long run phenomenon. When coupled with econometric estimates of households' willingness to trade off current and future consumption, as well as labor and leisure choices, this is a powerful framework for looking at the disaggregated household impacts of policies.

A second important strength of the framework developed by these authors is that the model endogenizes technological progress. As we all know, technological change is what drives economic growth in the long run, and it is also a key factor in determining the industrial composition of the economy. Yet most of models simply treat total factor productivity (TFP) growth as exogenous – often assumed to be equal across sectors. Rarely is productivity growth permitted to vary by sector, nor does it exhibit input-using or input-saving biases. Thus it is significant that the authors are able to draw on the recent work of Jin and Jorgenson (2010), which provides historical estimates of the price and non-price determinants of productivity growth over the 1960-2005 period, in order to predict future patterns of productivity growth. However, the benefits of this feature of the model are largely eliminated in this particular application, *where baseline economic growth is dictated exogenously* by government forecasts.

A third strength of this work is the disaggregation of households by region, family size, residence and head of household gender and race. This is a huge amount of detail. The model has 244 household types, constructed based on the latest Consumer Expenditure Survey which includes 150,000 households. As a consequence, the authors can capture the differential impact of climate policy due to differences in spending and earnings patterns which accurately reflect those in the population at large. Unfortunately, as we will see below, by treating the electric power sector as a national industry, *most of the benefits of this household disaggregation are lost in the climate policy application considered here.*

The fourth strength of this modeling framework is the fact that it is based on econometrically estimated consumer and producer responses to prices based on historical data. Such factor and product substitution is the economy's first line of defense against higher prices on energy intensive commodities. The greater the

The B.E. Journal of Economic Analysis & Policy, Vol. 10 [2010], Iss. 2 (Symposium), Art. 18

degree of substitution in the economy, the lower the cost of meeting a given GHG target, so having an accurate representation of these substitution possibilities is critical. In this regard, the authors note that the current model has much greater price responsiveness than earlier versions of IGEM. This is an important change and it *contradicts received wisdom in the energy economics literature*, which suggests that energy demand has become more inelastic in the past decade or two. This is a point that deserves further discussion.

Evaluation: In evaluating this paper, I take as my starting point the *Brookings Papers* publication by Jorgenson, Slesnick and Wilcoxen (JSW, 1992), which lays out the basic intertemporal framework. In his published discussion of that work, Paul Joskow (1992) raises a number of points which remain relevant to the current work. I have structured my remarks around these same points, assessing how much progress has been made since the 1992 publication.

Modeling of electric power: Here, Joskow argued that the econometric estimates used in 1992 were dated and left the model as a prisoner of an era when natural gas was not a viable option for utilities, nuclear power was not in the mix, and there were fewer air pollution restrictions. The estimates of energy demand, including those for the electric power industry, have now been updated, which is a great improvement. However, the fact that these are national level estimates creates great problems for the subsequent, regionally disaggregated incidence analysis. Electric power rates are determined by regional regulatory bodies, and they vary hugely across regions. While the authors have utilized the Consumer Expenditure Survey to obtain regionally differentiated electric power rates for their econometric analysis, they do not have a mechanism in the model to adjust these rates differentially in the face of a carbon tax.

Climate policy hits those states with high coal-fired power intensities particularly hard. Since IGEM has only one national electric power sector, the change in electric rates which follow from their cap and trade experiment is uniform nationally and does not reflect the differential coal intensity of particular regions. Yet, these differences are at the heart of the national debate over cap and trade legislation, and they drive much of the regional differences in impact in the paper by Rausch *et al.* (2010). I believe that it is possible to modify IGEM to capture such effects. The authors already have regionally differentiated households. All that is needed is to have regionally differentiated power sectors, each selling to the regional households. These power sectors could plausibly continue to be nationally owned, so no further disaggregation of earnings would be required. This task is much easier than fully disaggregating the national model and dealing with inter-regional trade.

Hertel: Comment on "The Distributional Impact of Climate Policy"

Base case: In his comments on the earlier JSW work, Joskow highlights the authors' baseline as being particularly suspect. In the present work, the authors have addressed this criticism head-on. They now adjust their economy-wide TFP growth rates to hit GDP targets proffered by US government agencies, and shared with other models of climate change policy. This is a welcome and important accommodation to those attempting to compare model results. However, energy consumption levels are also targeted. This is more questionable in my mind, as the authors might reasonably lay claim to having a better analytical framework than the one used to project US energy demand by the federal government.

No summary information on elasticities: In his published remarks, Joskow complained that nowhere in the paper are the key elasticities reported. Of course this is a problem with any CGE model, as they typically require a vast number of parameters. This is even more problematic in the case of the Jorgenson *et al.* model, since the authors utilize flexible functional forms that embody a full matrix of unrestricted cross-price elasticities for each sector/household type. The authors do present one table of elasticities in the paper. This table reports price and income elasticities of demand for the aggregated consumer demand categories, including leisure demand. The problem is that these are not the elasticities that matter for the particular policy at hand. The changes in household income, wages, capital services and nondurables composite are small, relative to the changes in energy prices.

What would be really valuable for the reader would be to see the energy price responsiveness of households and firms. Since there are many households in the model, and many sectors, some kind of summary measure is required. In my own experience, I have found general equilibrium demand elasticities to be quite useful (e.g., Hertel, 1997, chapter 3). The general equilibrium demand elasticities capture the model's entire response to a single price perturbation – induced by a small output tax for an individual sector – say coal. Since this perturbation induces not only a change in the demand for coal, but also a change in the demand for all other products, we observe an entire column of elasticities in response to the coal price perturbation. By undertaking one such price perturbation for each sector, one obtains a full matrix of own- and cross-price elasticities that summarize the economy's response to a marginal perturbation in the price of a given product at a given point in time. Producing such a matrix for the economy's 1 year, 5 year, 10 year and 50 year responses to energy price shocks would be enormously useful for model consumers.

Sensitivity to key parameters: A related concern expressed by Joskow was that he wanted to be able to "play around" with some key parameters to see how

The B.E. Journal of Economic Analysis & Policy, Vol. 10 [2010], Iss. 2 (Symposium), Art. 18

the model results varied as a function of such parametric sensitivity analysis. While such *unsystematic* sensitivity analysis is useful, Jorgenson *et al.* can actually do much better in this regard. Since their parameters have been econometrically estimated, they have access to the underlying probability distributions associated with these estimates. By sampling from these distributions, and resolving the model each time for a new set of parameters, the authors can obtain probability distributions for all of the key model outputs. With these in hand, they can provide the reader with confidence intervals on model results. (See Hertel *et al.*, 2007, for an illustration of this linkage between econometric estimation of model parameters and confidence intervals on model results.) This would greatly enhance both the scientific and policy credibility of their findings.

In light of the fact that the Jorgenson *et al.* model is fully intertemporal, the computing time required to solve the model thousands of times, as required of a Monte Carlo analysis, may be prohibitive. In this context, the authors may wish to consider the Gaussian Quadrature (GQ) approach proposed by DeVuyst and Preckel (1997). This analytical sampling technique limits the number of sample points required, and, as those authors demonstrate, can provide a good approximation to a full-blown Monte Carlo analysis of CGE model sensitivity to parameter uncertainty. This problem is also well-suited to distributed computing, since each draw from the parameter distribution is a separate solve.

Policy Realism: In their early work, JSW simply utilized a carbon tax to achieve GHG emissions stabilization at 1990 levels. In his Brookings Papers remarks, Joskow criticized this policy implementation as being unrealistic. In this more recent work, Jorgenson *et al.* have worked closely with the EPA to implement many more of the specific features of the policies. This is a great improvement. However, there remains a gaping hole in the realism of their representation of climate policy. The authors recycle climate policy revenue via "lump sum" transfer back to households *in proportion to full expenditure.* They argue that this has the advantage of being neutral across the wealth spectrum, so that the differential impact across wealth levels of the climate policy is solely due to changes in commodity and factor prices. However, while this approach to redistribution is analytically appealing, it is not the approach taken in the proposed climate legislation, whereby a substantial portion of the revenue is distributed to households on an equal amount per household basis. This dramatically changes the incidence profile of the policy. When revenue is recycled to households on a *per capita* or *per household* basis, the proportional gain is much larger for the poor, and the policy becomes progressive, as opposed to slightly regressive. Without implementing this alternative policy for comparison purposes, the authors risk gravely misleading policy makers – who are unlikely to read footnote 6 in this paper – regarding the likely incidence of climate policy in the US.

References:

DeVuyst, E. A., and P. V. Preckel (1997), "Sensitivity Analysis Revisited: A Quadrature-based Approach", *Journal of Policy Modeling* 19(2): 175–185.

Hertel, T. W. (ed.), (1997), *Global Trade Analysis: Modeling and Applications*, Cambridge, UK: Cambridge University Press.

Hertel, T. W., D. Hummels, M. Ivanic and R. Keeney (2007), "How Confident can we be in CGE Analyses of Free Trade Agreements?" *Economic Modelling* 24: 611-635.

Jin, H., and D. W. Jorgenson (2010), "Econometric Modeling of Technical Change," *Journal of Econometrics* 152(2), August: 205-219.

Jorgenson, D. W., D. T. Slesnick, and P. J. Wilcoxen (1992), "Carbon Taxes and Economic Welfare," *Brookings Papers on Economic Activity: Microeconomics 1992*, pp. 393-431.

Jorgenson, D. W., R. Goettle, M. S. Ho, D. T. Slesnick, and P. J. Wilcoxen (2010), "The Distributional Impact of Climate Policy", *The B.E. Journal of Economic Analysis & Policy* 10(2) (Symposium), Article 17.

Joskow, P. (1992), "Comments", on D.W. Jorgenson, D. T. Slesnick, and P. J. Wilcoxen (1992), "Carbon Taxes and Economic Welfare," *Brookings Papers on Economic Activity: Microeconomics 1992*, pp. 393-431.

Rausch, S., G. E. Metcalf, J. M. Reilly, and S. Paltsev (2010), "Distributional Implications of Alternative U.S. Greenhouse Gas Control Measures," *The B.E. Journal of Economic Analysis & Policy* 10(2) (Symposium), Article 1.

[15]

The B.E. Journal of Economic Analysis & Policy

Symposium

Volume 10, *Issue* 2	2010	*Article* 11

DISTRIBUTIONAL ASPECTS OF ENERGY AND CLIMATE POLICY

CIM-EARTH: Framework and Case Study

Joshua Elliott[*]   Ian Foster[†]   Kenneth Judd[‡]

Elisabeth Moyer[**]   Todd Munson[††]

[*]Computation Institute, University of Chicago and Argonne National Laboratory, jelliott@ci.uchicago.edu

[†]Computation Institute, University of Chicago and Argonne National Laboratory, foster@anl.gov

[‡]Hoover Institution, kennethjudd@mac.com

[**]University of Chicago, moyer@uchicago.edu

[††]Computation Institute, University of Chicago and Argonne National Laboratory, tmunson@mcs.anl.gov

Recommended Citation
Joshua Elliott, Ian Foster, Kenneth Judd, Elisabeth Moyer, and Todd Munson (2010) "CIM-EARTH: Framework and Case Study," *The B.E. Journal of Economic Analysis & Policy*: Vol. 10: Iss. 2 (Symposium), Article 11.
Available at: http://www.bepress.com/bejeap/vol10/iss2/art11

CIM-EARTH: Framework and Case Study*

Joshua Elliott, Ian Foster, Kenneth Judd, Elisabeth Moyer, and Todd Munson

Abstract

General equilibrium models have been used for decades to obtain insights into the economic implications of policies and decisions. Despite successes, however, these economic models have substantive limitations. Many of these limitations are due to computational and methodological constraints that can be overcome by leveraging recent advances in computer architecture, numerical methods, and economics research. Motivated by these considerations, we are developing a new modeling framework: the Community Integrated Model of Economic and Resource Trajectories for Humankind (CIM-EARTH). In this paper, we describe the key features of the CIM-EARTH framework and initial implementation, detail the model instance we use for studying the impacts of a carbon tax on international trade and the sensitivity of these impacts to assumptions on the rate of change in energy efficiency and labor productivity, and present results on the extent to which carbon leakage limits global reductions in emissions for some policy scenarios.

KEYWORDS: general equilibrium models, dynamic trajectories, carbon leakage

---

*We thank Margaret Loudermilk for her help in revising this paper and the referees for their thoughtful comments. This work was supported in part by grants from the MacArthur Foundation and the University of Chicago Energy Initiative and by the Office of Advanced Scientific Computing Research, Office of Science, U.S. Department of Energy, under Contract DE-AC02-06CH11357.

1   Introduction

Computable general equilibrium (CGE) models (Johansen, 1960; Robinson, 1991; Sue Wing, 2004) and their stochastic counterparts, dynamic stochastic general equilibrium models (del Negro and Schorfheide, 2003), form the backbone of policy analysis programs around the world and have been used for decades to obtain insights into the economic implications of policies (Bhattacharyya, 1996; Shoven and Whalley, 1984; de Melo, 1988). Hundreds of such models have been built (Devarajan and Robinson, 2002; Conrad, 2001) and used to explore policy-relevant questions such as the impacts on consumers of new tax policies or increases in fossil energy prices. These models also form a core component when studying the interaction between economic activity and the earth system with an integrated assessment model (Dowlatabadi and Morgan, 1993; Weyant, 2009).

Despite successes, however, these economic models have substantive limitations (Scrieciu, 2007): models may not incorporate the details required to answer questions of interest, cost estimates from different models often differ considerably (Vuuren et al., 2009; Weyant, 1999; de la Chesnaye and Weyant, 2006; Lee, 2006), and little quantification of the uncertainty inherent in these estimates is performed. Many limitations of current models are due to computational and methodological constraints that can be overcome by leveraging advances in computer architecture, numerical methods, and economics research. For example, many contemporary models use simple mathematical formulations, numerical methods, and computer systems that restrict model size and complexity unnecessarily. More modern formulations and solvers and more powerful computer systems offer the potential to solve models several orders of magnitudes larger, while still providing solutions in a reasonable time. Thus, we can in principle create models that incorporate more details of importance to decision makers such as increased industrial, regional, or temporal resolution and capital and commodity vintages (Benhabib and Rustichini, 1991; Cadiou et al., 2003; Salo and Tahvonen, 2003). For example, understanding the distributional impacts of carbon emission policies (Fullerton, 2009; Fullerton and Rogers, 1993) requires overlapping generations for each income group (Auerbach and Kotlikoff, 1987). Having these models, we can then study the interactions between policies that vary by region and characterize important aspects of uncertainty such as the sensitivity of model outputs to baseline assumptions.

Motivated by these considerations, we are developing a new modeling framework: the Community Integrated Model of Economic and Resource Trajectories for Humankind (CIM-EARTH). Our goal is to facilitate and en-

The B.E. Journal of Economic Analysis & Policy, Vol. 10 [2010], Iss. 2 (Symposium), Art. 11

courage the creation, execution, and testing of new economic models with significantly greater fidelity and sophistication than is the norm today. We envision the framework as combining (a) high-level programming that permits the convenient formulation of a wide range of models, (b) a flexible implementation that permits the efficient solution of these models using advanced numerical methods and high-performance computer systems, and (c) a suite of tools for parameter estimation and model evaluation.

We seek not only to provide access to better economic formulations and numerical methods but also to encourage the development and use of open models. Transparent policy studies require that software and data be *accessible* and *understandable*. Thus, we distribute our framework under an open-source license that permits others to read the software, modify it, and redistribute the modifications. Equally important, we structure the code in a way that makes its meaning readily apparent. In addition, we design our software to be *modifiable* and *extensible*, so as to facilitate the reuse of methodologies and tools; a model generated by one researcher can be tested by others with different data, compared to other models, and extended in new directions. In this way, the barriers to entry for newcomers are greatly reduced, increasing the diversity and quality of the ideas explored.

In this paper, we describe the key features of the CIM-EARTH framework (Section 2), detail the particular model instance we use for studying the impacts of a carbon tax on international trade and the sensitivity of these impacts to the assumptions used to construct a baseline scenario (Section 3), and present results on the extent to which carbon leakage limits global reductions in emissions for a handful of policy scenarios (Section 4). We conclude in Section 5.

The focus of this paper is the tools and techniques employed, rather than a detailed policy analysis. In particular, Section 3.2 develops an ensemble of baseline scenarios for sensitivity studies using the energy efficiency and labor productivity parameters that are highly relevant when analyzing emission abatement policies, Section 4.1 describes our method for measuring embedded carbon in goods and services for use in estimating the carbon content of international trade flows, Section 4.2 introduces a matrix method for displaying international carbon flows in order to visualize the impacts of climate policies, and Section 4.3 provides results over our ensemble of baseline scenarios. These tools and techniques can be adapted for more detailed studies of carbon leakage and for other policy studies. Scalar versions of the model used in the case study are available from `www.cim-earth.org`.

2   CIM-EARTH Framework

To develop an accessible, understandable, modifiable, and extensible framework, our overall architecture uses a modular design; proven numerical libraries such as PATH (Dirkse and Ferris, 1995; Ferris and Munson, 1999, 2001), TAO (Benson et al., 2010), and PETSc (Balay et al., 1997); and a high-level specification language. We discuss here those parts of the CIM-EARTH framework that are concerned with specifying and solving CGE models.

CGE models determine prices and quantities for commodities such that supply equals demand (Ballard et al., 1985; Ginsburgh and Keyzer, 1997; Scarf and Shoven, 1984). Such models may feature the following:

- Many *industries* that each hire labor, rent capital, and buy inputs to produce an output. Each industry chooses a feasible production schedule to maximize its profit.
- Many *consumers* that choose what to buy and how much to work subject to the constraint that expenditures cannot exceed income. Each consumer chooses a feasible consumption schedule to maximize his utility function.
- Many *markets* where industries and consumers trade and where wage rates and commodities prices are set to clear the markets. In particular, if the price of a commodity is positive, then supply must equal demand.

Model instances are specified by defining the type of model (deterministic or stochastic, myopic or forward looking); the size of the model (number of regions, industries, consumers, and time periods); the details for the industries and consumers (production and utility functions and their nested structure), their parametrization (elasticities of substitution) and calibration data (expenditures and tax data for the base year); dynamic trajectories (land and labor endowments, energy efficiency, and capital accumulation); and any coupling with other system components.

The initial version of the CIM-EARTH framework has been implemented in the AMPL modeling language (Fourer et al., 2003). This language is convenient for expressing large optimization and complementarity problems using sets and algebraic constraints, provides access to a variety of commercial and academic numerical methods, and automatically computes the derivative information required by these methods when calculating a solution. We are currently developing a next-generation system that uses a domain-specific language to simplify model specification and can utilize parallel computers when solving large models.

The B.E. Journal of Economic Analysis & Policy, Vol. 10 [2010], Iss. 2 (Symposium), Art. 11

Figure 1: **Basic nesting for production function.**

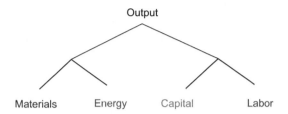

The primary challenge in developing such models is estimating the production and utility functions that characterize the physical and economic processes constraining the supply and demand decisions of industries and consumers. For our CGE models, we use nested constant elasticity of substitution (CES) production and utility functions in calibrated share form (Boehringer et al., 2003),

$$\mathbf{y} = \left(\sum_i \theta_i \left(\gamma_i \mathbf{x}_i \right)^{\frac{\sigma-1}{\sigma}} \right)^{\frac{\sigma}{\sigma-1}},$$

where $\mathbf{y}$ is the ratio between the output of the industry to a base-year value, $\mathbf{x}_i$ are the ratios of the inputs to their base-year values, γ_i are efficiency parameters that determine how effectively these factors can be used, θ_i are the share parameters with $\theta_i > 0$ and $\sum_i \theta_i = 1$, and σ controls the degree to which the inputs can be substituted for one another. The special cases of Leontief ($\sigma=0$) and Cobb-Douglas ($\sigma=1$) functions are supported by our framework.

The nesting structure is depicted graphically by a tree, with each node representing a production function with its own elasticity of substitution that aggregates the inputs from below into a bundle. The root node represents the total output from the production process. Figure 1 shows a simple case. In the CIM-EARTH framework, we add intermediate variables for the internal nodes, encode the individual functions by specifying the inputs and output, and reconstruct the tree from this information. Since the nesting structure is typically the same for each industry independent of the region in which it resides, we provide facilities to convey this information and reduce the amount of required coding.

Tables are used to convey the parametrization and calibration data. These data include expenditures on inputs and tax information. The share parameters are automatically computed given the nesting structure of the

4

production functions and the expenditure data for the base year. Also included is support for ad valorem and excise taxes, import and export duties, and endogenous tax rates such as those encountered in cap-and-trade policies.

Once the model structure and data are provided, we enter a preprocessing phase to check consistency and make any necessary modifications. Consistency checks include testing the nesting structure to ensure it is a tree. Modifications are made to the tree structure, for example, to eliminate inputs that have zero expenditures or minuscule shares. The modifications are applied iteratively so that if all the branches of a particular node are eliminated, that node is also eliminated. Such modifications are necessary to ensure that the nesting structure matches the expenditure data.

After preprocessing is complete, we have a set of constrained optimization problems for the industries and consumers and market clearing conditions. Because the optimization problems solved by the industries and consumers are convex in their own variables and satisfy a constraint qualification, we can replace each with an equivalent complementarity problem obtained from the first-order optimality conditions by adding Lagrange multipliers on the constraints. These optimality conditions in combination with the market clearing conditions form a square complementarity problem.

The simplest dynamic CGE models are *myopic*, meaning that industries and consumers look only at their current state and do not consider the future. In this case, after the preprocessing step, we loop over time and solve a complementarity problem for each time step with fixed trajectories for land and labor endowments, efficiency parameters, and emission factors. The capital stocks are dynamically updated at the end of each time period based on depreciation and investment. Summary reports are written to user-defined files once the complementarity problem for each time step is solved.

The complementarity problem solved at each time step is automatically generated by the framework and is emitted in a scalar form so that it can be inspected. The complementarity problem is solved by applying a generalized Newton method such as PATH (Dirkse and Ferris, 1995; Ferris and Munson, 1999, 2001). PATH is a sophisticated implementation of a Josephy-Newton method that solves a linear complementarity problem at each iteration using a variant of Lemke's method to obtain a direction and then searches along this direction to obtain sufficient decrease for the merit function.

The B.E. Journal of Economic Analysis & Policy, Vol. 10 [2010], Iss. 2 (Symposium), Art. 11

3 Model Instance

We next provide a detailed discussion of the model instance implemented in the CIM-EARTH framework used for studying the impacts of a carbon tax on international trade and the sensitivity of these impacts to the assumptions used to construct a baseline scenario. In particular, we specify the structure of the production functions and the data used to calibrate them in Section 3.1 and the exogenous trajectories for important economic drivers that define an ensemble baseline scenarios in Section 3.2.

3.1 Structure

Table 1 shows the regions, industries, and factors in the model instance used for our study. The regions are labeled with the aggregation used for reporting purposes. For each industry we indicate the structure of the production functions: (A) agriculture, (E) extraction of fossil fuels, (M) manufacturing, (N) electricity generation, (P) petroleum refining, and (S) service industries. This industry aggregation was chosen to contain more detailed resolution in the energy-intensive industries and in the industries that provide transport services to importers for moving commodities around the world, since these industries would be most affected by a carbon tax or cap-and-trade program.

 The production functions in each region have the nested structure summarized in Figure 2 and are loosely based on those used in the EPPA model (Babiker et al., 2001). As before, each node represents a CES function aggregating the production factors below it. The structure of the production functions for the importers of each commodity in each region is also provided. In addition, there is a capital goods industry in each region that aggregates materials using a single Leontief production function. The capital goods industries do not demand fossil fuels, refined petroleum, electricity, or production factors. These capital goods produced are demanded only by consumers in their role as investors. All these industries are subject to ad valorem and excise taxes. We use elasticities of substitution taken from the CGE literature for the industries and consumers (Balistreri et al., 2003; Liu et al., 2004; Webster et al., 2008; Sokolov et al., 2009) and the GTAP database for the base-year revenues, expenditures, and tax data. In particular, the share parameters are calibrated with the GTAP version 7 database of global expenditure values with a 2004 base year (Gopalakrishnan and Walmsley, 2008).

 Trade among regions is handled through importers of each commodity in each region. Importers are modeled like other industries using the nested

Elliott et al.: CIM-EARTH: Framework and Case Study

Table 1: **Regions, industries, and factors for the CGE model used for this study.**

Regions	Industries	Factors
United States (USA)	Agriculture and Forestry (A)	Capital
Western Europe (EUR)	Coal Extraction (E)	Labor
Rest of Europe (EUR)	Gas Extraction (E)	Land
Russia, Georgia, and Asiastan (RUS)	Oil Extraction (E)	Nat. Resources
Japan (JAZ)	Cement (M)	
Oceania (JAZ)	Chemicals (M)	
Canada (CAN)	Nonferrous Metals (M)	
China, Mongolia, and Koreas (CHK)	Steel and Iron (M)	
Brazil (LAM)	Other Manufacturing (M)	
Mexico (LAM)	Electricity (N)	
Rest of Latin America (LAM)	Petroleum Refining (P)	
Middle East and North Africa (ROW)	Air Transport (S)	
Rest of Africa (ROW)	Land Transport (S)	
India (ROW)	Sea Transport (S)	
Rest of South Asia (ROW)	Government Services (S)	
Rest of Southeast Asia (ROW)	Other Services (S)	

Note: The regions are labeled with the aggregation used for reporting the results obtained. The industries are labeled by their production function structure: (A) agriculture, (E) extraction of fossil fuels, (M) manufacturing, (N) electricity generation, (P) petroleum refining, and (S) service industries.

CES production function shown in Figure 2. Table 2 shows the elasticities of substitution between domestic and imported commodities and the Armington international trade elasticities used for this study. We use a Leontief production function to aggregate between the imported commodity and the relevant total transport margin, so that the amount of transport demanded scales with the amount of the commodity imported. We use three types of transportation: land transportation, including freight by trucks and pipelines; air transportation; and sea transportation. Since importers do not care about the origination of these transport services, we model international transportation of each type as a homogeneous commodity with one global price. The model instance has separate homogeneous transportation service industries for air, land, and sea transportation. Each aggregates a single type of domestic transportation service from all regions into a single commodity using a Cobb-Douglas production function. These homogeneous transportation services are used only for international trade; domestic transportation services are included in the materials nest of the other production functions.

This model does not contain a government consumer; it contains only a government services industry, which include defense, social security, health care, and education. Industries and consumers demand these government ser-

Figure 2: **Structure of the production functions for the model instance used for this study.**

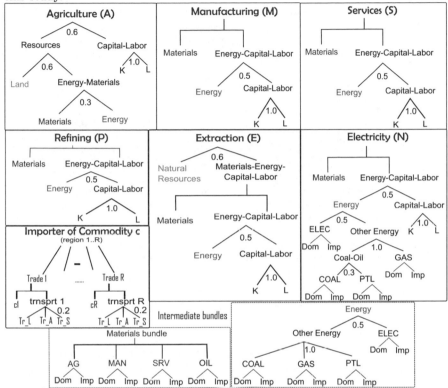

Note: Each node represents a production function. Nodes with vertical line inputs use Leontief functions; the other nodes are labeled with their elasticities of substitution. Table 2 shows the elasticities of substitution between domestic and imported commodities and the Armington international trade elasticities.

vices. The government services industry is treated like any other industry and is subject to ad valorem and excise taxes. All taxes collected by a region are returned to consumers in that region.

Capital is specific to each region in the model instance. Within each region, we use a perfectly fluid capital model with a 4% yearly depreciation rate. To spur investment in capital, we use the standard practice in myopic CGE models in which investment contributes to consumer utility with the

Elliott et al.: CIM-EARTH: Framework and Case Study

Table 2: **Elasticity of substitution parameters between domestic and imported commodities and the Armington international trade elasticities by industry for the CGE model used for this study.**

Industry	Elasticity of Substitution	
	Domestic/Import	Armington
Agriculture and Forestry (A)	2.7	5.6
Coal Extraction (E)	3.0	6.1
Gas Extraction (E)	17.2	34.4
Oil Extraction (E)	5.2	10.4
Cement (M)	2.9	5.8
Chemicals (M)	3.3	6.6
Nonferrous Metals (M)	4.2	8.4
Steel and Iron (M)	3.0	5.9
Other Manufacturing (M)	3.4	7.2
Electricity (N)	2.8	5.6
Petroleum Refining (P)	2.1	4.2
Air Transport (S)	1.9	3.8
Land Transport (S)	1.9	3.8
Sea Transport (S)	1.9	3.8
Government Services (S)	1.9	3.8
Other Services (S)	1.9	3.8

Note: The industries are labeled by their production function structure: (A) agriculture, (E) extraction of fossil fuels, (M) manufacturing, (N) electricity generation, (P) petroleum refining, and (S) service industries.

investment amount calibrated to historical data. In particular, the investment commodity the consumer buys is the output from the capital goods industry. Investment enters the consumer utility function in a Cobb-Douglas nest with the government services and consumption bundles, implying that fixed shares of consumer income in each year is used for government services, investment, and consumption. The change in the capital endowment in the next period relative to the amount in the base year is obtained from the dynamic equation

$$\mathbf{y}_{K,t+1} = (1 - \delta)\mathbf{y}_{K,t} + \frac{\bar{x}_{I,0}}{\bar{y}_{K,0}}\mathbf{x}_{I,t},$$

where $\mathbf{y}_{K,t}$ is the change in capital endowment ($\mathbf{y}_{K,0}{=}1$), $\mathbf{x}_{I,t}$ is the change in investment, and δ is the capital depreciation rate. The ratio of the base-year investment quantity $\bar{x}_{I,0}$ to the base-year capital stock $\bar{y}_{K,0}$ is available from the GTAP database.

The B.E. Journal of Economic Analysis & Policy, Vol. 10 [2010], Iss. 2 (Symposium), Art. 11

3.2 Ensemble of Baseline Scenarios

We construct an ensemble of trajectories for important economic drivers such as labor productivity and energy efficiency by extrapolation from historical data that are input into the model instance. By running the model instance for each set of forecasts without making any policy changes, we obtain an ensemble of baseline scenarios that can be compared to existing baseline scenarios from the literature. Moreover, by exploring policy scenarios over a range of baseline scenarios, we can determine the sensitivity of a policy to the assumptions used to generate the baseline scenario.

Our approach differs from much of the carbon leakage literature that typically starts from a reference baseline scenario, chooses a single set of trajectories to replicate it, and then determines the change in outcome for a variety of policy scenarios, often without discussion of the scientific underpinnings of the baseline scenario or how it has been integrated into the model. While the trajectory of CO_2 emissions, for example, may match the EIA forecast, the parameters tuned to achieve this result and thus the assumptions made are not described. This lack of documentation makes it difficult to compare results to the literature, since the results are reported relative to a single hypothetical baseline scenario for which the assumptions are not defined.

We now detail the construction of our ensemble of baseline scenarios, which are parametrized by national aggregate energy efficiency and labor productivity parameters. The space is reduced to two dimensions by assuming perfect correlations for energy efficiency and labor productivity across regions. We then compare the results from our baseline scenarios to forecasts of emissions from the literature.

3.2.1 Energy Efficiency

We incorporate an energy efficiency parameter into each industry production function to model the efficiency by which their input energy is used. The inverse of regional industrial energy intensity is used as a proxy for the energy efficiency of industry. Historical industry gross domestic product is obtained from the UN database of national accounts, and historical industry energy use is obtained from the IEA World Energy Balance database. These data are used to calculate the yearly rate of change in industrial energy intensity. Rates for all regions in our model instance are available from 1972 to 2007. We truncate the data set to eliminate the two largest positive and negative rates of change to eliminate strong variations from one-time political events or economic crashes.

The median baseline scenario forecasts a constant rate of change in energy intensity after 2008 equal to a weighted geometric mean of the historical rates. Minimum and maximum values on the forecast rate of change in energy intensity for the other baseline scenarios are obtained using the standard deviation of the historical rate data with a skewness factor. The skewness factor is related to the slope of a linear regression model of the historical rate data. For regions with a negative slope in the linear regression model, the rate of change in energy intensity for the forecast years is skewed lower. Positive slopes are treated similarly. In particular, we use

$$
\begin{aligned}
\min &= \mu - \tfrac{\sigma}{4} + 5\beta \\
\max &= \mu + \tfrac{\sigma}{4} + 5\beta
\end{aligned}
\tag{1}
$$

where μ is the geometric mean of the historical data, σ is the standard deviation, and β is the slope of the linear regression model. The only exception is for the "Rest of Europe" region, where rapid increases in energy efficiency in recent years from technological improvements and economic shifts spurred by membership in the European Union would forecast an energy efficiency level surpassing the gross energy efficiency levels forecast for the more developed parts of Europe, the United States, and Japan by 2030–2040. Since these efficiency levels seem highly unlikely, the skewness factor is set to zero for this region. The minimum, median, and maximum values for the forecast rate of change in energy intensity for each region is found in Table 3. Intermediate values are obtained by linear interpolation. Negative values for the rate of change in energy intensity imply increased energy efficiency.

3.2.2 Population Growth and Labor Productivity

The other economic drivers we consider are population growth and labor productivity, which are combined to estimate the labor endowment in each region. We use gross population data from 1950 to 2008 with forecasts to 2050 from the 2008 United Nations population database (United Nations, 2008) and historical economic activity rates from 1980 to 2006 from the International Labor Organization (International Labor Organization, 2010) with projections to 2020 to determine the economically active segment of the population.

Labor productivity is chosen to match forecasts extrapolated from historical trends using data from the International Labor Organization Database of Key Indicators of the Labor Market (International Labor Organization, 2009). This database contains data for most countries spanning 1980 to 2005. For simplicity, we currently base labor productivity on the index of gross domestic product per person employed, even though productivity indices are

The B.E. Journal of Economic Analysis & Policy, Vol. 10 [2010], Iss. 2 (Symposium), Art. 11

Table 3: **Maximum, median, and minimum values for the average percent rate of change in energy intensity and labor productivity.**

Region	Energy Intensity			Labor Productivity		
	max	median	min	max	median	min
United States	-1.65	-2.51	-3.17	1.93	1.76	1.59
Western Europe	-1.74	-2.09	-2.46	1.58	1.45	1.31
Rest of Europe	-1.39	-2.41	-3.43	3.83	3.26	2.70
Russia, Georgia, and Asiastan	-1.08	-1.88	-3.71	2.64	1.63	0.62
Japan	-0.97	-1.80	-2.19	1.94	1.69	1.45
Oceania	0.28	-0.35	-0.98	1.82	1.54	1.25
Canada	-0.75	-1.31	-2.10	1.57	1.33	1.08
China, Mongolia, and Koreas	-1.93	-2.54	-3.63	7.62	6.89	6.16
Brazil	1.02	0.67	0.32	1.27	0.27	0.09
Mexico	-1.48	-2.07	-3.38	1.34	0.38	0.25
Rest of Latin America	-0.17	-0.44	-1.68	1.28	0.78	0.27
Mid East and North Africa	0.83	0.07	-1.36	1.75	1.32	0.88
Rest of Africa	-0.77	-1.09	-1.98	1.24	0.76	0.29
India	-1.53	-1.92	-2.99	4.45	4.01	3.56
Rest of South Asia	0.09	-1.03	-1.78	2.78	2.50	2.21
Rest of Southeast Asia	0.30	-0.74	-1.73	4.31	3.82	3.33

Note: The median value is the linearly weighted geometric mean and the min and max values are defined in (1).

available at sectoral resolution covering agriculture, forestry and fishing, manufacturing, trade, and transportation and communication for many countries. Forecasts of the rate of change in labor productivity are constructed in a manner similar to the energy efficiency parameter. In particular, our median baseline scenario assumes a constant rate of change in labor productivity equal to a weighted geometric mean of the historical rates. The skewness parameter is set to zero for all regions except Mexico and Brazil when determining the minimum and maximum values because it either has a negligible impact or produces unrealistic maximum values. For Mexico and Brazil, the skewness parameter is set to the slope of the linear regression model to prevent the minimum rate of change in the labor productivity parameter from becoming negative. While the rate of change in labor productivity could be negative for some regions over the next 25 years in some time periods, a sustained negative rate of change in labor productivity is not realistic. The minimum, median, and maximum values for the forecast labor productivity intensity for each region are found in Table 3.

Elliott et al.: CIM-EARTH: Framework and Case Study

Figure 3:   **Percent rate of change in global CO$_2$ emissions.**

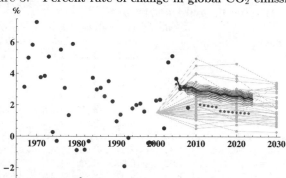

Note: The comparison of historical data (large blue dots), the 2005 EIA forecast (small red dots), SRES scenarios (light gray connected dots), and our ensemble of baseline scenarios (dark solid lines) is plotted as yearly rates of change.

Figure 4:   **Global gross CO$_2$ emissions.**

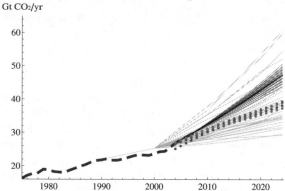

Note: The comparison of historical data (dashed blue line), 2005–2009 EIA forecasts (small red dots), SRES scenarios (light gray solid lines), and our ensemble of baseline scenarios (dark solid lines) is plotted as gross annual emissions.

3.2.3   Comparison to Emission Forecasts

By assuming perfect correlation among regions, we generate an ensemble of baseline scenarios containing 25 members by taking the cross product of five energy intensity and five labor productivity levels. In one scenario, for example, each region uses the minimum value for their energy intensity parameter and the maximum value for their labor productivity parameter. Figures 3–8 compare the emissions generated by our model instance for each element of

The B.E. Journal of Economic Analysis & Policy, Vol. 10 [2010], Iss. 2 (Symposium), Art. 11

Figure 5: **Percent rate of change in USA CO_2 emissions.**

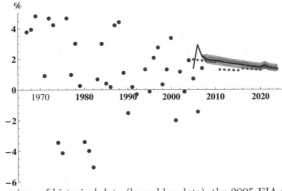

Note: The comparison of historical data (large blue dots), the 2005 EIA forecast (small red dots), and our ensemble of baseline scenarios (dark solid lines) is plotted as yearly rates of change.

Figure 6: **USA gross CO_2 emissions.**

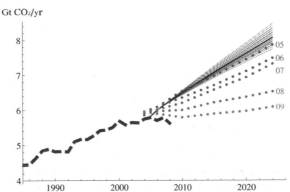

Note: The comparison of historical data (dashed blue line), the 2005–2009 EIA forecasts (small red dots), and our ensemble of baseline scenarios (dark solid lines) is plotted as gross annual emissions. The EIA forecasts are labeled by the year they were released to highlight the direction of the significant changes over the last 5 years.

the baseline ensemble to historical CDIAC data (Boden et al., 2009), 2005–2009 EIA reference case forecasts (United States Energy Information Agency, 2009), and 40 SRES scenarios from the IPCC AR4 (Nakicenovic et al., 2000). The unit for the reported gross emissions is billions of tonnes (Gt) CO_2.

Our model is calibrated from data up to and including 2005 and we have made no effort to account for the recent global recession. In particular,

14

Elliott et al.: CIM-EARTH: Framework and Case Study

Figure 7: **Percent rate of change in CHK CO₂ emissions.**

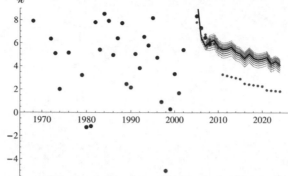

Note: The comparison of historical data (large blue dots), the 2005 EIA forecast (small red dots), and our ensemble of baseline scenarios (dark solid lines) is plotted as yearly rates of change.

Figure 8: **CHK gross CO₂ emissions.**

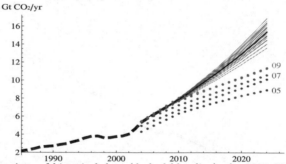

Note: The comparison of historical data (dashed blue line), the 2005–2009 EIA forecasts (small red dots), and our ensemble of baseline scenarios (dark solid lines) is plotted as gross annual emissions. The EIA forecasts are labeled by the year they were released to highlight the direction of the significant changes over the last 5 years.

the trajectory for the USA is similar to the 2005 EIA baseline both in the yearly rate of change (Figure 5) and in gross emissions (Figure 6). The EIA has adjusted their baseline USA forecasts substantially over the last 5 years. In particular, the slope of the EIA forecasts has been significantly modified. While we appreciate that the global recession has produced several years of flat or reduced emissions, we see no structural changes to the economy that would cause the rate of emissions growth to stay low once economic growth returns to previous levels, and so have not built this effect into the baseline ensemble.

The B.E. Journal of Economic Analysis & Policy, Vol. 10 [2010], Iss. 2 (Symposium), Art. 11

The difference between the rate of change and gross emissions trajectories for the global aggregate in our baseline ensemble and the baseline forecasts produced by the EIA are due almost entirely to the divergence between forecasts for CHK beyond 2010. Since our parameters are rooted in extrapolation from the historical record, our trajectories miss the dramatic slowing in emissions for CHK forecast by the EIA. Our baseline trends show a decline in the yearly rate of change in emissions for CHK after 2011 which is consistent with the EIA forecasts. The significant drop in the rate of change in CHK emissions in 2010, however, is not represented. In particular, the rate of change in emissions for CHK in the EIA forecasts drops from 6% for 2009 to just more than 3% for 2010 and continues to decline beyond 2010.

4 Case Study

In our case study, we examine the impacts of a carbon tax on international trade, the extent to which carbon leakage limits global reductions in emissions, the impact of border tax adjustments on reducing carbon leakage, and the sensitivity of these impacts to the assumptions used to generate the baseline scenario. To determine the extent of the carbon leakage for a particular policy, we must first measure the emissions embedded in the traded commodities. Our approach to carbon accounting is detailed in Section 4.1. We report results from four policy scenarios in Section 4.2 using a matrix to visualize international emission flows and then evaluate the dependence of the emission forecasts to the underlying baseline scenario assumptions in Section 4.3.

4.1 Carbon Accounting

We measure the embedded emissions in each commodity by assuming conservation of emissions. In particular, the emissions content of the output for an industry is the sum of the emissions content of the constituent inputs and the emissions generated during the production process from burning fossil fuels. Alternative methods for determining emissions content can be found in Davis and Caldeira (2010), for example.

Specifically, conservation of emissions is stated as

$$E_{j,t} y_{j,t} = \sum_i \left(E_{i,t} + e_i^j \right) x_{i,t}^j \tag{2}$$

where $E_{j,t}$ is the emissions content per unit of commodity j at time t, $y_{j,t}$ is the quantity of commodity j output by the industry, e_i^j is an emissions factor that determines the emissions generated per unit from input commodity i during the production process, and $x^j_{i,t}$ is the quantity of input commodity i used in the production of commodity j at time t. The emissions directly generated by industry j at time t is the summation $\sum_i e^j_i x^j_{i,t}$, while global emissions is $\sum_j \sum_i e^j_i x^j_{i,t}$, which equals the sum of the embedded emissions in all the commodities consumed when the markets clear.

The emissions factors are positive when the input is a fossil fuel that is burned and zero for all other inputs. The emission factors are industry specific to account for regional and industrial differentiation in the types of input commodities and their emission rates. For example, the steel industry uses a large amount of coking coal with a high carbon content, while the electricity generated by coal-fired power plants typically comes from lignite with a low carbon content. Further, some industries use fossil inputs in the generation of their outputs but do not burn them and hence have no new emissions generated from them. In particular, the chemicals and plastics industries use natural gas in their production processes, but do not burn the natural gas.

For a simple example, assume we have four commodities, coal, electricity, steel, and automobiles, where coal is burned to generate electricity, electricity is used to produce steel and automobiles, steel is used to produce automobiles, and consumers demand the automobiles. Given 10 units of each commodity, we would then solve the system of equations

$$
\begin{aligned}
10E_{coal} &= 0 \\
10E_{elec} &= 10\left(E_{coal} + e_{coal}\right) \\
10E_{steel} &= 7E_{elec} \\
10E_{auto} &= 3E_{elec} + 10E_{steel},
\end{aligned}
$$

where the emissions factor for coal, e_{coal}, is known and we assume that the production of coal does not generate any emissions. All emissions in this example are generated from burning coal to produce electricity and total $10e_{coal}$ emission units. The solution to the system of equation is

$$
\begin{aligned}
E_{coal} &= 0 \\
E_{elec} &= e_{coal} \\
E_{steel} &= 0.7e_{coal} \\
E_{auto} &= e_{coal}.
\end{aligned}
$$

Since consumers demand all the automobiles produced, the emissions content of the automobiles demanded, $10E_{auto}$, equals the total emissions generated.

The B.E. Journal of Economic Analysis & Policy, Vol. 10 [2010], Iss. 2 (Symposium), Art. 11

CGE models do not typically determine the actual quantities supplied and demanded, but rather calculate the change in quantity relative to a base year. Therefore, we rewrite (2) as

$$E_{j,t}\bar{y}_j \mathbf{y}_{j,t} = \sum_i (E_{i,t} + e_i^j)\bar{x}_i^j \mathbf{x}_{i,t}^j,$$

where $\bar{y}_j$ is the base-year volume of commodity j output by the industry and $\mathbf{y}_{j,t}$ is the change in output relative to the base year at time t with $y_{j,t} = \bar{y}_j \mathbf{y}_{j,t}$, and $\bar{x}_i^j$ is the base-year volume of commodity i input for the production of commodity j and $\mathbf{x}_{i,t}^j$ is the change in demand relative to the base year for those inputs at time t with $x_{j,t} = \bar{x}_j \mathbf{x}_{j,t}$. The emissions generated by the producers from each input in the base year, $\bar{f}_i^j = e_i^j \bar{x}_i^j$, is obtained from the energy volume information in the GTAP-E database (Burniaux and Truong, 2002).

Moreover, the base-year quantities in the emissions expression, $E_i \bar{x}_i^j$, are generally unavailable. Hence, we compute the total emissions for the industry rather than compute the emissions content per commodity unit. In particular, we make the substitution

$$F_{j,t} = E_{j,t}\bar{y}_j$$

to obtain the equivalent system

$$F_{j,t}\mathbf{y}_{j,t} = \sum_i \left(F_{i,t}\frac{\bar{x}_i^j}{\bar{y}_i} + \bar{f}_i^j\right)\mathbf{x}_{i,t}^j.$$

In those cases where we know the base year volume data, we directly compute the ratio of x_i^j to $\bar{y}_i$. In all other cases, we compute the ratio from available expenditure data,

$$\frac{\bar{x}_i^j}{\bar{y}_i} = \frac{\bar{p}_i\bar{x}_i^j}{\bar{p}_i\bar{y}_i} = \frac{\bar{e}_i^j}{\bar{r}_i} \equiv \Phi_i^j,$$

where $\bar{p}_i$ is the base-year price of commodity i. The expenditure and revenue data for each industry, $\bar{e}_i^j$ and $\bar{r}_i$, respectively, are known from the base-year calibration data. In particular, Φ_i^j is the fraction of commodity i used by industry j to produce their output. If the volume and expenditure data are consistent, then the ratios computed from either method will be identical. For

Elliott et al.: CIM-EARTH: Framework and Case Study

a calibrated model where the markets clear in the base year, $\sum_j \Phi_i^j = 1$ for all commodities i. We then obtain the system of equations

$$F_{j,t}\mathbf{y}_{j,t} = \sum_i \left(F_{i,t}\Phi_i^j + \bar{f}_i^j \right) \mathbf{x}_{i,t}^j. \tag{3}$$

Working though the simple example above and assuming the model is calibrated with $\mathbf{y}=1$ and $\mathbf{x}=1$, we obtain the equivalent system

$$\begin{aligned}
F_{coal} &= 0 \\
F_{elec} &= F_{coal} + 10e_{coal} \\
F_{steel} &= 0.7 F_{elec} \\
F_{auto} &= 0.3 F_{elec} + F_{steel}.
\end{aligned}$$

The solution to this system is

$$\begin{aligned}
F_{coal} &= 0 \\
F_{elec} &= 10e_{coal} \\
F_{steel} &= 7e_{coal} \\
F_{auto} &= 10e_{coal}.
\end{aligned}$$

Since consumers demand all the automobiles produced and the model is calibrated, the embedded emissions in the automobiles demanded are $F_{auto} = 10e_{coal} = 10E_{auto}$, which is consistent with the earlier calculation.

We estimate the emissions content F for each industry for given Φ, $\bar{f}$, $\mathbf{x}$, and $\mathbf{y}$ by solving the linear system of equations (3). These amounts are then used to determine the carbon taxes on imports and refunds on exports for border tax adjustments. This system has more variables than equations because of the land, labor, and capital factors. In our computations, we ignore the emissions from these factors by fixing their embedded emissions to zero. We are then left with a square system of linear equations that can be solved.

When a commodity is imported, we tax the embedded carbon emissions. In terms of quantities, the tax collected by importer j on commodity i is

$$tE_i x_i^j,$$

where t is the tax rate per unit emissions. As before, we change variables so that all quantities are measured relative to a base year. In particular,

$$tE_i x_i^j = tE_i \bar{x}_i^j \mathbf{x}_i^j = tE_i \bar{y}_i \frac{\bar{x}_i^j}{\bar{y}_i} \mathbf{x}_i^j = tF_i \frac{\bar{p}_i \bar{x}_i^j}{\bar{p}_i \bar{y}_i} \mathbf{x}_i^j = tF_i \Phi_i^j \mathbf{x}_i^j.$$

We need compute only the total emissions in each industry, F_i; the emissions per unit commodity, E_i, are not necessary.

The B.E. Journal of Economic Analysis & Policy, Vol. 10 [2010], Iss. 2 (Symposium), Art. 11

4.2   Results for Policy Scenarios

The issue of carbon leakage has generated a significant literature and a variety of approaches to estimation have produced a wide range of leakage estimates. Babiker (2005), for example, uses the EPPA model to predict leakage in excess of 100% in one scenario based on an assumption of increasing returns to scale. There exist far fewer estimates of the effects of border tax adjustments. Babiker and Rutherford (2005) model the Kyoto Protocol and find substantial leakage and small effects from border tax adjustments.

We consider four policy scenarios in this study:

1. A reference scenario with no climate policy using the median baseline scenario described in Section 3.2 (REF).
2. A policy scenario using the median baseline scenario in which each Annex B country taxes carbon at 28.6 \$/t CO_2 (AB).
3. A policy scenario using the median baseline scenario in which each Annex B country taxes carbon and imposes a tariff on the estimated unpaid carbon content of imports from all non-Annex B countries at 28.6 \$/t CO_2 (AB-T).
4. A policy scenario using the median baseline scenario in which each Annex B country taxes carbon and assesses a border tax adjustment that taxes the estimated total carbon content of all imports and refunds the collected carbon taxes on all exports based on the total carbon content at 28.6 \$/t CO_2 (AB-BTA).

More policy scenarios can be found in Elliott et al. (2010). A carbon price of 28.6 \$/t CO_2 was chosen because it is close to the median value in proposed climate legislation. The two trade policy options AB-T and AB-BTA are examined for their effect on carbon leakage. For all policy scenarios, we solve the CGE model instance for the given year using the embedded emissions estimate from the previous year to determine the tariffs and border tax adjustments, since this approach mimics how the policies would be implemented. For the AB-T scenario where we compute only unpaid emissions content, we simply set $\bar{f}_i^j=0$ for the producers in the Annex B countries. We always use the full emissions data when computing the total emissions in the results.

To present results, we define a carbon flow matrix showing international emission flows as a result of international trade. We calculate the total emissions produced in each region from the fossil fuels burned, estimate the export emissions flows, and assign the remaining emissions to local consumption. We aggregate from 16 regions to 8 regions in the carbon flow tables for readability.

20

Elliott et al.: CIM-EARTH: Framework and Case Study

Table 4: **Fossil fuel CO_2 accounting in 2004 for the reference scenario.**

REF 2004	USA	EUR	RUS	JAZ	CAN	CHK	LAM	ROW	Prod.
		Annex B				Non Annex B			
USA	5.012	0.280	0.007	0.095	0.177	0.109	0.209	0.112	6.002
EUR	0.303	3.928	0.063	0.072	0.028	0.096	0.066	0.306	4.863
RUS	0.071	0.408	1.468	0.022	0.003	0.083	0.022	0.100	2.178
JAZ	0.084	0.082	0.003	1.146	0.008	0.160	0.012	0.098	1.593
CAN	0.248	0.033	0.001	0.009	0.223	0.012	0.008	0.010	0.543
CHK	0.577	0.587	0.032	0.390	0.050	3.679	0.103	0.478	5.897
LAM	0.293	0.122	0.006	0.018	0.016	0.036	0.956	0.040	1.487
ROW	0.300	0.657	0.031	0.289	0.020	0.376	0.055	3.199	4.928
Cons.	6.888	6.096	1.610	2.043	0.5260	4.5509	1.432	4.344	27.491

Note: All numbers in billions of tonnes (Gt) CO_2. The table shows carbon producers (or exporters) on the vertical and carbon consumers (or importers) on the horizontal. The diagonal gives domestic consumption. The right column labeled "Prod." gives the total carbon produced (emitted) in each region. The bottom row labeled "Cons." gives the total carbon consumed in each region (embedded in domestic goods and imports).

The countries in each aggregate region are shown in parenthesis in the first column of Table 1. Japan and Oceania, for example, are aggregated to JAZ.

The carbon flow matrix for the reference scenario in the 2004 base year is shown in Table 4. All numbers are reported in billions of tonnes (Gt) CO_2. The row sum in the right column of the table is the total emissions generated by producers in the region, while the column sum in the bottom row shows the total embedded emissions in the commodities demanded by the consumers in each region. The difference between the row sum and the column sum determines whether the region is a net importer or exporter of emissions. In particular, USA is a net importer of 0.885 Gt CO_2, while CHK is a net exporter of 1.347 Gt CO_2. The lower-right corner indicates global emissions of 27.491 Gt CO_2. For the 2004 base year, the global emissions are in good agreement with the emissions database produced by GTAP from the IEA energy database (Lee, 2009) and with the CDIAC National Fossil-Fuel CO_2 Emissions database Boden et al. (2009). The remainder of the matrix indicates the international emission flows. The diagonal value is the emissions generated in the given region that are not exported, while the off-diagonal values are the emissions embedded in imports and exports. Looking at the USA row, of the 6.002 Gt CO_2 produced, 5.012 Gt CO_2 are embedded in commodities demanded by the USA consumer, while 0.280 Gt CO_2 are exported by USA to EUR. From the USA column, of the 6.888 Gt CO_2 in commodities demanded by the USA consumer, 5.012 Gt CO_2 were generated domestically, while 0.577 Gt CO_2 are imported by USA from CHK.

The B.E. Journal of Economic Analysis & Policy, Vol. 10 [2010], Iss. 2 (Symposium), Art. 11

Table 5: **Percent change in emissions in 2020 for the AB policy scenario.**

AB vs. REF	Annex B					Non Annex B			
	USA	EUR	RUS	JAZ	CAN	CHK	LAM	ROW	Prod.
USA	**-27.2**	**-20.0**	-22.5	-27.0	-21.7	**-25.4**	**-30.0**	-29.6	**-26.8**
EUR	**-23.6**	**-23.3**	-19.6	-18.3	-17.7	-21.6	-23.4	**-28.2**	**-23.5**
RUS	**-38.0**	**-33.9**	**-29.4**	-34.6	-34.0	**-37.6**	-40.0	**-35.6**	**-31.5**
JAZ	-14.2	-14.4	-17.2	**-32.9**	-18.8	**-22.3**	-19.3	-25.0	**-28.8**
CAN	**-20.8**	-18.6	-16.2	-19.0	**-26.1**	-19.8	-20.1	-20.7	**-22.8**
CHK	1.1	1.8	2.0	3.0	2.0	**2.8**	2.3	1.3	**2.4**
LAM	**24.7**	14.0	47.7	4.3	25.9	3.0	**6.6**	5.4	**10.7**
ROW	8.0	**12.8**	18.5	**15.2**	8.4	**6.2**	9.6	**4.7**	**6.6**
Cons.	**-19.4**	**-15.1**	**-26.7**	**-15.6**	**-17.0**	0.3	-1.0	-0.1	**-9.9**

Note: This policy scenario has a constant 28.6 $/t CO_2 (USD per tonne carbon) tax levied in all Annex B countries starting in 2012, relative to the reference scenario. The largest gross changes ($|\Delta E| \geq 0.05$ Gt CO_2) are shown in bold, and the smallest ($|\Delta E| \leq 0.01$ Gt CO_2) are shown faded.

Table 5 shows the carbon flow matrix for the AB policy scenario with a carbon price of 28.6 $/t CO_2 (AB) relative to the reference scenario. The upper-left block of the matrix shows decreased trade among the Annex B regions relative to the reference scenario, while the lower-right block shows increased trade among the non-Annex B regions. The emissions embedded in exports from Annex B to non-Annex B countries shown in the upper-right block of the matrix decreases from a combination of the exporting nations switching to cleaner production processes and exporting less because the cleaner commodities are more expensive. Carbon leakage is indicated by the lower-left block of the matrix. In particular, imports to Annex B counties from non-Annex B countries increase since the commodities from non-Annex B countries are less expensive than the cleaner commodities produced domestically. As can be seen by comparing the change in total emissions produced by an Annex B region (the last column of the table) to the change in total consumed emissions (the last row in the table), the emissions from consumption for each Annex B region falls more slowly than their generation of emissions because of leakage. For example, USA reduces its generation of emissions in 2020 by 26.8% relative to the reference scenario with the AB policy, but only decreases its consumption of emissions by 19.4%. This leakage is the result of increased imports of dirty commodities from non-Annex B regions.

The addition of a tariff on embedded emissions in the AB-T policy scenario has a small, but not insubstantial effect on global emissions. Where the emissions are generated changes substantially from the AB policy scenario, as shown in Table 6. In particular, increased trade among the Annex B countries causes them to increase their emissions generated. For example, USA has a

Elliott et al.: CIM-EARTH: Framework and Case Study

Table 6: **Percent change in emissions in 2020 for the AB-T policy scenario.**

AB-T vs. REF	Annex B					Non Annex B			Prod.
	USA	EUR	RUS	JAZ	CAN	CHK	LAM	ROW	
USA	**-25.5**	**-18.1**	-20.7	-16.4	-19.2	**-35.8**	**-36.2**	**-37.7**	**-25.8**
EUR	-12.4	**-19.9**	-17.6	-14.9	-14.3	**-33.4**	-31.5	**-36.6**	**-21.0**
RUS	-27.8	**-30.0**	**-27.8**	-24.2	-30.9	**-49.9**	-51.6	**-45.1**	**-30.3**
JAZ	-12.8	-15.0	-17.0	**-25.8**	-18.1	**-35.5**	-28.0	-35.8	**-26.7**
CAN	-14.8	-18.6	-13.7	-16.7	**-22.3**	-32.1	-29.1	-30.7	**-19.8**
CHK	**-9.4**	**-10.5**	-12.1	**-11.1**	-11.9	**3.8**	13.6	**8.6**	**0.9**
LAM	-8.5	-4.0	10.6	-2.7	-4.6	3.6	**5.7**	7.3	**2.7**
ROW	-5.2	**-6.9**	-9.2	-8.1	-6.2	**8.0**	16.6	**4.6**	**2.8**
Cons.	**-21.1**	**-17.3**	**-26.5**	**-18.8**	**-18.6**	**0.5**	-0.8	0.2	**-10.7**

Note: A policy scenario with a constant 28.6 \$/t CO_2 (USD per tonne carbon) tax levied in all Annex B countries and on the unpaid emissions embedded in imports from non-Annex B countries starting in 2012, relative to the reference scenario.

Table 7: **Percent change in emissions in 2020 for the AB-BTA policy scenario.**

AB-BTA vs. REF	Annex B					Non Annex B			Prod.
	USA	EUR	RUS	JAZ	CAN	CHK	LAM	ROW	
USA	**-25.6**	**-19.0**	-21.4	-18.7	-20.4	**-23.9**	-16.9	-22.0	**-24.7**
EUR	-13.7	**-20.2**	-17.5	-15.9	-15.6	-25.8	-21.8	**-23.1**	**-20.1**
RUS	-31.8	**-32.6**	**-30.3**	-29.0	-33.5	-16.4	-18.9	-12.4	**-28.7**
JAZ	-13.4	-15.9	-16.5	**-26.3**	-18.8	**-25.8**	-20.4	-27.7	**-25.1**
CAN	-15.6	-19.1	-13.5	-18.2	**-23.1**	-24.7	-20.8	-22.1	**-19.6**
CHK	**-8.0**	**-8.9**	-10.0	**-9.9**	-10.3	**3.1**	9.4	**5.6**	**0.3**
LAM	-9.9	-2.7	15.5	-1.3	-2.8	-1.0	**4.0**	0.6	**0.5**
ROW	-4.1	-5.8	-6.4	-7.1	-4.5	1.8	7.9	**3.2**	**1.2**
Cons.	**-20.8**	**-17.4**	**-28.5**	**-18.4**	**-18.6**	**0.5**	0.2	0.5	**-10.8**

Note: A policy scenario with a constant 28.6 \$/t CO_2 (USD per tonne carbon) tax levied in all Annex B countries and on the total emissions embedded in all imports with refunds on all exports for the carbon taxes levied starting in 2012, relative to the reference scenario.

26.8% reduction in emissions produced in the AB scenario relative to the reference scenario, but only a 25.8% reduction in the AB-T scenario. Trade among the non-Annex B countries generally increases. However, the off-diagonal blocks show decreased trade between Annex B and non-Annex B countries, including further reductions to exports from Annex B to non-Annex B regions shown in the upper-right block of the matrix. The net result is a small reduction in global emissions. Moreover, we can see a narrowing of the change in total emissions produced by an Annex B region to the change in total consumed emissions compared to the AB policy scenario, indicating less carbon leakage.

The B.E. Journal of Economic Analysis & Policy, Vol. 10 [2010], Iss. 2 (Symposium), Art. 11

Results from adding border tax adjustments in the Annex B countries are shown in Table 7. The refunds on the collected carbon taxes on all exports from Annex B countries results in increased production and more exports to non-Annex B countries, but reduces the amount of trade between non-Annex B countries and hence their emissions. The result is a small reduction in total global emissions. The producers in Annex B countries are better off because exports to non-Annex B countries shown in the top-right block of the matrix are reduced less under this scenario than the AB-T policy scenario. For example, USA has a 35.8% reduction in exports to CHK relative to the reference scenario under the AB-T policy scenario, but only a 23.9% reduction with the AB-BTA policy scenario. We also see a narrowing of the change in total emissions produced by an Annex B region to the change in total consumed emissions compared to the AB-T policy scenario, indicating further reductions in carbon leakage.

4.3 Results for Ensemble of Baseline Scenarios

The forecasts generated by any model instance depend on both the policy scenario and the baseline assumptions used to construct the dynamic trajectories for labor productivity and energy intensity. We now study the sensitivity of the emissions forecasts from our model to the baseline assumptions. We are most interested in gross emissions since they are used to define policy targets in international agreements and are useful for measuring carbon leakage by comparing the overall reduction in emissions from Annex B countries to subsequent increases in emissions for non-Annex B countries.

To complete this analysis, we produced forecasts of emissions with our model instance using the parameters in each of the 25 baseline scenarios outlined in Section 3.2. Figure 9 compares the reduction in USA emissions (upper plots), the reduction in total emissions from Annex B countries (middle plots), and the total increase in emissions from non-Annex B countries (lower plots) in 2020 across the range of baseline scenarios for the AB policy scenario. Each point in the figure corresponds to a set of baseline assumptions. The point in the upper left plot with coordinates $(7.87, -2.11)$ corresponds to the median baseline scenario. The first coordinate, 7.87 Gt CO_2, is the gross emissions forecast for the reference policy. The second coordinate, -2.11 Gt CO_2, indicates the change in gross emissions for the AB policy scenario relative to the reference policy with the same baseline. The sum of the two coordinates, 5.76, is the gross emissions forecast for the AB policy scenario with the median baseline scenario. The right plots are similar except the second coordinate is the percent change in emissions relative to the reference policy.

24

Elliott et al.: CIM-EARTH: Framework and Case Study

Figure 9: **Policy implications of AB scenario.**

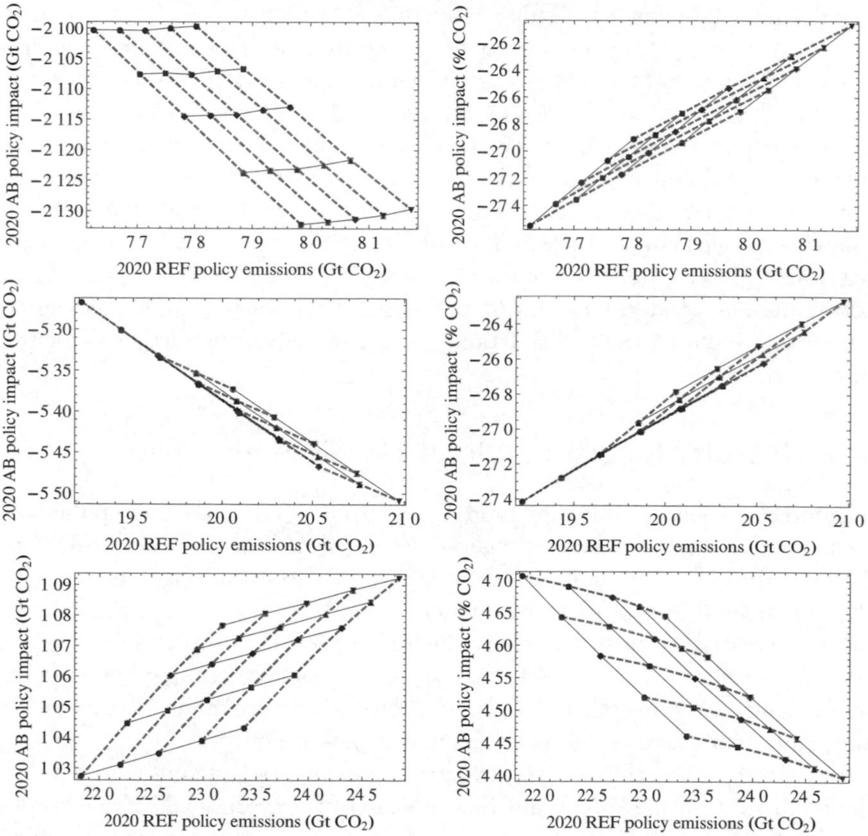

Note: Each point in the plots corresponds to a baseline scenario. The coordinates are the forecasts for the reference policy and the AB policy scenario. Along the horizontal, each baseline scenario is ranked by the gross CO_2 emissions in 2020 for the reference policy. On the vertical, we plot the impact of the AB policy scenario. The left plots measures gross emissions reductions, while the right plots measures percent reductions. The upper, middle, and lower plots are for the USA, Annex B counties, and non-Annex B countries, respectively. Points connected by solid black lines have a fixed value of the labor productivity parameter and varying energy intensity; points connected by red dashed lines have fix energy intensity.

The points are connected with solid black lines to indicate baseline scenarios with the same value of the labor productivity parameter and varying energy intensity. The points connected with red dashed lines have the same value of the energy intensity parameter and varying labor productivity. The top black line in the upper-left plot corresponds to baseline scenarios with the minimum value for the rate of change in labor productivity, while the bottom black line is the baseline scenarios with the maximum value. The left dashed line is the scenarios with the lowest rate of change in energy intensity, while the right dashed line is the highest. The slope of the lines indicates the sensitivity of the metric to different rates of change in energy intensity (black lines) or labor productivity (red dashed lines). Flat lines, such as the black lines found in the upper-left plot of the gross reduction in emissions for the AB policy scenario relative to the reference policy for the USA, indicate low sensitivity, while more vertical lines indicate higher sensitivity.

The relative impact of the policy on gross emissions in the Annex B countries appears to be more sensitive to the rate of change in energy intensity than to the rate of change in labor productivity. The USA is a notable exception, where gross emissions reductions are more sensitive to the rate of change in labor productivity than to the rate of change in energy intensity. Measured in percent reduction relative to the reference policy, the ensemble of baseline scenarios has a sensitivity range for gross emissions in the Annex B countries of about 1.3%. Emissions increases in the non-Annex B regions are more sensitive to changes in labor productivity than to energy intensity.

Overall, the median baseline scenario shows a decrease of 5.4 Gt CO_2 across the Annex B counties and an increase in emissions of 1.07 Gt CO_2 across the non-Annex B countries, implying almost 20% carbon leakage. Unsurprisingly, baseline scenarios with higher emissions lead to more carbon leakage, since Annex B regions are forced to pay more carbon taxes for production. In percent terms, the carbon leakage is less sensitive to different rates of change in labor productivity than to different rates of change in energy intensity.

In the USA, the ensemble of baseline scenarios shows a gross 2020 emissions range of 7.6–8.1 Gt CO_2 or about 6% of the median value, with a gross emissions reduction range of 2.10–2.13 Gt CO_2 or about 1.5% of the median value. In aggregate, the range across all Annex B countries is almost 10% of median gross emissions and the gross emissions reduction is around 4.5% of the median value. The results across the non-Annex B countries are similar with a range of almost 13% of the median value for gross emissions and a range of about 5.5% of the median value for gross emissions reductions. Given that these results are over a fairly compact ensemble of baseline scenarios when compared to the span of the EIA baseline forecasts over the past five

Elliott et al.: CIM-EARTH: Framework and Case Study

years shown in Figure 6, measuring policy impacts with high accuracy against historical targets, such as the USA emissions target under the nonbinding international agreement from the December 2009 Copenhagen meeting of a 17% reduction from 2005 emissions levels by 2020, would be difficult.

5   Conclusion

In this paper, we introduced the CIM-EARTH framework for specifying CGE models, detailed a model instance used to study the impact of carbon emission policies, and presented results for a handful of policies and the sensitivity of those results to baseline assumptions on the rate of change in both the labor productivity and energy intensity trajectories. Additionally, we described a method for measuring embedded carbon in commodities, used it to estimate carbon content in international trade flows, and introduced a matrix method to display international emissions flows to highlight the major mitigation impacts of climate policies.

Many extensions to the modeling framework are either under development or planned. In particular, we are planning to introduce fully dynamic CGE models and to augment the set of building blocks available to assemble model instances. These building blocks include capital and product vintages (Benhabib and Rustichini, 1991; Cadiou et al., 2003; Salo and Tahvonen, 2003) and overlapping consumer generations (Auerbach and Kotlikoff, 1987). We also plan to add private learning, research and development, and technology adoption (Boucekkine and Pommeret, 2004; Futagami and Iwaisako, 2007; Hritonenko, 2008; Zou, 2006) and to include household production functions, nonseparable utility functions, and heterogeneous beliefs.

Our framework allows us to systematically explore the sensitivity of the simulation results to a wide range of input uncertainties including the elasticities of substitution, the base-year expenditure and revenue data, and the assumptions used to construct the dynamic trajectories. To explore the sensitivity of policy-relevant metrics to the dynamic trajectories, we constructed an ensemble of baseline scenarios. Our results indicate that the gross reduction in emissions relative to the reference policy is more sensitive to energy intensity than to labor productivity across the Annex B countries. The exception is the USA, which is much more sensitive to labor productivity than to energy efficiency. The non-Annex B countries are more sensitive to labor productivity. Given the range of results across our relatively small ensemble of baseline scenarios, measuring policy impacts against historical targets, such as a 17% emissions reduction from 2005 levels by 2020 in the USA, with high accuracy may already be problematic.

The B.E. Journal of Economic Analysis & Policy, Vol. 10 [2010], Iss. 2 (Symposium), Art. 11

References

Auerbach, A. J. and L. J. Kotlikoff (1987): *Dynamic Fiscal Policy*, Cambridge, U.K.: Cambridge University Press.

Babiker, M. H. (2005): "Climate Change Policy, Market Structure, and Carbon Leakage," *Journal of International Economics*, 65, 421–445.

Babiker, M. H., J. Reilly, M. Mayer, R. S. Eckaus, I. S. Wing, and R. C. Hyman (2001): "The MIT Emissions Prediction and Policy Analysis (EPPA) Model: Revisions, Sensitivities, and Comparisons of Results," Technical Report 71, MIT Joint Program Report Series, URL http://globalchange.mit.edu/pubs/abstract.php?publication_id=643

Babiker, M. H. and T. F. Rutherford (2005): "The Economic Effects of Border Measures in Subglobal Climate Agreements," *The Energy Journal*, 26, 99–126.

Balay, S., W. D. Gropp, L. C. McInnes, and B. F. Smith (1997): "Efficient Management of Parallelism in Object Oriented Numerical Software Libraries," in E. Arge, A. M. Bruaset, and H. P. Langtangen, eds., *Modern Software Tools in Scientific Computing*, Birkhauser Press, 163–202.

Balistreri, E. J., C. A. McDaniel, and E. V. Wong (2003): "An Estimation of U.S. Industry-Level Capital-Labor Substitution," Computational Economics 0303001, EconWPA, URL http://ideas.repec.org/p/wpa/wuwpco/0303001.html

Ballard, C., D. Fullerton, J. B. Shoven, and J. Whalley (1985): *A General Equilibrium Model for Tax Policy Evaluation*, National Bureau of Economic Research Monograph, University of Chicago Press.

Benhabib, J. and A. Rustichini (1991): "Vintage Capital, Investment, and Growth," *Journal of Economic Theory*, 55, 323–339.

Benson, S., L. C. McInnes, J. Moré, T. Munson, and J. Sarich (2010): *Toolkit for Advanced Optimization (TAO) Web Page*, URL http://www.mcs.anl.gov/tao.

Bhattacharyya, S. C. (1996): "Applied General Equilibrium Models for Energy Studies: A Survey," *Energy Economics*, 18, 145–164.

Boden, T., G. Marland, and R. Andres (2009): *Global, Regional, and National Fossil-Fuel CO_2 Emissions*, Carbon Dioxide Information Analysis Center, Oak Ridge National Laboratory, U.S. Department of Energy, Oak Ridge, Tenn. U.S.A., URL http://cdiac.ornl.gov/trends/emis/tre_glob.html, doi 10.3334/CDIAC/00001_V2010.

Boehringer, C., T. F. Rutherford, and W. Wiegard (2003): "Computable General Equilibrium Analysis: Opening a Black Box," Discussion Paper 03-56, ZEW, URL ftp://ftp.zew.de/pub/zew-docs/dp/dp0356.pdf.

Elliott et al.: CIM-EARTH: Framework and Case Study

Boucekkine, R. and A. Pommeret (2004): "Energy Saving Technical Progress and Optimal Capital Stock: The Role of Embodiment," *Economic Modelling*, 21, 429–444.

Burniaux, J.-M. and T. Truong (2002): "GTAP-E: An Energy-Environmental Version of the GTAP Model," GTAP Technical Papers 923, Center for Global Trade Analysis, Department of Agricultural Economics, Purdue University, URL http://ideas.repec.org/p/gta/techpp/923.html.

Cadiou, L., S. Dées, and J.-P. Laffargue (2003): "A Computational General Equilibrium Model with Vintage Capital," *Journal of Economic Dynamics and Control*, 27, 1961–1991.

Conrad, K. (2001): "Computable General equilibrium Models in Environmental and Resource Economics," IVS discussion paper series 601, Institut für Volkswirtschaft und Statistik, University of Mannheim, URL http://ideas.repec.org/p/mea/ivswpa/601.html

Davis, S. J. and K. Caldeira (2010): "Consumption-based Accounting of CO_2 Emissions," *Proceedings of the National Academy of Sciences*, doi 10.1073/pnas.0906974107.

de la Chesnaye, F. C. and J. P. Weyant, eds. (2006): *Multi-Greenhouse Gas Mitigation and Climate Policy*, The Energy Journal, Special Issue.

de Melo, J. (1988): "CGE Models for the Analysis of Trade Policy in Developing Countries," Policy Research Working Paper Series 3, The World Bank, URL http://ideas.repec.org/p/wbk/wbrwps/3.html

del Negro, M. and F. Schorfheide (2003): "Take Your Model Bowling: Forecasting with General Equilibrium Models," *Economic Review*, 35–50.

Devarajan, S. and S. Robinson (2002): "The Influence of Computable General Equilibrium Models on Policy," TMD discussion papers 98, International Food Policy Research Institute (IFPRI), URL http://ideas.repec.org/p/fpr/tmddps/98.html

Dirkse, S. P. and M. C. Ferris (1995): "The PATH Solver: A Non-Monotone Stabilization Scheme for Mixed Complementarity Problems," *Optimization Methods and Software*, 5, 123–156.

Dowlatabadi, H. and M. G. Morgan (1993): "Integrated Assessment of Climate Change," *Science*, 259, 1813–1932.

Elliott, J., I. Foster, S. Kortum, T. Munson, F. Pérez Cervantes, and D. Weisbach (2010): "Trade and Carbon Taxes," *American Economic Review: Papers and Proceedings*, 100, 465–469.

Ferris, M. C. and T. Munson (1999): "Interfaces to PATH 3.0: Design, Implementation and Usage," *Computational Optimization and Applications*, 12, 207–227.

The B.E. Journal of Economic Analysis & Policy, Vol. 10 [2010], Iss. 2 (Symposium), Art. 11

Ferris, M. C. and T. Munson (2001): *GAMS/PATH User Guide: Version 4.6*, Department of Computer Sciences, University of Wisconsin, Madison, URL `ftp://ftp.cs.wisc.edu/math-prog/solvers/path/pathlib/doc/gams_user..ps`

Fourer, R., D. M. Gay, and B. W. Kernighan (2003): *AMPL: A Modeling Language for Mathematical Programming*, Pacific Grove, California: Brooks/Cole–Thomson Learning, second edition.

Fullerton, D., ed. (2009): *Distributional Effects of Environmental and Energy Policy*, Surrey, U.K.: Ashgate Publishing.

Fullerton, D. and D. L. Rogers (1993): *Who Bears the Lifetime Tax Burden?*, Washington, D.C.: Brookings Institution Press.

Futagami, K. and T. Iwaisako (2007): "Dynamic Analysis of Patent Policy in an Endogenous Growth Model," *Journal of Economic Theory*, 132, 306–334.

Ginsburgh, V. and M. Keyzer (1997): *The Structure of Applied General Equilibrium Models*, Cambridge, MA: The MIT Press.

Gopalakrishnan, B. N. and T. L. Walmsley, eds. (2008): *Global Trade, Assistance, and Production: The GTAP 7 Data Base*, Purdue University: Global Trade Analysis Center, Department of Agricultural Economics, URL `https://www.gtap.agecon.purdue.edu/databases/v7/v7_doco.asp`.

Hritonenko, N. (2008): "Modeling of Optimal Investment in Science and Technology," *Nonlinear Analysis: Hybrid Systems*, 2, 220–230.

International Labor Organization (2009): *Key Indicators of the Labour Market*, International Labour Organization Economic and Labour Market Analysis Department, URL `http://www.ilo.org/empelm/`.

International Labor Organization (2010): *Yearbook of Labour Statistics Database*, International Labor Organization Department of Statistics, URL `http://laborsta.ilo.org`.

Johansen, L. (1960): *A Multisectoral Study of Economic Growth*, Amsterdam: North Holland.

Lee, H.-L. (2009): *An Emissions Data Base for Integrated Assessment of Climate Change Policy Using GTAP*, URL `https://www.gtap.agecon.purdue.edu/resources/download/4470.pdf`

Lee, N. (2006): "Bridging the Gap Between Theory and Practice in Integrated Assessment," *Environmental Impact Assessment Review*, 26, 57–78.

Liu, J., C. Arndt, and T. Hertel (2004): "Parameter Estimation and Measures of Fit in A Global, General Equilibrium Model," *Journal of Economic Integration*, 19, 626–649.

Nakicenovic, N., J. Alcamo, G. Davis, B. de Vries, J. Fenhann, S. Gaffin, K. Gregory, A. Grübler, T. Jung, and T. K. et al. (2000): *Intergovernmental Panel on Climate Change Special Report on Emissions Scenarios*, Cambridge, U.K.: Cambridge University Press.

Robinson, S. (1991): "Macroeconomics, Financial Variables, and Computable General Equilibrium Models," *World Development*, 19, 1509–1525.

Salo, S. and O. Tahvonen (2003): "On the Economics of Forest Vintages," *Journal of Economic Dynamics and Control*, 27, 1411–1435.

Scarf, H. E. and J. B. Shoven (1984): *Applied General Equilibrium Analysis*, Cambridge, U.K.: Cambridge University Press.

Scrieciu, S. S. (2007): "The Inherent Dangers of Using Computable General Equilibrium Models as a Single Integrated Modelling Framework for Sustainability Impact Assessment: A Critical Note on Bohringer and Loschel (2006)," *Ecological Economics*, 60, 678–684.

Shoven, J. B. and J. Whalley (1984): "Applied General-Equilibrium Models of Taxation and International Trade: An Introduction and Survey," *Journal of Economic Literature*, 22, 1007–1051.

Sokolov, A. P., P. H. Stone, C. E. Forest, R. Prinn, M. C. Sarofim, M. Webster, S. Paltsev, C. A. Schlosser, D. Kicklighter, S. Dutkiewicz, J. Reilly, C. Wang, B. Felzer, J. M. Melillo, and H. D. Jacoby (2009): "Probabilistic Forecast for Twenty-First-Century Climate Based on Uncertainties in Emissions (Without Policy) and Climate Parameters," *Journal of Climate*, 22, 5175–5204.

Sue Wing, I. (2004): "Computable General Equilibrium Models and Their Use in Economy-Wide Policy Analysis," Technical Note 6, Joint Program on the Science and Policy of Global Change, URL http://globalchange.mit.edu/files/document/MITJPSPGC_TechNote6.pdf.

United Nations (2008): *World Population Prospects: The 2008 Revision Population Database*, United Nations Population Division, URL http://esa.un.org/unpp

United States Energy Information Agency (2009): *International Energy Outlook*, United States Energy Information Agency, report DOE/EIA-0484, URL http://www.eia.doe.gov/.

Vuuren, D., J. Lowe, E. Stehfest, L. Gohar, A. Hof, C. Hope, R. Warren, M. Meinshausen, and G.-K. Plattner (2009): "How Well do IAMs Model Climate Change?" *IOP Conference Series: Earth and Environmental Science*, 6, 2005.

The B.E. Journal of Economic Analysis & Policy, Vol. 10 [2010], Iss. 2 (Symposium), Art. 11

Webster, M., S. Paltsev, J. Parsons, J. Reilly, and H. Jacoby (2008): "Uncertainty in Greenhouse Emissions and Costs of Atmospheric Stabilization," Technical Report 165, MIT Joint Program Report Series, URL http://globalchange.mit.edu/files/document/MITJPSPGC_Rpt165.pdf.

Weyant, J. (2009): "A Perspective on Integrated Assessment," *Climatic Change*, 95, 317–323.

Weyant, J. P. (1999): *Energy and Environmental Policy Modeling*, Boston, MA: Kluwer Academic.

Zou, B. (2006): "Vintage Technology, Optimal Investment, and Technology Adoption," *Economic Modeling*, 23, 515–533.

[16]

The B.E. Journal of Economic Analysis & Policy

Symposium

Volume 10, Issue 2	2010	Article 12

DISTRIBUTIONAL ASPECTS OF ENERGY AND CLIMATE POLICY

Comment on "CIM-EARTH: Framework and Case Study"

Don Fullerton[*]

*University of Illinois at Urbana-Champaign, dfullert@illinois.edu

Recommended Citation
Don Fullerton (2010) "Comment on "CIM-EARTH: Framework and Case Study"," *The B.E. Journal of Economic Analysis & Policy*: Vol. 10: Iss. 2 (Symposium), Article 12.
Available at: http://www.bepress.com/bejeap/vol10/iss2/art12

Fullerton: Comment on "CIM-EARTH: Framework and Case Study"

Given the first fifteen years of my career spent working on various computable general equilibrium (CGE) models of the U.S. economy and tax system before turning to smaller models of environmental policy, it was with considerable *déjà vu* that I read the very interesting and comprehensive paper by Elliott, Foster, Judd, Moyer, and Munson (EFJMM, 2010). They see some gaps in the current crop of CGE climate policy models and have started a new one called CIM-EARTH with multiple inputs, outputs, regions, consumers, parameters, and simulations of alternative policies. One important goal is to make the programs available and understandable to others via open software. The paper reviewed here is a sort of progress report on the current status of the model. It is highly disaggregated but does not yet have dynamics over time with savings behavior based on forward-looking households.

The example used to demonstrate the model is the simulation of a tax equal to \$28.6 per ton of CO_2 in Annex B countries of the Kyoto Protocol. One variation has a tariff to reflect the carbon content of imports, and another variation has full border tax adjustments (BTA) with tariff on imports and rebate of tax upon export. For the median baseline scenario, they find that other nations increase their emissions by 20% of the amount that Annex B nations cut emissions. This rate of leakage is consistent with other literature.

The real contribution of the paper, however, is to introduce the new model and show how trade shifts around the carbon among 16 regions. They also devote careful attention to the problem of baseline sensitivity. Five different baseline rates of labor productivity growth and five different baseline rates of energy efficiency growth are used to generate 25 different possible baselines *without* a carbon tax. Each is used to see the effects of three policies. These 75 different simulations each generate matrices of results of all sorts, particularly the percent change in emissions in production from each region and the change in carbon consumption in each region. These matrices show the leakage from each Annex B nation to each other nation.

These authors come smack up against the standard problem of how to make use of a CGE model and how to interpret results. With so many model features, and so many matrices of results, how can we know which model features are the primary "drivers" of each result? Ultimately, we have no way to know for sure, though the authors do a heroic and helpful job of explanation.

This point warrants further discussion, because it is the major problem of CGE models, and the main advantage of using a small analytical model. If the model is simple enough to solve for expressions that show the result of interest as a function of all parameters and the exogenous tax change, then the reader can see exactly how the assumptions affect results. In a CGE model, readers cannot see how the assumptions affect results. They only get to see the effects of assumptions that the authors choose to vary. In this case, the authors choose to vary two assumptions about the baseline, to generate 25 different baselines, each with their own results. But we do not see the effects of thousands of *other* assumptions, such as alternative aggregations, elasticity parameters, functional forms, abatement technologies, imperfect competition, factor mobility, and a host of possible market imperfections *not* discussed in the paper.

I understand, because I've been there before. At one presentation of my CGE results a few years ago, a member of the audience stood up to say: "How can we trust any of your results? Since you are just making up the assumptions, you are essentially making up your results. Garbage in, garbage out!" I did not think that criticism was fair then, nor do I now. After all, every economic model uses assumptions to derive results! The problem here is that the process is more opaque. Many readers of this new literature will have that criticism, even if they don't voice it, so let me help suggest ways to avoid those pitfalls.

First of all, the goals in academia are completely different from the goals of policy discussions in government. They may want actual forecasts of future economic outcomes in much detail for every specific industry and every kind of person. They think the model should be realistic. In contrast, a good academic economist knows that realism is to be avoided! We want to strip the model of every complication that is not absolutely essential for the question of interest. Bigger is not better. It's worse, because it obscures what's important.

In 1991, Tyler Fox and I published a short note in *Economics Letters* called "The Irrelevance of Detail in a Computable General Equilibrium Model." In it, we reconsider the calculation of deadweight loss (*DWL*) from divergent capital tax rates in 18 sectors of a complicated CGE model of the U.S. economy. To create a simpler model, we remove the 12 consumer groups, the 18×18 input-output matrix, the conversion of 18 producer goods into 15 consumer goods, the industry for government enterprises, all other taxes and transfers, the entire foreign sector, all savings decisions, and the labor/leisure choice. With all those changes to the model, the measured annual *DWL* fell from \$10.912 to \$10.183 billion (in 1973\$). The point is to keep the detail that's relevant, and remove the rest. Consider the simple *partial* equilibrium formula of Harberger (1964):

$$DWL = 0.5 \sum_{i=1}^{N} \varepsilon t_i^2 X_i P_i$$

For years, we've already known that the *DWL* depends on the elasticities ε, the set of tax rates t_i, and the value of each sector ($X_i P_i$). The rest just doesn't matter. We vary the key elasticities and find *DWL* from \$6.7 billion to \$16.9 billion.

Similarly, EFJMM could remove a lot of detail and get essentially the same 20% leakage rate. We just need to consider which features of the model *are* important for deriving that result; but that's a job that needs to be done anyway! No results should be reported at all without knowing which features of the model

Fullerton: Comment on "CIM-EARTH: Framework and Case Study"

are driving the key results. For that reason, it may be good practice to use both analytical and numerical models, the former to see which assumptions are key, and the latter to put quantitative meat on the bones.

Moreover, results depend on assumptions *other* than the measures of elasticities, production shares, tax rates, and other features in the model; they depend on choices about features *not* in the model, choices not stated! This model assumes consumers maximize utility subject to a budget, and firms maximize profits subject to a production function; results would differ under other behaviors, in imperfect competition, or with other forms of mobility.

Here is an example that applies to leakage, which this paper finds to be +20%. This model also has a fixed amount of capital in each region, and it has a particular form of nesting in production. The nature of the model makes leakage necessarily positive, so the goal here is just to measure its size. In contrast, Fullerton and Karney (2010) find that leakage can be negative, but the mechanism requires two model features not present here: international capital mobility, and firms that can substitute out of carbon dioxide emissions into more use of abatement capital. The reason is that the firm taxed on its carbon emissions then demands more capital for abatement and thereby draws capital away from the other country, which shrinks production and emissions in the other country.

The authors (EFJMM) are fully aware of these difficulties. They are working on simpler toy models that better illustrate the key economic relationships driving results. And they have other long working papers with both analytical and computational results for comparison. My purpose here is just to help readers understand what they are up against!

References

Elliott, Joshua, Ian Foster, Kenneth Judd, Elisabeth Moyer, and Todd Munson (2010), "CIM-EARTH: Framework and Case Study", *The B.E. Journal of Economic Analysis & Policy*, 10(2).

Fox, Tyler, and Don Fullerton (1991), "The Irrelevance of Detail in a Computable General Equilibrium Model", *Economics Letters*, 36, 67-70.

Fullerton, Don, and Dan Karney (2010), "Negative Leakage," working paper, University of Illinois at Urbana-Champaign.

Harberger, Arnold C. (1964), "The Measurement of Waste," *American Economic Review*, 54, May, 58-76.

[17]

The B.E. Journal of Economic Analysis & Policy

Symposium

Volume 10, Issue 2	2010	Article 13

DISTRIBUTIONAL ASPECTS OF ENERGY AND CLIMATE POLICY

The Global Effects of Subglobal Climate Policies

Christoph Boehringer*   Carolyn Fischer[†]

Knut Einar Rosendahl[‡]

*Oldenburg University, christoph.boehringer@uni-oldenburg.de

[†]Resources for the Future, fischer@rff.org

[‡]Statistics Norway, knut.einar.rosendahl@ssb.no

Recommended Citation

Christoph Boehringer, Carolyn Fischer, and Knut Einar Rosendahl (2010) "The Global Effects of Subglobal Climate Policies," *The B.E. Journal of Economic Analysis & Policy*: Vol. 10: Iss. 2 (Symposium), Article 13.

Available at: http://www.bepress.com/bejeap/vol10/iss2/art13

The Global Effects of Subglobal Climate Policies*

Christoph Boehringer, Carolyn Fischer, and Knut Einar Rosendahl

Abstract

Individual countries are in the process of legislating responses to the challenges posed by climate change. The prospect of rising carbon prices raises concerns in these nations about the effects on the competitiveness of their own energy-intensive industries and the potential for carbon leakage, particularly leakage to emerging economies that lack comparable regulation. In response, certain developed countries are proposing controversial trade-related measures and allowance allocation designs to complement their climate policies. Missing from much of the debate on trade-related measures is a broader understanding of how climate policies implemented unilaterally (or subglobally) affect all countries in the global trading system. Arguably, the largest impacts are from the targeted carbon pricing itself, which generates macroeconomic effects, terms-of-trade changes, and shifts in global energy demand and prices; it also changes the relative prices of certain energy-intensive goods. This paper studies how climate policies implemented in certain major economies (the European Union and the United States) affect the global distribution of economic and environmental outcomes and how these outcomes may be altered by complementary policies aimed at addressing carbon leakage.

KEYWORDS: cap-and-trade, emissions leakage, border carbon adjustments, output-based allocation, general equilibrium model

---

*Carolyn Fischer (corresponding author) is a senior fellow at Resources for the Future, 1616 P Street, Washington, D.C. 20036; email: fischer@rff.org. Christoph Böhringer is a professor and holds the Chair of Economic Policy in the Department of Economics at the University of Oldenburg. Knut Einar Rosendahl is a senior research fellow at Statistics Norway in Oslo. The authors are indebted to Rod Ludema, Don Fullerton, and an anonymous referee for helpful comments. Support from the Research Council of Norway Clean Energy for the Future (RENERGI) program, the German Research Foundation (BO 1713/5-1), and the Mistra Foundation's Environment and Trade in a World of Interdependence (ENTWINED) program is gratefully acknowledged.

Individual countries, particularly members of the Organisation for Economic Co-operation and Development (OECD), are in the process of legislating responses to the challenges posed by climate change. The prospect of rising carbon prices raises concerns in these nations about the effects on the competitiveness of their own energy-intensive industries and the potential for carbon leakage, which is conventionally defined as the change in foreign emissions as a share of the domestic emissions reductions. Of particular concern is leakage to emerging economies that lack comparable regulation. In response, certain OECD nations are proposing trade-related measures to complement their climate policies. However, these measures are controversial. Some analysts believe they may harm industries in developing countries while minimally mitigating total global carbon emissions. Others have been more acute, stating that these trade policy measures are disguised restrictions to trade, intended primarily to protect the competitiveness of domestic industries in OECD countries rather than the integrity of emissions reductions.

Missing from much of the debate on trade-related measures is a broader understanding of how climate policies implemented unilaterally (or subglobally) affect all countries in the global trading system. Arguably, the largest consequences are from the targeted carbon pricing itself, which generates macroeconomic effects, terms-of-trade changes, and shifts in global energy demand and prices; it also changes the relative prices of certain energy-intensive goods. And in addition to trade-related measures, other climate policy design options can affect the distribution of outcomes around the world. Using a computable general equilibrium (CGE) model of global trade and emissions, we examine how climate policies implemented in two major economies, the European Union and the United States, affect the global distribution of economic and environmental outcomes, and how these outcomes may be altered by complementary policies aimed at addressing carbon leakage.

Background

Competitiveness and emissions leakage issues have been at the fore of climate policy discussions in all the major economies implementing or proposing to implement significant emissions cap-and-trade programs, including the United States, the European Union, Australia, and New Zealand. The European Union has so far used preferential allocation of grandfathered allowances to energy-intensive manufacturing to allay concerns about losing profits to foreign competitors. The U.S., Australian, and New Zealand proposals employ another form of free allocation—output-based rebating—to offset most of the carbon cost increases to their energy-intensive, trade-exposed sectors.

The B.E. Journal of Economic Analysis & Policy, Vol. 10 [2010], Iss. 2 (Symposium), Art. 13

An important feature of output-based rebating (OBR) is that unlike with ordinary grandfathering, the allocations are updated based on recent measures of economic activity, namely production. Additional production then garners additional allowances, the value of which functions as a subsidy to production. The approach is similar to using tradable performance standards, in that above-average (or above-standard) emitters face a net liability, while below-average emitters get a net subsidy. Although grandfathered allowances are an unconditional transfer that does not affect operating costs, OBR lowers marginal costs, which are stronger determinants of competitiveness than fixed costs, at least in the short run. The U.S. 2009 American Clean Energy and Security Act (H.R. 2454, or ACESA) proposes that the per unit allocation for eligible sectors be 100 percent of average emissions (both direct and indirect), up to a maximum of 15 percent of the total cap. Australia's Carbon Pollution Reduction Scheme would offer the most energy-intensive activities allowances equal to 60 to 90 percent of historical average emissions (direct and indirect), phasing down gradually over time. The European Union plans a similar benchmarking exercise for its next period, but the results will be used to determine grandfathered allowances; still, the fact that allowances will be granted to new entrants and forfeited for significant reductions in capacity means that the allowances are not truly unconditional and in the long run may have properties somewhat similar to OBR.

OBR keeps the playing field level by keeping domestic costs from rising. Thus, while the emissions price ensures that producers still have economic incentives to reduce their emissions intensity, the subsidy discourages them from reducing emissions by decreasing production (Böhringer and Lange 2005; Fischer 2001). This latter effect creates an important tradeoff: without imposing higher prices on "dirty" products that reflect their full carbon footprint, downstream consumers have less incentive to use alternative products or conservation measures to reduce emissions. Although eligible sectors may benefit and leakage may be reduced, those forgone domestic reductions must be made up elsewhere, driving up the emissions price and overall costs to meet the national cap (Fischer and Fox 2007). Therefore, OBR might be justified on efficiency grounds only as a second-best unilateral strategy when leakage rates are substantial (Böhringer et al. 1998a, 1998b).

An alternative policy measure is border carbon adjustments (BCA), which require importers to purchase emissions allowances in proportion to the emissions embodied in the foreign production of the goods. This method levels part of the playing field by bringing the price of imported goods up to the level of those at home, retaining the incentives for consumers to find and innovate low-carbon alternatives. The other side of border adjustment would give rebates to exported goods, based on a benchmark of their emissions intensity; the rebates keep domestic goods competitive on world markets, without offering the subsidy at home. Full border adjustment would combine adjustments for imports and exports, effectively implementing destination-based carbon pricing. However, most policy proposals focus only on import adjustments.

ACESA proposes to transition from OBR to import adjustments for eligible sectors starting in 2020. The legislation exempts imports from countries that are undertaking comparable steps to mitigate greenhouse gas emissions, as well as parties to multilateral climate agreements or sectoral agreements with the United States; least developed nations are also exempt. The idea of border adjustment of carbon pricing has advocates in Europe (e.g., Godard 2007; Grubb and Neuhoff 2006). Although the European Union has no specific plans to use BCA, it is retaining that option for the future, if insufficient international cooperation emerges.

Border adjustments have their own controversies, however. In theory, trade measures against carbon leakage could help support a new multilateral climate agreement (Karp and Zhao 2008). But some analysts voice apprehension that unilateral trade measures could poison future climate negotiations (Houser et al. 2008) or trade relations (ICTSD 2008). For example, the U.S. Trade Representative has vowed to resist any EU attempt to impose climate-linked tariffs on U.S. products. In addition to the political questions, many scholars have asked whether and how border adjustments, particularly unilateral ones, might be compatible with World Trade Organization (WTO) obligations.[1]

Free allocations may have their own potential conflicts with WTO obligations in the Agreement on Subsidies and Countervailing Measures (Charnovitz et al. 2009). Although they can also distort trade, they do not seem to raise the hackles of the trade community in the same way as do import adjustments, perhaps in part because member countries recognize that to implement any serious climate regulation, such allocations will be necessary to gain domestic political acceptance. Thus, these leakage-oriented policies cannot really be evaluated apart from the climate policy which with they are paired. Indeed, in terms of economic impact, domestic carbon cap-and-trade programs are likely to have much stronger effects on trade partners than would countervailing policies.

---

[1] See, for example, Charnovitz et al. 2009; Pauwelyn 2007; Bagwati and Mavroidis (2007), de Cendra (2006), and Kommerskollegium 2004; and a summary in Fischer and Fox 2009.

The B.E. Journal of Economic Analysis & Policy, Vol. 10 [2010], Iss. 2 (Symposium), Art. 13

Literature

Several recent studies have analyzed the effects of unilateral or subglobal climate policies in combination with trade measures and allocation schemes. Some of these studies focus just on specific energy-intensive, trade-exposed sectors, like copper, steel, or cement.[2] Although these partial equilibrium studies ignore important global, general equilibrium effects, they provide some useful insights for interpreting results from general equilibrium approaches. Fischer and Fox (2009) compare different border adjustment options with output-based rebating. They find that while all such policies improve domestic competitiveness for a given sector, none necessarily reduce global emissions, since some emissions are being repatriated along with output. The results depend on the relative elasticities of substitution, as well as relative emissions rates between home and foreign goods. They also note important general equilibrium effects of the climate policies themselves, driven by global energy price changes as well as relative price changes for manufacturing: countervailing policies merely affect the latter, but the full extent of emissions leakage is much more sensitive to the former.

Multicountry, multisector computable general equilibrium (CGE) models are typically used to study the global effects of climate policy, and their specifications can have important implications for leakage and policy outcomes. Many of these studies concentrate on the economic effects on the implementing parties, as well as the implications for global emissions. Peterson and Schleich (2007) investigate border carbon adjustment options for Annex B (industrialized) countries, concentrating on the calculation of the carbon content for imports, which affects the stringency of the border adjustment, and on the breadth of their application across sectors. They find that border adjustment increases the welfare losses for unilaterally abating regions, in part by driving up carbon prices and shifting burdens to less intensively traded sectors. Fischer and Fox (2007, 2009) compare designs for domestic rebate (output-based allocation) programs within a unilateral U.S. climate policy. Their model also considers interactions with labor tax distortions, and they show that output-based rebating (designed appropriately) can generate lower leakage and higher welfare than grandfathering and, in some circumstances, even auctioning.

Burniaux and Martins (2000) show that average (economy-wide) leakage is highly sensitive to the parameterization of fossil fuel supply curves. Average leakage rates in various CGE studies range from 10 to 30 percent (Babiker and

---

2 See, for example, Gielen and Moriguchi 2002; Demailly and Quirion 2006, 2008a; Ponssard and Walker 2008; Fischer and Fox 2009.

Rutherford 2005), although some models report leakage rates below 10 percent for a coalition of Annex I countries, those with reduction commitments under the Kyoto Protocol (Burniaux et al. 2009; Mattoo et al. 2009) and other models find rates above 100 percent for oligopolistic market structures with increasing returns to scale (Babiker 2005). For individual sectors, however, calculated leakage rates can be much higher than the average leakage rates (Paltsev 2001; Fischer and Fox 2009; Ho et al. 2008).

Böhringer et al. (1998a) show that leakage rates are also highly sensitive to the specification of international trade in the CGE model, with important implications for the efficiency effects of output-based allocation of emissions allowances. If products of the same variety produced in different regions are traded as homogeneous goods, leakage rates are rather high and a policy of output-based allocation turns out to be pareto-superior to auctioned permits (or likewise uniform emissions taxes). If, however, these traded goods are treated as qualitatively different, leakage rates are rather low and the better unilateral climate policy applies auctioned permits rather than output-based allocation.

Babiker and Rutherford (2005) consider a coalition of Kyoto ratifiers pursuing their emissions targets; the reference scenario, where coalition members implement Kyoto with no border adjustment, is compared with scenarios with such adjustment measures as import tariffs, export rebates, exemption of energy-intensive industries, and voluntary export restraints by noncoalition countries. They find that most coalition members are better off with tariffs rather than rebates for mitigating their own welfare losses. Exemptions are the most costly to the coalition members but the most effective at reducing carbon leakage. Major noncoalition members, like China, India, and Brazil, are found also to benefit from the adjustment policies, with the exception of the import tariff policy.

Mattoo et al. (2009) also look at the effects of border carbon adjustment options implemented by a coalition of industrialized countries. In their analysis, Mattoo et al. find that import taxes confer the largest welfare losses on lower- and middle-income countries, particularly when imposing countries fully adjust for emissions intensities in the country of origin. Mattoo et al. argue that border adjustments based on domestic or best-available technology emissions rates are able to offset most of the competitiveness impacts with less detrimental effects on developing countries. They also downplay the leakage effects but note that these are sensitive to the major parameter assumptions.

The purpose of this paper is to explore more deeply the effects of unilateral climate policies in the United States and the European Union on welfare, competitiveness, and carbon leakage in different parts of the world. We consider climate policies that differ with respect to their treatment of the energy-intensive sectors (EIS) that both are sensitive to climate policies and have significant international trade volumes.

The B.E. Journal of Economic Analysis & Policy, Vol. 10 [2010], Iss. 2 (Symposium), Art. 13

The policies we consider are (i) full auctioning of all allowances, (ii) output-based allocation to EIS, (iii) import tariffs for EIS goods based on their embodied carbon (using direct emissions in export country), (iv) export rebate to EIS sectors, and (v) full border adjustments—that is, a combination of (iii) and (iv).

Model structure and parameterization

To quantify the economic and emissions effects of unilateral carbon abatement strategies, we build on a generic multiregion, multisector CGE model of global trade and energy use established by Böhringer and Rutherford for the economy-wide analysis of climate policy issues (see Böhringer and Rutherford 2010 for a recent application of the static model versions and its detailed algebraic description). A multiregion setting is essential for analyzing the economic consequences of climate policy regimes: in a world that is increasingly integrated through trade, policy interference in larger, open economies not only causes adjustment of domestic production and consumption patterns but also influences international prices via changes in exports and imports. The changes in international prices—that is, the terms-of-trade—imply secondary effects that can significantly alter the effects of the primary domestic policy. In addition to the consistent representation of trade links, a detailed tracking of energy flows as the main source for emissions of carbon dioxide (CO_2) is a prerequisite for the assessment of climate policies.

The static CGE model used for our numerical analysis features a representative agent in each region that receives income from three primary factors: labor, capital, and fossil fuel resources (coal, gas, and crude oil). Labor and capital are intersectorally mobile within a region but immobile between regions. Fossil fuel resources are specific to fossil fuel production sectors in each region. Production of commodities other than primary fossil fuels is captured by three-level constant elasticity of substitution (CES) cost functions describing the price-dependent use of capital, labor, energy, and material in production. At the top level, a CES composite of intermediate material demands trades off with an aggregate of energy, capital, and labor subject to a constant elasticity of substitution. At the second level, a CES function describes the substitution possibilities between intermediate demand for the energy aggregate and a value-added composite of labor and capital. At the third level, capital and labor substitution possibilities within the value-added composite are captured by a CES function, whereas different energy inputs (coal, gas, oil, and electricity) enter the energy composite subject to a constant elasticity of substitution. In the production of fossil fuels, all inputs, except for the sector-specific fossil fuel resource, are aggregated in fixed proportions. This aggregate trades off with the sector-specific

6

fossil fuel resource at a constant elasticity of substitution. The latter is calibrated to be consistent with supply elasticities of 1 for crude oil and natural gas and 4 for coal. These elasticities are in line with other studies (e.g., Aune et al. 2008) and reflect differences in the market structure and production flexibility across the fuels.

Final consumption demand in each region is determined by the representative agent who maximizes utility subject to a budget constraint with fixed investment (i.e., given demand for the savings good) and exogenous government provision of public goods and services. Total income of the representative household consists of net factor income and tax revenues. Consumption demand of the representative agent is given as a CES composite that combines consumption of nonelectric energy and a composite of other consumption goods. Substitution patterns within the nonelectric energy bundle are reflected by means of a CES function; other consumption goods trade off with each other at a unitary elasticity of substitution (i.e., a Cobb-Douglas relationship).

Bilateral trade is specified following the Armington (1969) approach of product heterogeneity—that is, domestic and foreign goods are distinguished by origin.[3] All goods used on the domestic market in intermediate and final demand correspond to a CES composite that combines the domestically produced good and the imported good from other regions differentiated by demand category *g*. That is, the composition of the Armington good differs across sectors and final demand components. Domestic production is split between input to the formation of the Armington good and export to other regions subject to a constant elasticity of transformation. The balance-of-payment constraint, which is warranted through flexible exchange rates, incorporates the base-year trade deficit or surplus for each region.

Figure 1 provides tree diagrams for the nesting structure in production and consumption underlying the actual model specification (CES stands for constant elasticity of substitution, and CET stands for constant elasticity of transformation).

---

[3] The only exception is crude oil, where we assume product homogeneity.

The B.E. Journal of Economic Analysis & Policy, Vol. 10 [2010], Iss. 2 (Symposium), Art. 13

Figure 1. Nesting structure in production and consumption

1a. Production (except for fossil fuels)

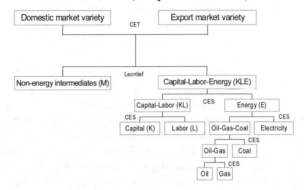

1b. Fossil fuel production

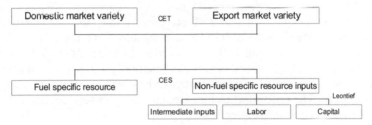

1c. Armington good production

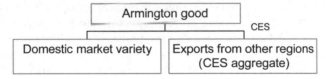

1d. Final consumption

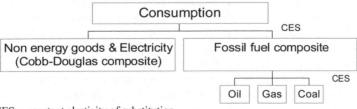

CES = constant elasticity of substitution
CET = constant elasticity of transformation

The model builds on the most recent Global Trade Analysis Project (GTAP) data set with detailed accounts of regional production, regional consumption, and bilateral trade flows as well as energy flows and CO_2 emissions, all for the base year 2004 (Badri and Walmsley 2008). As is customary in applied general equilibrium analysis, base year data together with exogenous elasticities determine the free parameters of the functional forms. Elasticities in international trade are based on empirical estimates reported in the GTAP database.

As to sectoral and regional model resolution, the GTAP database is aggregated toward a composite data set that accounts for the specific requirements of international climate policy analysis. At the sectoral level, the model captures details on sector-specific differences in factor intensities, degrees of factor substitutability, and price elasticities of output demand to trace the structural change in production induced by policy interference. The energy goods identified in the model are coal, crude oil, natural gas, refined oil products, and electricity. This disaggregation is essential for distinguishing energy goods by CO_2 intensity and the degree of substitutability.

The model then incorporates CO_2-intensive (energy-intensive) commodities with significant shares of international trade that are potentially most affected by unilateral climate policies and therefore are considered for border adjustment measures: chemicals (CRP); nonmetallic minerals (NMM), a category that includes cement, glass, and ceramic production; pulp, paper, and print (PPP); iron and steel (I_S); and nonferrous metals (NFM), the category including copper and aluminium. The remaining sectors include transport services and a composite of all other industries and services.

With respect to the regional disaggregation, the model covers the industrialized and developing regions that are major players in international climate negotiations and at the same time intertwined through bilateral trade links: the United States, the European Union, Canada, Japan, Russia, China, India, Brazil, Mexico, and Australia–New Zealand (Aust.NZ). The model also encompasses the Organization of Oil Exporting Countries (OPEC) and aggregate regions for other Asia (Oth.ASIA), other America (Oth.AMER), other Africa (Oth.AFR), other Europe (Oth.EUR), and other former Soviet Union (Oth.FSU).

The calibration is a deterministic procedure and does not allow for a statistical test of the model specification, other than the replication of the initial benchmark. The policy simulations compute counterfactual equilibria to provide information on the policy-induced changes on major economic variables. Recognizing the reliance on exogenous elasticity values and a single base-year observation, as well as the much greater complexity of actual policies, one should not interpret the results as forecasts of outcomes, but rather use them for understanding essential trade-offs, directions of impact, and general rather than precise measures of potential magnitude. Further sensitivity analysis can enhance confidence in policy recommendations.[4] For example, Burniaux and Martins (2000), as well as Babiker and Rutherford (2005), have conducted extensive multidimensional sensitivity analysis in general equilibrium models of carbon leakage, finding that the results are particularly sensitive to assumptions about fossil fuel supply elasticities, but manufactured product differentiation and capital mobility are less influential. In our model, we find that our qualitative conclusions remain robust to a variety of specifications for the fossil fuel supplies, although the magnitudes of the effects do vary.

Policy scenarios

Our reference scenario is one without climate policies, the historical outcome of the base year of the model, 2004. Note that this was before the EU Emission Trading Scheme (ETS) was implemented, and before the Kyoto Protocol entered into force. Thus, climate policies are almost absent internationally, and importantly, no cap exists on emissions in Annex B countries.

The policy scenarios are presented in Table 1. Scenarios 1A–1E assume that the United States unilaterally reduces its domestic CO_2 emissions. Common to all five scenarios is that all sectors of the economy face the same price on CO_2 emissions through an economy-wide cap-and-trade system. This price will, however, differ across the five scenarios.

In Scenario 1A ("AUCTION"), U.S. emissions are reduced by 20 percent compared with the base-year level. Furthermore, all quotas are auctioned off, and no other policies are implemented. As we do not consider revenue recycling, this scenario is equivalent to grandfathered permits, citizen dividends, or the like.

Scenarios 1B–1E all have special treatments for energy-intensive sectors or industries (EIS) that are considered to be significantly trade exposed (the aforementioned sectors of iron and steel; chemical products; nonmetallic mineral products; paper, pulp, and print; and nonferrous metals). To abstract from the need to consider environmental benefits of climate policies, we construct Scenarios 1B–1E to have the same global emissions reduction as in Scenario 1A, which turns out to be 4 percent (see Table 1). Because the special treatment of energy-intensive sectors will tend to reduce carbon leakage (i.e., increase emissions abroad), the emissions reduction in the United States will be slightly lower than 20 percent in Scenarios 1B–1E (cf. Table 2).

Scenario 1B ("OUTPUT") represents the combination of the economy-wide cap with output-based rebating to the EIS. It assumes that quotas are

---

[4] See Böhringer et al. (2003).

allocated free of charge to these industries in proportion to their production level. The allocation rate is adjusted such that the total allocation to energy-intensive industries equals their share of base-year emissions, adjusted for the reduction in total emissions (defined by the cap).

Scenarios 1C–1E assume that the cap-and-trade system is complemented by different border adjustment policies directed toward energy-intensive industries. In Scenario 1C ("REBATE"), export of EIS goods is rebated in proportion to export levels such that the total value of rebates equals the total costs of CO_2 quotas needed to produce these exported goods. In Scenario 1D ("TARIFF"), a tariff is imposed on imports of energy-intensive goods. The tariff is set equal to the embodied carbon in the EIS good[5] times the price of carbon in the domestic cap-and-trade system. The tariff is differentiated across regions such that regions with relatively high emissions in producing an EIS good face a relatively high tariff. Scenario 1E ("BTAX") combines the export rebate and the import tariff—that is, a combination of Scenarios 1C and 1D.

Scenarios 2A–2E are similar to Scenarios 1A–1E except that the European Union instead of the United States reduces domestic CO_2 emissions (by 20 percent in Scenario 2A). Note that the global emissions reduction will be lower in these scenarios than in Scenarios 1A–1E because EU emissions are considerably below U.S. emissions in the reference scenario.[6] Note also that the economy-wide cap differs from the actual EU ETS, which covers only energy and energy-intensive manufacturing sectors.

Scenario 3A assumes that both the United States and the European Union impose a 20 percent reduction in their CO_2 emissions, whereas Scenarios 3B–3E have the same global emissions as Scenario 3A, achieved by ratcheting the U.S. and EU targets proportionately. Obviously, global emissions reductions are bigger than in Scenarios 1A–1E and 2A–2E. We do not have international emissions trading in these scenarios, so carbon prices differ between the United States and the European Union. In the border adjustment cases, we assume that the export rebate is not given for exports from the United States to the European Union (or vice versa), and similarly, import tariffs are not imposed on imports from the European Union and the United States.

---

[5] We only consider the direct emissions embodied in the EIS good. For example, indirect emissions from using electricity produced by coal power are not included here. The tariff we consider can of course give incentives to switch from using fossil fuels to using electricity, without necessarily reducing total emissions from producing these goods.

[6] As a consequence, the environmental benefits of Scenarios 2A–2E will be lower than in Scenarios 1A–1E, which should be kept in mind when comparing the effects of U.S. and EU policies.

The B.E. Journal of Economic Analysis & Policy, Vol. 10 [2010], Iss. 2 (Symposium), Art. 13

Table 1 summarizes the scenarios. All scenarios in the first column ("A") assume a 20 percent cut in domestic CO_2 emissions (U.S. and/or EU), and all scenarios in the same row have identical cuts in global CO_2-emissions.

Table 1. Policy scenarios

	A (Auction)	B (Output)	C (Rebate)	D (Tariff)	E (Btax)	Global emissions reduction
1	U.S.	U.S.	U.S.	U.S.	U.S.	4.0%
2	EU	EU	EU	EU	EU	2.4%
3	U.S. and EU	U.S. and EU	U.S. and EU	U.S. and EU	U.S. and EU	6.6%

Simulation results

We first discuss the economic welfare effects at home and abroad of the primary policies of reducing domestic CO_2 emissions in the United States and/or in the European Union. Economic welfare impacts are reported as Hicksian equivalent variation in income, which denotes the amount necessary to add to (or subtract from) the benchmark income of the representative consumer so that she enjoys a utility level equal to the one in the counterfactual policy scenario on the basis of *ex ante* relative prices. In our analysis we do not attempt to measure the economic benefits of emissions reductions. Because we omit environmental benefits, to avoid confusion, we subsequently use the term *consumption* instead of *economic welfare*. To put it differently, while keeping global emissions reductions constant within each of our three scenario categories, we provide a consistent cost-effectiveness analysis across policy variants A–E. Subsequently, we compare the effects of combining the emissions caps with countervailing policies and discuss the implications for carbon leakage and competitiveness.

Figure 2 presents the consumption effects, by region, of the three auctioned emissions cap scenarios: the United States alone (1A), the European Union alone (2A), and both economies (3A). We first notice that the costs of these targets are substantially higher in the European Union than in the United States. This result may emerge from two important differences. First, the United States has cheaper abatement options than the European Union, with respect to both energy efficiency improvements and fuel switching in the electricity sector. The resulting carbon prices are \$14.70/ton CO_2 for the U.S. unilateral auction policy, compared with \$33.30/ton for the European Union. Both prices rise slightly (to \$15.40 and \$34.10, respectively) when the cap-and-trade programs are jointly implemented, since emissions reductions in either of the two regions will tend to increase emissions in the other region (cf. the discussion of carbon leakage below). Second, the countries have different trade intensities and terms-of-trade effects. For example, we also see that the consumption reductions in the European Union are lower when the United States also cuts its emissions (3A vs. 2A), whereas the United States faces slightly bigger consumption reductions when the European Union also cuts its emissions (3A vs. 1A). Natural gas prices are stimulated by the climate policies, whereas coal and crude oil prices are depressed.[7] The European Union is a net importer of all three fossil fuels, whereas the United States was a net exporter of coal in 2004.

Next, if only the United States or the European Union reduces CO_2 emissions, the effects on total consumption can be quite different for other trade partners. In particular, Russia is made better off by emissions reductions in the European Union because of higher gas export revenues, but it is severely harmed by reductions in the United States because of lower oil export revenues. Brazil and Canada are worse off when the European Union reduces emissions but unaffected or better off when the United States reduces emissions. China, Other Asia, Other America, and Other Africa are also harmed much more by EU policies than by U.S. policies. This is somewhat surprising, given that the *absolute* level of emissions reductions is lowest when only the European Union cuts emissions by 20 percent. One explanation is that climate policy is more costly in the European Union than in the United States, and lower incomes reduce import demand and hence export prices in other regions (remember also that the EU economies are more trade intensive than the U.S. economy).

Indeed, several other regions are worse off than the United States and the European Union when both these two regions reduce their CO_2 emissions by 20 percent. In particular, OPEC experiences a joint loss in consumption of 1.3 percent, mainly because of lower crude oil prices. Other oil exporting regions or countries, like Russia and other former Soviet Union republics, Mexico, and Africa, all see consumption losses even larger than the U.S. consumption losses. On the other hand, countries that import crude oil and coal, like Japan and especially India, benefit from lower international crude and coal prices. Canada actually gains when the United States, its close neighbor, takes on emissions reductions. Canada increases its output and net export of both natural gas and energy-intensive goods, and it suffers a consumption loss only when the European Union alone reduces emissions. Global consumption is actually reduced more in

---

[7] In these three scenarios, U.S. and EU coal prices fall 15–1 percent, crude prices fall 1–2 percent, and natural gas prices increase 6–8 percent.

The B.E. Journal of Economic Analysis & Policy, Vol. 10 [2010], Iss. 2 (Symposium), Art. 13

relative terms than U.S. consumption when both the United States and the European Union cut back their emissions.

Figure 2. Consumption effects of 20% cut in CO_2 emissions by United States, European Union, or both, under auctioning (Scenarios 1A, 2A, and 3A)

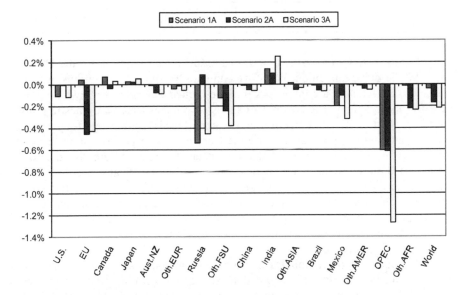

Next, we consider the effects of the countervailing policies. Table 2 shows the domestic emissions reductions in the different scenarios. As already indicated, the complementary policies imply less domestic reductions in order to reach the same global emissions reductions. We also notice that the biggest changes are in the European Union. This greater sensitivity is likely driven by the fact that EIS emissions represent a larger share of the cap in the European Union than in the United States. Improved competitiveness of EIS sectors through OUTPUT or BTAX, which affects foreign emissions, then requires fewer domestic emissions reductions to meet the global target (cf. also the discussion of leakage below).

Boehringer et al.: The Global Effects of Subglobal Climate Policies

Table 2. Emissions reductions, by policy scenario

	Domestic emissions reductions (percentage)					Global reduction
	A (Auction)	B (Output)	C (Rebate)	D (Tariff)	E (Btax)	
1 (U.S.)	20.0	19.8	19.9	19.8	19.7	4.0
2 (EU)	20.0	19.3	19.5	18.6	18.2	2.4
3 (U.S. and EU)	20.0	19.7	19.8	19.3	19.1	6.6

Figure 3 shows the costs, in terms of reduced consumption, for the United States and the European Union. When we compare the five policy options, we see that the cost differences are rather small in the United States but somewhat larger in the European Union. The two most expensive policies (from a domestic point of view) in both regions are auction and export rebates. Import tariffs, possibly complemented with export rebates, are the least expensive policy option in both economies. This assumes, however, that import tariffs can be differentiated across regions based on differences in embodied carbon. If the tariffs also accounted for indirect emissions (cf. footnote 5), additional simulations show that the costs would be even lower. Because the environmental benefits are held constant across the five policy options, we may conclude that both the United States and the European Union are better off in terms of welfare with special treatment of energy-intensive industries.

The fact that other policies dominate an auctioned cap reflects to some degree the auctioned cap's increased domestic emissions reductions. However, even if the *domestic* emissions reductions are kept constant, at 20 percent, both output-based allocation and import tariffs reduce the costs slightly. This is due to some beneficial terms-of-trade effects from protecting EIS in these two large economies. Interestingly, the U.S. carbon price is almost insensitive to the adjustment policies, but the EU carbon price *decreases* 4 percent in the OUTPUT scenario and 13 percent with BTAX. Thus, the effect of increased competitiveness for EIS sectors, which would tend to increase the carbon price needed to reach a fixed domestic target, is dominated by the effect of lower domestic emissions reductions.

The B.E. Journal of Economic Analysis & Policy, Vol. 10 [2010], Iss. 2 (Symposium), Art. 13

Figure 3. U.S. and EU costs (reduced consumption)
of cut in CO$_2$ emissions

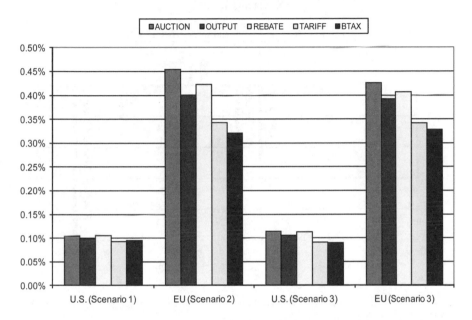

As shown in Figure 4, the differences among policies are moderate for most regions and small for the world as a whole. However, the consumption losses for crude oil and coal exporters are generally highest when the United States and the European Union also impose import tariffs, in which case the demand for fossil fuels outside these two regions is lower than in the other scenarios, leading to even lower fossil fuel prices (see the discussion of leakage below). Import tariffs also hurt exporters of energy-intensive goods, notably China, India, and Canada. We also observe from the figure that total costs for the world are lower with special treatment of energy-intensive industries, which presumably is due to generally higher emissions intensities for these sectors outside the United States and (especially) the European Union.

**Figure 4. Consumption effects of cut in CO_2 emissions
in both United States and European Union
(Scenarios 3A–3E)[*]**

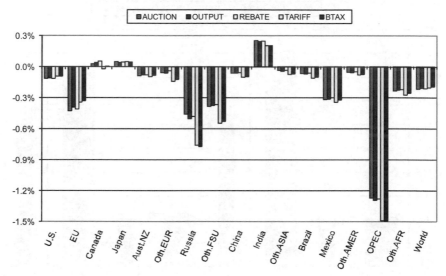

[*] The results for the United States and the European Union are the same as the last 10 bars in Figure 3.

The importance of international price effects of climate policy in the United States and the European Union is readily apparent from Figure 5, which shows the change in the Laspeyres index for the different regions (in Scenarios 3A–3E). The Laspeyres index of the terms of trade measures the ratio of the price index of exports to the price index of imports, in which prices are weighted by the baseline quantities of exports and imports. For the regions without emissions constraints, we notice that their consumption effects, as shown in Figure 4, can to a large degree be traced back to changes in the international prices triggered by the U.S. and EU emissions policies. For the United States and the European Union, the two figures illustrate that beneficial terms-of-trade effects contribute significantly to reducing the overall costs of climate policies in these regions. In particular, the percentage reduction in U.S. consumption is clearly below the percentage increase in the country's Laspeyres index. Not surprisingly, Figure 5 shows that import tariffs give more advantageous terms-of-trade effects than other policies for both the United States and the European Union.

The B.E. Journal of Economic Analysis & Policy, Vol. 10 [2010], Iss. 2 (Symposium), Art. 13

**Figure 5. Terms-of-trade effects (Laspeyres index)
of cut in both U.S. and EU CO₂ emissions
(Scenarios 3A–3E)**

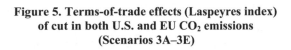

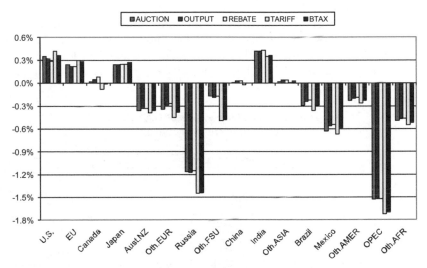

As already mentioned, subglobal climate policies in one or more regions typically lead to higher overall carbon emissions elsewhere. Leakage occurs not only through the international energy markets, as the drop in demand in the abating countries lowers global prices of fossil fuels, but also through the markets for energy-intensive goods, as the costs of producing such goods in the abating countries rise.

Figure 6 shows the size of global leakage in the policy scenarios. First, we observe that carbon leakage is highest when the European Union reduces its emissions. The leakage rate is then up to 28 percent, compared with up to 10 percent when the United States reduces its emissions. One reason for this difference is that the European Union is a more open economy than the United States, meaning that imports and exports constitute a larger share of the EU economy. This is true both for energy-intensive goods and for fossil fuels, of which the European Union is a much bigger importer (relative to own consumption) than the United States. This fact matters in particular for coal and gas, where transport costs are important, leading to differentiated prices around the world. Another reason for higher leakage with EU policies is that this region's energy-intensive industries are less carbon-intensive than the same industries in the United States. Thus, relocation of industrial activities away from the abating region has more adverse effects on global emissions when the European Union imposes climate policies.

When both regions reduce emissions, the leakage rates are closer to the U.S. policy scenarios. This is partly because the United States has substantially higher emissions than the European Union, and partly because some of the leakages in Scenarios 1 and 2 take place in the European Union and the United States, respectively (see below). Thus, when both regions reduce their emissions, overall leakage tends to fall (which is obviously the case generally with increases in the regional coverage of subglobal action).

Figure 6. Global leakage effects of U.S. and/or EU
CO_2 emissions reductions

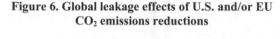

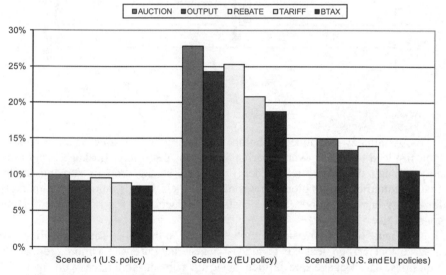

Further, we see from Figure 6 that leakage differs little among the five alternative U.S. policies. This is not the case with the EU policies, however, where the differences in leakage are somewhat bigger across policy scenarios. The explanation is, again, that energy-intensive industries in the European Union are less carbon intensive than in the United States, and thus EU policies that prevent relocation of these industries to other regions produce less leakage. Moreover, the U.S. electricity sector has bigger potential for technology switching toward less carbon intensive technologies, and thus the relative emissions reduction in the power sector is bigger than in the European Union, which has relatively bigger emissions reductions in the energy-intensive industries. As a

The B.E. Journal of Economic Analysis & Policy, Vol. 10 [2010], Iss. 2 (Symposium), Art. 13

consequence, special treatment of these industries tends to have bigger effects on leakage in the European Union. Full border adjustment policies are most effective in reducing leakage, with the import tariff being more important than the export rebate. This is partly because the import tariff is the only instrument that can differentiate among regions, so that production of energy-intensive goods in regions with very carbon-intensive production is particularly penalized. Another reason is that even though output-based allocation, export rebates, and import tariffs all improve the competitiveness of domestic industries, the two former policies are subsidies to domestic production, whereas the latter policy is a tax on foreign production.

The figure further illustrates that border adjustment policies or output-based allocation can reduce carbon leakage to only a certain extent (by 33 percent, or nine percentage points, at most in these scenarios, when compared with the auction scenarios). The reason is that most of the carbon leakage in our study takes place via the international markets for fossil fuels: that is, lower consumption of oil and coal in the United States and the European Union leads to lower prices of these fuels, which in turn leads to increased demand for these fuels in other regions. The alternative policies are hardly able to deal with this sort of leakage (although import tariffs can to some degree, because foreign production is taxed).

Figure 7 shows how global leakage is distributed across regions in Scenarios 3A–3E. The share of global leakage that takes place in other OECD countries is 23 to 29 percent.[8] When only the United States reduces emissions, this share is in fact about 50 percent, and the European Union accounts for one-quarter of global leakage. On the other hand, EU policies have small leakage effects in the United States, and the OECD share of global leakage is then 25 to 30 percent. This difference is again due to the greater openness of the EU economy. A substantial share of leakage takes place in Russia. When we add Russia to the OECD countries, we approximate Annex B. The leakage to Annex B countries in Scenarios 3A–3E amounts to around 40 percent of global leakage.

Outside OECD and Russia, a large share of leakage occurs in Asian and African countries, whereas leakage to American countries is more moderate. African countries (outside OPEC) see significant increases in production of electricity and energy-intensive goods. The relative increases in such production in Asian countries are somewhat smaller, but the size of these countries is much bigger and hence the share of global leakage is larger.

---

[8] This is seen by summing the leakage figures for Canada, Japan, Aust.NZ, and Oth.EUR and dividing by total leakage (shown in the last bars of Figure 5).

**Figure 7. Leakage effects of U.S. and EU
CO₂ emissions reductions
(Scenarios 3A–3E)**

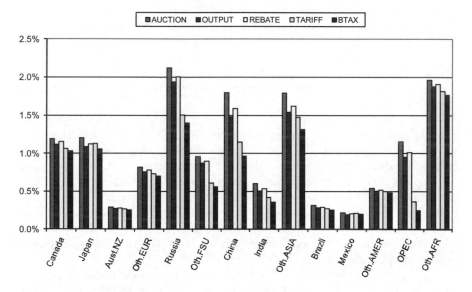

If we look at the different policies, we observe that border adjustment policies and output-based allocation hardly reduce leakage to other OECD countries, and also to Latin American countries. The largest leakage effects are in Asian countries like China, former Soviet Union republics like Russia, and OPEC. These countries significantly increase their production and export of energy-intensive goods (particularly chemicals, iron and steel, and nonferrous metals) in the auction scenario, but less so under alternative policies (especially import tariffs).

The effects of unilateral climate policies on production of energy-intensive goods in the United States and the European Union are shown in Figure 8 and Figure 9. Production of chemical products, nonferrous metals, and (in the European Union) iron and steel are the most heavily affected, with reductions around 3 percent in the auction scenarios. Production of paper, pulp, and print has generally lower CO₂ intensities than the other energy-intensive industries. Mineral production like cement has substantial emissions of CO₂ but is less traded than the other goods because of higher transport costs.

Both output-based allocation and border adjustment policies dampen the production decrease. The policies have largest effects in the European Union

The B.E. Journal of Economic Analysis & Policy, Vol. 10 [2010], Iss. 2 (Symposium), Art. 13

Both output-based allocation and border adjustment policies dampen the production decrease. The policies have largest effects in the European Union (Figure 9), consistent with the discussion of leakage, above. Full border adjustment policies (import tariffs and export rebate) are most effective in most cases, especially in the European Union. Overall, output-based allocation has about the same effect as import tariffs alone. All the policies have rather small effects on production of nonferrous metals (and paper, pulp, and print). The main explanation is that in the abating countries, these industries' heavy use of electricity becomes more costly, which the complementary policies do not target. If the import tariffs also take into account the indirect emissions (i.e., from electricity production) embodied in the products, production of nonferrous metals in the European Union will actually *increase* (not shown in the figure).

**Figure 8. Effects of U.S. climate policies on U.S. production
(Scenarios 1A–1E)**

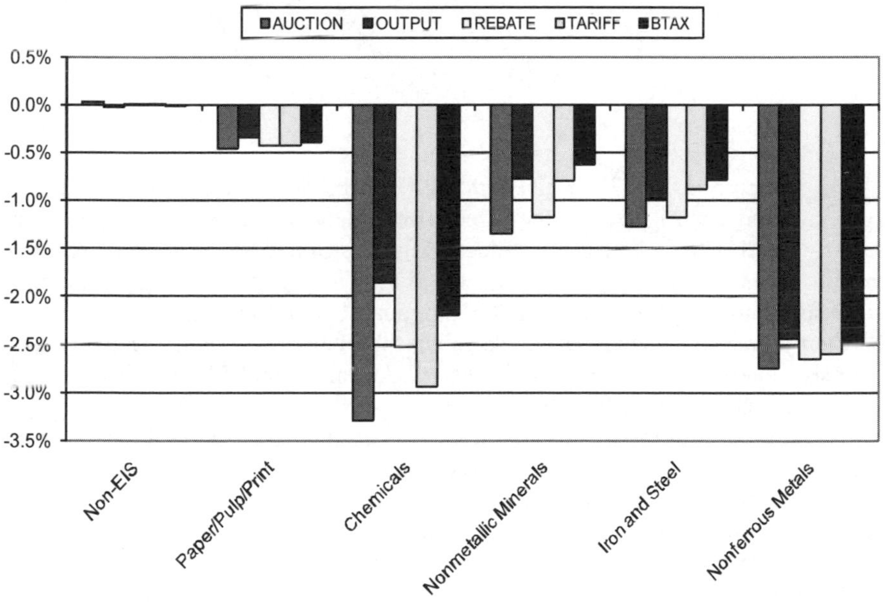

**Figure 9. Effects of EU climate policies on EU production
(Scenarios 2A–2E)**

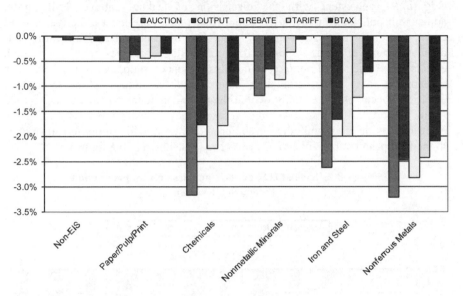

In Scenarios 3A–3E, when both the United States and the European Union cut emissions by 20 percent, the effects on production in the two regions are slightly smaller. In particular, the reductions in the United States are smaller when the European Union also cuts emissions. As noted above, U.S. climate policies have substantial consequences for the European Union, including its production of energy-intensive goods, but not so much the other way around. Thus, energy-intensive industries in the United States benefit more from EU policies than EU industries benefit from U.S. policies.

In the rest of the world, production in EIS increases across the board with the caps, while the OtherManuf&Serv sector (i.e., non-EIS, nonenergy, and nontransport) experiences production decreases in all countries except OPEC. The adjustment policies in part shift production in the other regions back toward non-EIS from EIS. Overall, however, aggregate production remains higher than with no climate policy action (Figure 10). By this metric, then, nonimplementing regions reap competitiveness benefits from the restrictions in the European Union and United States. Of course, in terms of consumption rather than production, they remain worse off, as shown earlier.

The B.E. Journal of Economic Analysis & Policy, Vol. 10 [2010], Iss. 2 (Symposium), Art. 13

**Figure 10. Percentage change in total production, by region
(Scenarios 3A–3E)**

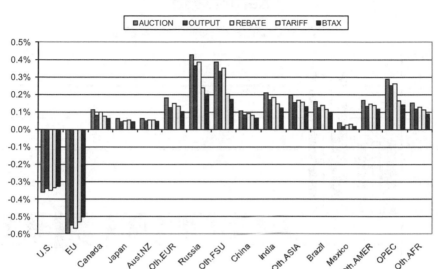

The changes in prices of energy intensive goods in these countries tend to run counter to the production changes. Carbon pricing typically raises the domestic prices of energy-intensive products in the regulating countries; with (only) auctioning in the United States and the European Union, the prices of chemical, metal, and mineral products in these two regions increase by 1.2 to 2.4 percent. Output-based allocation, acting as an implicit output subsidy to these sectors, lowers the price increase by one-quarter to two-thirds. On the other hand, border adjustment policies have only small effects on these prices. For nonregulating trade partners, however, prices of these goods fall as global demand decreases and global energy prices adjust. Figure 11 gives as an example the relative price effects in the iron and steel sectors across a range of regions under different policy choices by the United States and European Union (Scenarios 3A–3E); the results for other EIS are similar. The regions included represent most of the major trading partners in EIS goods, covering the "BASIC" countries—Brazil, South Africa, India, and China—as well as Russia and Japan, in addition to the United States and European Union. For the nonregulating countries, auctioned allowances have the smallest effect on global steel prices. Output-based allocation lowers prices further, in nonregulating as well as regulating countries. The border adjustment scenarios have the largest downward effects on prices in partner countries, but the effects in the regulating countries are minimal.

Boehringer et al.: The Global Effects of Subglobal Climate Policies

**Figure 11. Price effects of U.S. and EU climate policies
on iron and steel sectors, across trading partners
(Scenarios 3A–3E)**

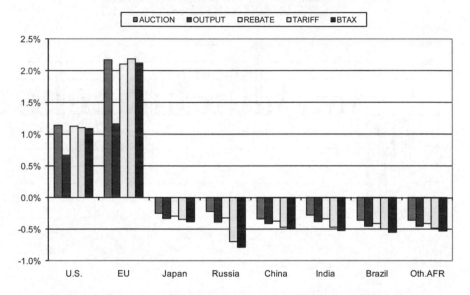

Figures 12–15 illustrate the effects of CO_2 emissions targets in both the United States and the European Union on exports of iron and steel, chemicals, nonferrous metals, and all energy-intensive goods, respectively, from these same countries.[9] To some degree, the results mirror the changes in production. In all auction cases, exports from these other countries increase, meaning that the changes in competitiveness outweigh the global demand reductions due to carbon pricing in the European Union and United States. All of the countervailing policies reduce these gains in exports in nonregulating countries to some extent and lessen the losses in the regulating countries. In no case do the European and United States experience increases in exports compared with no policy.

---

[9] We focus on these products because their aggregation corresponds most closely to the energy-intensive products. Exports of paper, pulp, and print are only slightly changed by these policies, in part because the category is dominated by print.

The B.E. Journal of Economic Analysis & Policy, Vol. 10 [2010], Iss. 2 (Symposium), Art. 13

We see some significant differences across regions and products. With full auction and no border adjustment policies, Asian countries see the highest relative increase in exports of chemical products and the lowest increase in exports of nonferrous metals. Other Africa sees the highest relative increase in exports of iron and steel products.

Output-based allocations and export rebates have quite similar effects across regions, with the former being somewhat more influential. These policies have small effects on exports of nonferrous metals, however, since we are not adjusting for the indirect emissions that are relatively important in these industries.

**Figure 12. Changes in exports of iron and steel
under U.S. and EU climate policies, by trading partner
(Scenarios 3A–3E)**

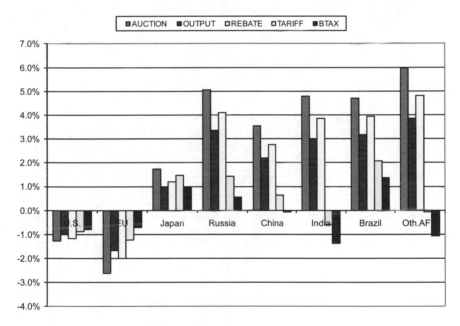

Boehringer et al.: The Global Effects of Subglobal Climate Policies

**Figure 13. Changes in exports of chemicals
under U.S. and EU climate policies, by trading partner
(Scenarios 3A–3E)**

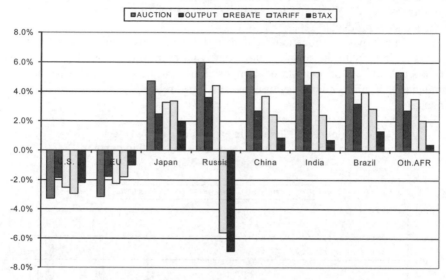

**Figure 14. Changes in exports of nonferrous metals
under U.S. and EU climate policies, by trading partner
(Scenarios 3A–3E)**

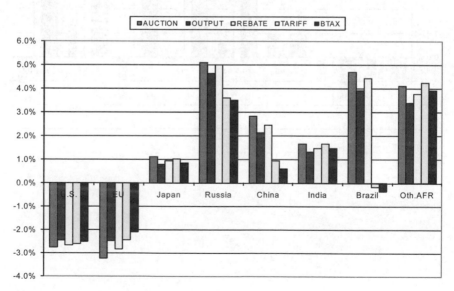

The B.E. Journal of Economic Analysis & Policy, Vol. 10 [2010], Iss. 2 (Symposium), Art. 13

**Figure 15. Changes in exports of all EIS products
under U.S. and EU climate policies, by trading partner
(Scenarios 3A–3E)**

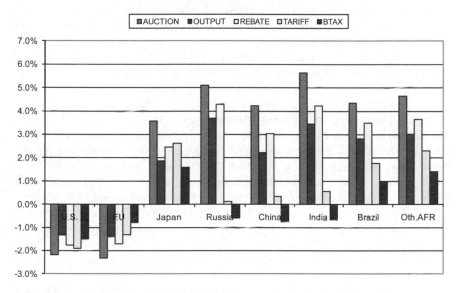

Import tariffs (and full border adjustment) have stronger effects on exports from these regions; still, in most cases, their exports are still higher with import tariffs in the United States and the European Union than without any climate policy whatsoever. The exceptions in this country sample are Russia, driven by the chemicals sector, and China and India, driven by nonmetallic minerals (in which exports experience a 10–12 percent drop with border adjustments). Import tariffs have much more differentiated effects across regions and sectors. The explanation is, of course, that the tariff treats imports from different regions differently, based on the embodied carbon in that region. Carbon intensities vary substantially across regions for all energy-intensive products. For instance, the carbon intensity of Brazilian production of nonmetallic minerals is several times smaller than the corresponding carbon intensity in China. And Japan has the smallest carbon intensity in production of iron and steel and nonferrous metals of the regions presented in these figures. The large variation in carbon intensities occurs partly because the five sectors in reality consist of a much larger number of subsectors producing different products with different carbon intensities. In addition, different regions use different technologies with different inputs of energy to produce the same product.

Conclusions

We find that the welfare effects of subglobal climate policies are significant not only for the countries undertaking them but also for their trade partners, who experience changes in demand for their products as well as changes in global energy prices. In some cases, the welfare losses (in terms of percentage changes in consumption) can be even larger abroad than at home, particularly for fossil fuel producers. When the United States is reducing its emissions, however, a few trade partners gain, including Japan, India, and Canada.

Policies intended to avoid leakage have little effect on welfare overall—even in the countries implementing them—because they mostly just shift global production in certain energy-intensive goods. Although these additional welfare changes are small, implementing countries do benefit from adjustment policies, while most nonimplementing countries (particularly developing ones) would prefer no adjustment; the net effect of antileakage policies is a slight reduction in the global costs of achieving a given level of emissions reductions.

Regarding leakage, we find that a significant share of leakage occurs via changes in global energy prices. Hence, none of the countervailing policies reduce leakage rates very much—at most by 22 percent, in the case of full border adjustments. Furthermore, 40 percent of leakage from U.S. and EU climate policies can be attributed to other Annex B nations.

Policies like output-based rebates and border adjustments do have significant effects on the energy-intensive sectors to which they are applied. Domestic production changes are mitigated and foreign exports are also reduced. Still, for the most part, domestic production is lower and foreign production and exports are higher than without any climate policy intervention. One exception in our modelling results is that full border adjustment can cause exports from other countries to decrease below baseline levels, particularly in the nonmetallic minerals sector.

The narrow debate about border adjustments may prove to be a tempest in a teapot that is being tossed around in a much larger tempest. Clearly, the main effects on global welfare, emissions, and leakage arise from the primary climate policies themselves. Developed countries should understand that most developing nations do not actually gain economically from the former's efforts to reduce greenhouse gas emissions. At the same time, developing countries should recognize that their sectors targeted specifically by antileakage policies do not necessarily lose, compared with a world without any climate policies. Ultimately, it is in all countries' interest to mitigate climate change as comprehensively and cost-effectively as possible, and the larger question is whether unilateral antileakage policies can help in the transition to concerted global action.

The B.E. Journal of Economic Analysis & Policy, Vol. 10 [2010], Iss. 2 (Symposium), Art. 13

References

Armington, P.S. 1969. A Theory of Demand for Producers Distinguished by Place of Production. *IMF Staff Papers* 16(1): 159–78.

Aune, F.R., R. Golombek, S.A.C. Kittelsen, and K.E. Rosendahl. 2008. *Liberalizing European Energy Markets: An Economic Analysis.* Cheltenham, UK: Edward Elgar Publishing.

Babiker, M.H. 2005. Climate Change Policy, Market Structure, and Carbon Leakage. *Journal of International Economics* 65(2): 421–45.

Babiker, M.H. and T.F. Rutherford. 2005. The Economic Effects of Border Measures in Subglobal Climate Agreements. *Energy Journal* 26(4): 99–126.

Badri, N.G., and T.L. Walmsley. 2008. *Global Trade, Assistance, and Production: The GTAP 7 Data Base.* West Lafayette, Indiana: Center for Global Trade Analysis, Purdue University.

Bagwati, J., and P.C. Mavroidis. 2007. Is Action against U.S. Exports for Failure to Sign Kyoto Protocol WTO-Legal? *World Trade Review* 6: 299–310.

Biermann, F., and R. Brohm. 2005. Implementing the Kyoto Protocol Without the United States: The Strategic Role of Energy Tax Adjustments at the Border, Global Governance. *Climate Policy* 4: 289–302.

Böhringer, C., and A. Lange. 2005. On the Design of Optimal Grandfathering Schemes for Emission Allowances. *European Economic Review* 49: 2041–55.

Böhringer, C., and T.F. Rutherford. 2010. The Costs of Compliance: A CGE Assessment of Canada's Policy Options under the Kyoto Protocol. *World Economy* 33(2): 177–211.

Böhringer, C., M. Ferris, and T.F. Rutherford. 1998a. Alternative CO_2 Abatement Strategies for the European Union. In *Climate Change, Transport and Environmental Policy*, edited by J.B. Braden and S. Proost. Northampton, Massachusetts: Edward Elgar Publishing, 16–47.

Böhringer, C., T.F. Rutherford, and A. Voss. 1998b. Global CO_2 Emissions and Unilateral Action: Policy Implications of Induced Trade Effects. *International Journal of Global Energy Issues* 11(1-4): 18–22.

Böhringer, C., T.F. Rutherford, and W. Wiegard. 2003. Computable General Equilibrium Analysis: Opening a Black Box. ZEW Discussion Paper 03-56. Mannheim: Center for European Economic Research.

Burniaux, J., and J.O. Martins. 2000. Carbon Emission Leakages: A General Equilibrium View. OECD Economics Department Working Paper 242. OECD Publishing. doi:10.1787/410535403555.

Burniaux, J.-M., J. Chateau, R. Dellink, R. Duval, and S. Jamet. 2009. The Economics of Climate Change Mitigation: How to Build the Necessary Global Action in a Cost-Effective Manner. OECD Economics Department Working Paper 201. OECD Publishing.

de Cendra, J. 2006. Can Emissions Trading Schemes Be Coupled with Border Tax Adjustments? An Analysis vis-à-vis WTO Law. *Review of European Community and International Environmental Law (RECIEL)* 15(2): 131–145. ISSN 0962 8797.

Charnovitz, S., G.C. Hufbauer, and J. Kim. 2009. *Global Warming and the World Trading System.* Washington, D.C.: Peterson Institute for International Economics.

Demailly, D., and P. Quirion. 2006. CO_2 Abatement, Competitiveness and Leakage in the European Cement Industry under the EU ETS: Grandfathering versus Output-Based Allocation. *Climate Policy* 6(1): 93–113.

Demailly, D., and P. Quirion. 2008a. European Emission Trading Scheme and Competitiveness: A Case Study on the Iron and Steel Industry. *Energy Economics* 30(4): 2009–27.

Fischer, C. 2001. Rebating Environmental Policy Revenues: Output-Based Allocations and Tradable Performance Standards. Discussion Paper 01-22. Washington, D.C.: Resources for the Future.

Fischer, C., and A.K. Fox. 2007. Output-Based Allocation of Emissions Permits for Mitigating Tax and Trade Interactions. *Land Economics* 83: 575–99.

The B.E. Journal of Economic Analysis & Policy, Vol. 10 [2010], Iss. 2 (Symposium), Art. 13

Fischer, C., and A.K. Fox. 2009. Combining Rebates with Carbon Taxes: Optimal Strategies for Coping with Emissions Leakage and Tax Interactions. RFF Discussion Paper 09-12. Washington, D.C.: Resources for the Future.

Gielen, D., and Y. Moriguchi. 2002. CO_2 in the Iron and Steel Industry: An Analysis of Japanese Emission Reduction Potentials. *Energy Policy* 30: 849–63.

Godard, O. 2007. Unilateral European Post-Kyoto Climate Policy and Economic Adjustment at EU Borders. EDF—Ecole Polytechnique: Cahier No. DDX 07-15.

Grubb, M., and K. Neuhoff. 2006. Allocation and Competitiveness in the EU Emissions Trading Scheme: Policy Overview. *Climate Policy* 6(1): 7–30.

Ho, M., R.D. Morgenstern, and J.-S. Shih. 2008. Impact of Carbon Price Policies on U.S. Industry. Discussion Paper 08-37. Washington, D.C.: Resources for the Future.

Houser, T., R. Bradley, B. Childs, J. Werksman, and R. Heilmayr. 2008. *Leveling the Carbon Playing Field: International Competition and U.S. Climate Policy Design*. Washington, D.C.: Peterson Institute for International Economics and World Resources Institute.

International Centre for Trade and Sustainable Development (ICTSD). 2008. Climate Change: Schwab Opposes Potential Trade Measures. *Bridges Trade BioRes* 8(4)· March 7.

Karp, L.S., and J. Zhao. 2008. A Proposal for the Design of the Successor to the Kyoto Protocol. Discussion Paper 2008-03. Cambridge, Massachusetts: Harvard Project on International Climate Agreements. September.

Kommerskollegium. 2004. Climate and Trade Rules: Harmony or Conflict? Stockholm: National Board of Trade.

Mattoo, A., A. Subramanian, D. Van der Mensbrugghe, and J. He. 2009. Reconciling Climate Change and Trade Policy. Policy Research Working Paper 5123 (November 1). Washington, D.C.: World Bank.

Paltsev, S. 2001. The Kyoto Protocol: Regional and Sectoral Contributions to the Carbon Leakage. *Energy Journal* 22(4): 53–79.

Pauwelyn, J. 2007. U.S. Federal Climate Policy and Competitiveness Concerns: The Limits and Options of International Trade Law. Nicholas Institute for Environmental Policy Solutions. Working Paper 07-02. Durham, North Carolina: Duke University.

Peterson, E.B., and J. Schleich. 2007. Economic and Environmental Effects of Border Tax Adjustments. Sustainability and Innovation Working Paper S1/2007. Karlsruhe, Germany: Fraunhofer Institute Systems and Innovation Research.

Ponssard, J.P., and N. Walker. 2008. EU Emissions Trading and the Cement Sector: A Spatial Competition Analysis. Manuscript. Paris: Laboratoire d'Économétrie, CNRS et École Polytechnique.

[18]

The B.E. Journal of Economic Analysis & Policy

Symposium

Volume 10, Issue 2	2010	Article 14

DISTRIBUTIONAL ASPECTS OF ENERGY AND CLIMATE POLICY

Comment on "The Global Effects of Subglobal Climate Policies"

Rodney D. Ludema*

*Georgetown University, ludemar@georgetown.edu

Ludema: Comment on "The Global Effects of Subglobal Climate Policies"

Since the earliest days of the climate change debate, supporters and opponents of carbon reduction policies have agreed on one thing: a country that unilaterally limits its carbon emissions shoulders the economic burden of that policy, while the rest of the world shares in the environmental benefit. Climate policy supporters have used this argument to push for broad-based international agreements, such as the Kyoto Protocol, and only after the failure of such agreements to garner political support in the U.S. have they come to embrace unilateral action as a worthwhile intermediate strategy. Opponents, on the other hand, have used this argument to denounce both unilateral climate policies and international agreements, as the latter inevitably fall short of universal participation.

To the extent that international trade has entered into this debate, it has been used to bolster the conventional wisdom. International trade adds to a country's burden of unilateral climate policy, it is argued, by putting domestic industries at a cost disadvantage relative to their foreign competitors. Trade also causes carbon leakage, which undermines the benefits of unilateral climate policy. These trade-related concerns have led to various proposals for incorporating trade policies into the unilateral climate policy packages being considered in both the U.S. and the European Union.

Seen against this backdrop, the paper by Boehringer, Fischer and Rosendahl (2010) has significant implications. Basically, they show that the conventional wisdom about unilateral climate policy is almost certainly wrong; moreover, it is wrong precisely because of international trade. In an open economy setting, when large countries like the U.S. and the EU enact economy-wide cap-and-trade systems, the burden of these policies does *not* fall entirely on the countries enacting them but are distributed, albeit unevenly, across the world. The reason is that the policies affect world prices. According to their model, world prices of coal and oil fall, the world price of natural gas rises and world prices of energy-intensive products mostly fall. These and a host of other world price changes combine to improve the terms of trade of U.S. and EU on balance. That is, on average the prices of their exports rise relative to their imports (see their Figure 5). For the same reason, the terms of trade deteriorate for nearly every other region of the world, save Canada, Japan and India. The biggest losers tend to be the major oil exporters, OPEC and Russia. The fact that much of the economic burden of U.S. and EU climate policies is passed on to the rest of the world through world prices means that the "go it alone" strategy currently being pursued by these countries is not nearly as suicidal as we thought.[1]

---

[1] Copeland and Taylor (2005) discuss other effects of changes in the terms of trade, including income effects that may induce the other country to cut emissions (negative leakage).

The B.E. Journal of Economic Analysis & Policy, Vol. 10 [2010], Iss. 2 (Symposium), Art. 14

However, this observation should also affect how we think about the use of trade policy to "level the playing field" and reduce carbon leakage. In my view, this is where the paper could be clearer. The paper uses a multi-region, multi-sector computable general equilibrium (CGE) model, which is great for tracking down all of the complex interactions between markets but somewhat obscures the first-order effects of the trade policies. In what follows, I attempt some clarifications.

To begin with, as long as carbon leakage does not fully offset the effect of a unilateral carbon cap,[2] the cap can be adjusted to achieve almost any desired level of global emissions. Since the environmental benefit of a cap depends solely on its global impact (and not the distribution of emissions across countries), the only remaining question for welfare is what are the economic costs of achieving a target level of global emissions via a unilateral cap. The paper acknowledges this point implicitly by holding constant the level of global emissions across the various policy scenarios it considers. The first policy scenario (1A) involves a 20% emissions cut by the U.S., which leads to a global emissions cut of around 4%. In the other policy scenarios (1B-1E), trade policies are added, but the level of the U.S. carbon cap is adjusted so as achieve the same 4% global emissions cut. The reason the paper does this is so that it can make welfare comparisons across policy scenarios without undertaking the difficult task of actually of measuring the welfare benefit from emissions reductions. The added advantage, however, is that it demonstrates a key point, which is that the differences among these policies lie not in their ability to affect global carbon emissions but in how the economic burden is allocated throughout the world. The only practical value of estimating carbon leakage, therefore, is to determine the level of the carbon cap needed to achieve the target level of global emissions.

Inexplicably, the paper does not apply this logic across countries. The U.S. and the EU are compared on the basis of the same 20% unilateral emissions cut, instead of the same global emissions cut. The EU policy results in only a 2.4% global emissions reduction. Similarly, when both cut emissions by 20% the result is a 6.6% global reduction. Because the emissions levels differ in these cases, there is no way to make welfare comparisons between them. It would have been useful to know, for example, how the cost to the U.S. of achieving a 4% reduction in global emissions differs when done in concert with the EU than when done unilaterally. Sadly, this question is not addressed.

---

[2] The model produces positive, but not fully offsetting, carbon leakage. Negative leakage is also possible, however. In addition to the income effect emphasized by Copeland and Taylor (2005), Fullerton and Karney (2010) show that negative leakage can occur through a substitution effect: a carbon cap causes home producers to substitute from carbon into capital for abatement, and thus bid capital away from unregulated countries causing a reduction in their output.

Ludema: Comment on "The Global Effects of Subglobal Climate Policies"

The paper addresses the question of burden sharing both across sectors of the economy and across countries in the aggregate, with the former being germane to the level-playing-field issue. To gather some intuition, it is helps to begin with a simple partial equilibrium model of an energy-intensive product like steel. A U.S. cap-and-trade system without border adjustments can be viewed as roughly equivalent to a production tax. It raises the marginal cost and thus price charged by U.S. steel producers. In response to this price hike, consumers of steel (e.g., automakers) throughout the world begin to switch to foreign sources of steel, which drives up world steel prices. In the end, U.S. steel producers lose, foreign steel producers win, and steel consumers worldwide lose, as they have to pay higher prices both in the U.S. and abroad. Now consider the effect of border adjustment policies -- the rebate of carbon permits on exports and the imposition of a carbon tariff on imports. This effectively converts the production tax into a tax on U.S. steel consumers. Instead of raising the world price of steel, a consumption tax would tend to lower it, because it reduces U.S. steel demand and by extension world steel demand. In the end, U.S. steel consumers lose, foreign steel consumers win, as they get lower prices without paying the tax, while steel producers throughout the world lose from the lower producer prices. Thus, while the two policies reduce world carbon emissions by the same amount by definition, they differ in their allocation of the burden between the world's producers and consumers. Without border adjustments, U.S. steel producers lose relative to foreign steel producers, but with border adjustments, U.S. steel consumers lose relative to foreign steel consumers. Thus, it is not question of whether border adjustments level the playing field. It is a question of whose playing field to level, that of the steel producers or steel users?

Their Figures 9 and 10 bear out most of the above story. Relative to the no-border-adjustments case, border adjustments increase U.S. steel production, reduce foreign steel production, and lower foreign steel prices. However, the figures also highlight an important general equilibrium element that is missing from our partial equilibrium story. A U.S. carbon cap (with or without border adjustments) reduces demand for energy, not just in the steel sector but in all sectors, which drives down energy prices worldwide. This lowers marginal costs of all steel producers. While this effect does not entirely offset the effect of the tax in the U.S., it causes steel production to rise, and steel prices to fall in the rest of the world, even without border adjustments.

Finally, let us consider aggregate effects. Whether or not U.S. welfare as a whole increases under border adjustment depends on the effect of the policies on the terms of trade. We have already seen that border adjustments tend to drive down the world prices of energy-intensive goods. Thus, if the U.S. is a net importer of these goods, border adjustments improve the U.S. terms of trade. This improvement is in addition to the terms-of-trade improvement already present in

The B.E. Journal of Economic Analysis & Policy, Vol. 10 [2010], Iss. 2 (Symposium), Art. 14

other sectors (e.g., the decreased price of oil) under cap-and-trade without border adjustment, and thus it further pushes the burden of U.S. climate policy onto the rest of the world.

These points can be seen clearly by comparing Figures 4 and 5. We see first that the consumption gains and losses from the combined U.S. and EU carbon caps are directly related to the effects those policies have on each country's terms of trade. We see second that the combined carbon caps cause major improvements in the U.S. and EU terms of trade and major deteriorations in the terms of trade of energy exporters, in particular. Third, see that the border adjustment policies further improve the U.S. and EU terms of trade, mainly at the expense of these same exporters.

One can begin to see why most observers think border adjustments are likely to get a chilly reception in the WTO. While the WTO contains exceptions that allow countries to use trade policy to achieve environmental ends, these rights are tightly circumscribed. In particular, the policy must not be a form of "disguised" protectionism. Given that 1) the environmental objective can be achieved without resort to border adjustments, 2) the only domestic objective served by border adjustments is to appease energy-intensive producers at the expense of consumers, and 3) the international effect of such adjustments is to further shift the burden of the carbon cap onto foreigners by deteriorating their terms of trade, a WTO panel would have little choice but to reject these policies if ever it were called upon to review them.

Summing up, the paper by Boehringer, Fischer and Rosendahl conveys several important messages about relationships between international trade and climate policy. Far from weakening the case for a U.S. or EU unilateral carbon cap, the presence of international trade strengthens it, because trade diffuses the economic burden throughout the world through changes in the terms of trade. Carbon leakage does not change this conclusion in the slightest. A carbon cap imposes costs on domestic producers, as well as consumers, of energy-intensive products, although this effect is mitigated by declines in world energy prices. Border adjustments can help to level the international playing field in energy-intensive sectors, like steel, but only at the expense of tilting the international playing field against other domestic producers, like automakers, that use such products as inputs -- making for a complex political calculus. Moreover, because they are not necessary to achieve the environmental goal, border adjustments could be viewed as a disguised protectionism under WTO rules.

It is generally inefficient to rely on unilateral action in the presence of international policy externalities, be they positive (carbon reduction) or negative (adverse terms of trade changes). Thus, the case for coordinated global action on climate change remains sound. The point illuminated by this paper is that, if a large energy-importing country imposes a unilateral carbon cap, the positive externality is offset to some degree by the negative one. This raises an intriguing question of whether it is possible to find a collection of energy-importing countries, such that coordinated carbon caps among these countries could result in carbon reduction and terms-of-trade externalities that are exactly offsetting. Such a collection might offer a way to obtain an efficient outcome without having to enlist the cooperation of all countries in the world. While answering this question would require modeling the welfare benefit of carbon reduction, the CGE model of this paper could handle the rest.

References

Boehringer, Christoph, Carolyn Fischer, and Knut Einar Rosendahl (2010), "The Global Effects of Subglobal Climate Policies," *The B.E. Journal of Economic Analysis & Policy* 10(2).

Copeland, Brian R., and M. Scott Taylor (2005), "Free Trade and Global Warming: A Trade Theory View of the Kyoto Protocol," *Journal of Environmental Economics and Management*, 49: 205-234.

Fullerton, Don, and Dan Karney (2010), "Negative Leakage," working paper, University of Illinois at Urbana-Champaign.

[19]

The B.E. Journal of Economic Analysis & Policy

Symposium

Volume 10, *Issue* 2	2010	*Article* 3

DISTRIBUTIONAL ASPECTS OF ENERGY AND CLIMATE POLICY

Equity, Heterogeneity and International Environmental Agreements

Charles D. Kolstad*

*University of California, Santa Barbara, kolstad@econ.ucsb.edu

Recommended Citation
Charles D. Kolstad (2010) "Equity, Heterogeneity and International Environmental Agreements," *The B.E. Journal of Economic Analysis & Policy*: Vol. 10: Iss. 2 (Symposium), Article 3.
Available at: http://www.bepress.com/bejeap/vol10/iss2/art3

Equity, Heterogeneity and International Environmental Agreements*

Charles D. Kolstad

Abstract

Much of the literature on international environmental agreements (IEA) considers the case of identical countries. There is a much smaller literature concerning the more complex but more realistic case of country heterogeneity. This paper involves modifying the standard static homogeneous country model of international environmental agreements. In particular, we consider two types of countries, differing in size as well as in marginal damage from pollution. Although the IEA does not have a unique size in this case, we do introduce two equilibrium refinements and explore the implications for coalition size. The two refinements include one based on efficiency and one based on equity.

*Charles D. Kolstad, Department of Economics and Bren School of Environmental Science & Management, University of California, Santa Barbara; University Fellow, Resources for the Future; Research Associate, National Bureau of Economic Research. Comments from Scott Barrett, Don Fullerton, Ross Mohr, an anonymous referee and participants at an RFF/University of Chicago workshop in January 2010 are much appreciated.

Kolstad: Equity, Heterogeneity and International Environmental Agreements

The 2009 stalemate at COP-15 in Copenhagen regarding taking action on climate change illustrates the difficulties in getting nearly 200 countries to agree to act on anything, let alone on an issue as complex and significant as climate change.[1]

The challenge faced by the participants at Copenhagen was to forge an international environmental agreement (IEA) to reduce greenhouse gas emissions leading to climate change. What makes the problem particularly tough is that countries cannot be forced to participate but must do so voluntarily. And because a reduction in greenhouse gases is truly a global public good, while abatement is privately costly, it is far more attractive to free-ride on any agreement than to participate as an abating party. From an incentive point of view, the only stable agreement is one for which every participant finds it individually rational to participate; that is, participation makes the country better off than free-riding.

The issue of the formation of IEAs has also been of concern to the academic community for some time, both in political science (eg, Young 1994) and economics (eg, Barrett 2003). The main question asked by this literature is what characteristics of the problem lead to strong or weak IEAs? A corollary to this question is what structural features can be incorporated into IEAs to improve their performance?

Although the political science literature is more nuanced than the economics literature in generating understanding of IEAs, the economics literature has provided many insights into how IEAs work. One of the first papers in this literature is due to Scott Barrett (1994). In that paper, he develops a simple game-theoretic model and argues that IEAs are unable to improve very much on the status quo of no agreement: either agreements involving very many parties are not stable or, if large agreements can be formed, they do not improve welfare much relative to the case of no agreement. Other questions/issues include the effect of uncertainty on IEA formation[2] and the use of commitment mechanisms to strengthen agreements.[3]

This paper raises the question of the effect of heterogeneity of countries on the formation of agreements. Simply put, is it easier for a meaningful agreement to form when countries are similar or when they are not? Do differences among countries retard or enhance the formation of IEAs? Does heterogeneity increase abatement and thus welfare? Implications are important for improved design of an IEA.

---

[1] COP-15 is the fifteenth "Conference of the Parties" to the Framework Convention on Climate Change treaty.

[2] On uncertainty, see Na and Shin (1998), Helm (1998), Kolstad (2005), Ulph (2004) and Kolstad and Ulph (2008).

[3] On commitment, see Carraro and Siniscalco (1993) and Barrett (1997b).

The B.E. Journal of Economic Analysis & Policy, Vol. 10 [2010], Iss. 2 (Symposium), Art. 3

Intuition suggests that if countries are very similar, it may be easier to reach agreement to solve a common problem, and thus homogeneity can increase efficiency. It turns out that this intuition is correct.

I. BACKGROUND

In reviewing the literature on International Environmental Agreements (IEAs), we first consider the general question of determinants of stable IEAs. We then turn to the issue of heterogeneity of participants, the issue of concern in this paper. Finally, we move away from pure theory, examining the experimental evidence on IEA formation.

A.   IEA Size

International environmental agreements are the subject of a significant economic literature, primarily post-1990.[4] Most of the literature focuses on self-enforcing agreements; i.e., agreements that are structured so that they are effective and cohesive (or stable) without recourse to a larger context of international law for enforcement.[5] The most common, and simplest, notion of stability draws on the cartel stability literature (eg, d'Aspremont et al, 1983; Donsimoni et al, 1986), wherein a stable cartel is defined as a cartel for which no individual members has an incentive to leave nor any outsider to join. This turns out to be a very strong assumption in the sense that many potential cartels fail the test. Chander and Tulkens (1992, 1994) adopt an even stronger assumption that should any individual member of a voluntary agreement choose to leave, the entire agreement would be null and void. Between these two concepts is the notion of "farsighted stability" (Ecchia and Mariotti, 1997; Eyckmans, 2003). The idea here is that an agreement is stable if no country has an incentive to leave or join, but in evaluating those incentives, countries look beyond their act of joining or leaving to the credible additional actions that other countries may take.

Some of the earliest work (Hoel, 1992; Carraro and Siniscalco, 1993; Barrett, 1994) finds that such agreements are either unlikely to consist of very many participants or, if the agreement involves a large number of countries, then

---

[4] Wagner (2001) and Barrett (2003, 2007) provide comprehensive reviews of this literature. See also Finus (2001) and several chapters in the volume edited by Guesnerie and Tulkens (2009).

[5] The literature on IEA uses the term "stable" to refer to coalitions of countries that will tend to stay together and not break up. This is a somewhat unfortunate choice of words, since stability is generally a dynamic term referring to the tendency of an equilibrium or coalition to remain unchanged when conditions are perturbed slightly. That is not the meaning here. In the interests of clarity, we use the standard term "stability" here to describe coalitions that are cohesive, recognizing the less-than-satisfactory nature of the term.

the gains from cooperation must be low.[6] The basic idea is that the incentives for free-riding must be low or else most countries will choose to free-ride and not belong to the agreement. A low incentive to free-ride is the flip-side of a small gain from cooperating.

This conclusion is based on a simple model of N homogeneous countries, each with a marginal cost of abatement c and a marginal damage from pollution b. Each country acts as a payoff maximizer, choosing emissions, e_i, from the interval [0,1], which may be discrete or continuous. In the simplest form of this model, payoffs are linear, and each country either pollutes or abates, resulting in a payoff of Π_i:

$$\Pi_i = ce_i - bE, \text{ where } E = \sum e_i. \tag{1}$$

The countries play a two-stage game: in the first stage, countries announce whether or not they will join the IEA. In the second stage, the IEA and the fringe choose emission levels. In each stage, Nash play is typically assumed. Using the notion of cartel stability mentioned above, it is easy to show that c/b is the unique size of a stable IEA, or more precisely the unique size is the integer rounded up from c/b. The logic is that with an agreement size of c/b, it is optimal for the IEA to abate. But for an agreement size of $c/b-1$, it is optimal to pollute. Thus every member of the IEA views itself as pivotal in that if any single member defects, the entire IEA ceases abating.

Payoffs are normalized so that if all countries abate, the aggregate payoff is zero. With a stable IEA of size c/b (ignoring the integer issues), the aggregate payoff is $-b(N-c/b)^2$. In a noncooperative equilibrium, each country pollutes with an aggregate payoff of $cN-bN^2$. Thus the gain from an IEA is $c(N-c/b)$. This suggests that an IEA does the most good when c/b is small and the least good when c/b is large. But this also means that when an IEA can do the most good, the equilibrium size will be small; and conversely, when the equilibrium size is large, the gain from the IEA is smaller.

The literature varies on the structure just described. Some authors set the problem up as a Stackelberg equilibrium, wherein the IEA is the leader and the fringe members are followers. As mentioned above, the strategies may be discrete as illustrated or they may be continuous. Payoffs may be linear as illustrated or quadratic or even more general. Finally, the model above is fundamentally static – a one-shot game. An obvious extension is to consider a repeated game framework or other dynamic representations. Clearly international agreements on the environment are not reached in a static context; however the

---

[6] Many of the results in the literature rely on simulation models and are thus less in the nature of proofs than illustrations. Rubio and Ulph (2007) provide analytic proofs of some of these early results.

difficulty of developing credible dynamic models has limited the literature to static models for the most part (though exceptions are Finus, 2001, and Barrett, 2002).

The example above illustrates a fundamental issue in this entire literature. How big will an agreement be? What is the size of a "stable" agreement? The answers hinge on what holds an agreement together, or what keeps countries in the agreement. Assumptions can range from complex commitment procedures, to punishments for defecting, to simple self-interest without commitment (as in the example).

B. Heterogeneity

Another issue is the extent to which participating countries are homogeneous or heterogeneous. Most of the results in this literature assume homogeneity of participating countries. But "real" countries are not all the same. Countries may be large or small. Some countries see small marginal damage from climate change (in the sense of small additional damage from additional change in the climate), due to low population levels, large geographic scale or other geographic specifics. Other countries experience high marginal damages. Similarly, the marginal cost of abatement (per ton of greenhouse gases may differ dramatically from one country to another. However, many agreements involve trading mechanisms designed to equalize the marginal cost of abatement among countries; thus marginal costs may not be a major source of heterogeneity. But total costs and total emissions may vary significantly. In comparing the United States to Switzerland, a big difference is that the US emits far more. If the US and Switzerland reduce emissions to a given level of marginal costs, the absolute size of the reduction will be far greater than in Switzerland.

The literature is modest on the significance of heterogeneity. One of the earliest papers on this topic is Barrett (1997a). In that paper he focuses on how countries within a coalition may share the joint payoff from cooperation. He uses Shapley values to divide the joint payoff and examines a world in which there are two types of countries. Although he is able to develop some analytic results, for the most part he uses simulations to derive his results. Botteon and Carraro (2001) extend this framework to five types of countries, again using simulation. McGinty (2006) extends this to 20 countries, also using simulation. To illustrate the difficulty of developing analytic solutions to this problem, it is only recently that an analytic solution has been found for the heterogeneity problem with two types of countries, continuous strategies and quadratic payoffs (Fuentes-Albero and Rubio, 2010). Results of this literature are mixed, though a common conclusion of the simulation literature is that, in some cases, heterogeneity increases the effectiveness of international environmental agreements.

Barrett (2001, 2002) takes a different approach to this problem. Rather than view the problem as a static game, he views it as a repeated game and posits credible punishments that can be built into the agreement. The classic problem with enforcing an agreement with punishment is that it is often not in a country's self-interest to punish. By weakening the abatement undertaken by the coalition, Barrett is able to construct renegotiation-proof agreements that involve credible punishments. Furthermore, weakening the requirements of coalition members tends to expand the size of the coalition. Barrett (2001) specifically shows that heterogeneity of countries can reduce the free-riding problem and thus help support larger coalitions. In his model, heterogeneity facilitates commitment. And commitment is the big problem in self-enforcing agreements. One of his results is that if the countries are substantially different, then only countries with high marginal damage will form a coalition/IEA. But these countries have a collective incentive to bribe the low damage countries to join the IEA, increasing the effectiveness of the IEA. Transfers are a moot issue in the homogeneous country case. But in the heterogeneous country case, the surplus that accrues to some countries is more than enough to "bribe" other countries to participate. The author points out that this is illustrated by the Montreal Protocol. Rich countries stood to gain the most and could afford to pay developing countries to participate.

C. Experimental Evidence

Until recently, virtually all of the economics literature on IEAs has been theoretical in nature. After all, treaties are complex, and it is difficult to do econometrics with them. But an IEA is fundamentally an institution for coordinating voluntary contributions of countries to a global public good. And we know from the literature on voluntary contributions to public goods that theory is often at odds with empirical and experimental evidence (eg, Isaac and Walker, 1988).

Burger and Kolstad (2009) take an experimental approach to validating the theoretical models of IEA formation. Using a structure very similar to the model of Barrett (1994) and others, they find that one of the basic theoretical results does not hold experimentally. As discussed above, a basic theoretical result is that a smaller abatement cost relative to marginal damage from emissions (c/b), results in a smaller stable IEA and thus a lower aggregate level of abatement. Experiments find the opposite result, very much in line with the experimental work on voluntary provision of public goods. In fact, the parallel with the voluntary provision of public goods literature is striking; probably the same underlying forces are at work increasing cooperation relative to the theory. This finding calls into question the validity of theoretical models of IEA formation.

The B.E. Journal of Economic Analysis & Policy, Vol. 10 [2010], Iss. 2 (Symposium), Art. 3

Kosfeld et al (2009) examine the same problem, though they couch it as a public goods provision problem. Furthermore, they only examine the case of four countries/participants. Nevertheless, they confirm the result of Burger and Kolstad (2009) that as the abatement cost-marginal damage ratio increases the size of the stable IEA increases, in contradiction to theory. However, the authors posit inequality aversion as an explanation for the experimental results. This represents a new and innovative direction of research in bringing together theory with experimental results on IEA formation.

II. A MODEL OF AGREEMENTS

We are interested in comparing the situation where all countries are identical to the situation where countries differ. Countries may differ in terms of size, marginal abatement costs, and/or marginal damage. The model we examine is closely related to that of Barrett (2001) and Saha (2007).

The basic structure of the problem is the standard two-stage game model of an IEA, as discussed earlier in this paper.[7] In the first stage, countries decide whether or not to join an IEA or to stay in the fringe (the "membership game"). In the second stage, the IEA acts as a joint payoff maximizer and acts as a single decision-maker in choosing emissions and competing with the individual members of the fringe (the "emissions game"). Equilibrium is a Nash equilibrium in the membership game followed by a Nash equilibrium in the emissions game.

We will introduce an additional variable into this problem, size of the county, and also let the abatement costs and marginal damages vary. Through the size variable we are capturing two things. One is that abatement in large countries is expected to be greater than in small countries. Furthermore, if one thinks of marginal damage as fundamentally per capita, then aggregate marginal damage in a country is also proportional to size. Letting size of county i be S_i, payoff for country i is still as in Eqn. (1), with two modifications:

- e_i is either 0 or S_i (rather than 0 or 1)   (2a)
- b_i is proportional to per capita damage (β_i) and S_i: $b_i \approx \beta_i S_i$.   (2b)

In other words, marginal damage b_i can vary from country to country, both in per capita terms and in aggregate terms. Furthermore, although the marginal cost of abatement is equalized across countries (due to approaches like the Clean Development Mechanism), the aggregate amount of abatement depends on the size of the country.

---

[7] As discussed in the previous section, there are more complex models of international agreements. The one adopted here is pedagogically attractive and the most common in the literature.

To illustrate the richness of this representation, consider a few examples. The US is a large country but one with relatively modest per capita damage from climate change (though this is clearly debatable). Canada and Australia are much smaller countries, though with similar per capita damage. India is a large country with potentially large per capita damage from climate change (though moderated by low per capita income).

To be specific, consider $i=1,...,N$ countries, each of "size" S_i, emitting pollution (e_i) which contributes to the global commons ($E=\sum_j e_j$). For simplicity, assume each country makes a discrete choice regarding how much pollution to emit, which without loss of generality may be restricted to 0 or S_i: to abate ($e_i=0$) or to pollute ($e_i=S_i$). Marginal damage from aggregate emissions varies from one country to another but the marginal cost of abatement does not. As in Eqn. (1), each country's payoff is represented as a linear function of own emissions and aggregate emissions:

$$\Pi_i(e_i,E_{-i}) \quad \equiv ce_i - b_i E$$

$$= ce_i - b_i(e_i + E_{-i}) \tag{3}$$

where $E_{-i}=\sum_{j\neq i} e_j$. Thus $\gamma_i \equiv b_i/c$ is the benefit-cost ratio for emissions control – the ratio of own marginal environmental damage from emissions to the marginal cost of emissions control. Clearly we wish to focus on the case of $\gamma_i < 1$ ($b_i < c$), and we make that assumption; otherwise abatement is a dominant strategy for individual countries, and cooperation is unnecessary. Furthermore, the γ_i's cannot be too small; otherwise full cooperation (and efficiency) will not result in any abatement.

A. A Self-Enforcing IEA

As mentioned earlier, we represent the formation of an IEA as a two-stage game, consisting first of a membership game followed by an emissions game. The first-stage membership game is an announcement game in which countries decide whether or not to join the IEA.[8] In the second-stage emissions game, the membership of the IEA is given and countries decide how much to emit. In the emissions game, we assume the members of the IEA decide on emissions jointly, and the non-members (the fringe) decide individually. The coalition acts as a singleton and each member of the fringe acts in a Nash noncooperative manner; a Nash equilibrium results. Membership of the coalition cannot change in the emissions game.

---

[8] In the announcement game, each country announces "in" or "out." A Nash equilibrium is a set of announcements for which no country can do better by unilaterally changing its announcement.

We now assume two types of countries in our world, characterized by different sizes and marginal damages: S_j and b_j. It is straightforward to consider the case of M different types, though the presentation is messier. Type 1 countries have size S_1 and marginal damage b_1; type 2 countries have size S_2 and marginal damage b_2. Without loss of generality, assume $b_1/S_1 < b_2/S_2$. Further, assume N_j of type j countries ($N=N_1 + N_2$), but only n_j of type j countries in a coalition/agreement ($n_j \leq N_j$). Thus a coalition (n_1,n_2) consists of n_1 of country 1 and n_2 of country 2. Further, by assumption, full cooperation involves abatement by all countries. As we will see below, this is equivalent to $N_1\gamma_1 + N_2\gamma_2 \geq 1$.

Equilibrium can be determined using backward recursion—first find the equilibrium in the emissions game, conditional on the membership in the IEA, and then find the equilibrium in the membership game. In the emissions game, it is a dominant strategy for the fringe to pollute. Conditional on having (n_1,n_2) members of the coalition, the coalition sees the following coalition aggregate payoff (CAP):

Abate: $\qquad CAP = -n_1b_1[S_1(N_1-n_1)+ S_2(N_2-n_2)] - n_2b_2[S_1(N_1-n_1)+ S_2(N_2-n_2)]$

$$= -(n_1b_1 + n_2b_2)\ [S_1(N_1-n_1)+ S_2(N_2-n_2)] \qquad (4a)$$

Pollute: $\qquad CAP = (n_1S_1+n_2S_2)c - [S_1N_1 + S_2N_2](n_1b_1 + n_2b_2) \qquad (4b)$

$\Rightarrow$ Abate weakly preferred iff $n_1b_1 + n_2b_2 \geq c \qquad (4c)$

In the membership game, it is clear that an equilibrium in the announcement game is associated with $n_1\gamma_1 + n_2\gamma_2$ being as close to 1 as possible without being less than 1 ($n_1b_1 + n_2b_2$ is as close to c as possible without violating the inequality). Then all members of the coalition are pivotal in the sense that if any one defects, the coalition's decision rule on emissions switches from abate to emit, and thus the defector is worse off.

An interesting consequence of the condition in Eqn. (4c) is that size of a country does not matter (from the point of view of emissions abatement). What does matter is damages. In fact, Eqn (4c) can be viewed as a variant of the Samuelson condition for the efficient provision of a public good. The left hand side of the inequality is the aggregate marginal damage from one ton of emissions, and the right hand side is the marginal cost of abating one ton.

We will now make the assumption that in equilibrium, Eqn. (4c) holds with equality. This is approximately true but ignores the integer nature of n_1 and n_2. We will be interested in comparative statics and welfare, and thus we argue this simplification is reasonable. Thus the equilibrium condition is

Kolstad: Equity, Heterogeneity and International Environmental Agreements

$$n_1\gamma_1 + n_2\gamma_2 = 1 \Leftrightarrow n_1b_1 + n_2b_2 = c \tag{5}$$

Note that the problem has multiple equilibria as stated. Many combinations of n_1 and n_2 will satisfy Eqn. (5).

B. Equilibrium Refinement

Multiple equilibria usually means a degree of arbitrariness in narrowing the set of possible equilbria. The issue here is: for all of the possible coalition compositions, which is likely to emerge? Two clear candidates emerge for shrinking this set, selecting a narrower set of equilibria for further consideration – what we call a "refinement." One criterion for refinement is based on efficiency; the other is based on equity (consistent with Kosfeld et al, 2009). The efficiency criterion would involve the coalition that maximizes the joint payoff for coalition members. The equity criterion is a bit more complex.

In fact there are many ways of defining an equity criterion. One could modify the efficiency criterion so that coalition members also take into account the welfare of non-members. As Prof. Scott Barrett has pointed out in discussant comments on this paper, equity could involve the richest countries (per capita) making the greatest sacrifices. Rather than consider all possible ways of characterizing equity, we adopt one of the simplest representations: equating the payoff to each of the two groups of countries within the coalition (there may be other interpretations of what equity means).

Another issue, also raised by Prof. Barrett, is that if one is using an equity refinement for the second stage game (the emissions game), then it would be logical to use an equity criterion for the first stage game – the participation game. This is a valid point; however, we are not expanding the number of equilibria in this game, just choosing among several. Thus we are not changing the fundamental nature of the two-stage game.

The aggregate payoff for each of the two groups in the abating coalition (CAP_j) can easily be computed:

$$CAP_j = -n_jb_j [S_1(N_1-n_1) + S_2(N_2-n_2)], \text{ for } j=1 \text{ or } 2 \tag{6}$$

The two refinement conditions would then be:

Efficiency: Find (n_1,n_2) to max $[CAP_1 + CAP_2]$ (7a)
Equity: Find (n_1,n_2) such that $CAP_1 = CAP_2$ (7b)

The B.E. Journal of Economic Analysis & Policy, Vol. 10 [2010], Iss. 2 (Symposium), Art. 3

In solving for (n_1, n_2) that satisfy either of these two refinement conditions, we must also satisfy Eqn (5), and n_j must lie on the closed interval $[0, N_j]$. There may be more than one solution to each of these two conditions.

It is easy to see that efficiency involves maximizing $[n_1 S_1 + n_2 S_2]$, subject to Eqn. (5). Because the problem is linear, the maximum occurs at a corner when either n_1 or n_2 is as large as possible, depending on which is larger, S_1/b_1 or S_2/b_2. By assumption, $S_1/b_1 > S_2/b_2$, so $n_1 = \min[c/b_1, N_1]$. This implies n_2 is either zero, or whatever is necessary so that Eqn. (5) holds: $n_2 = \max[0, (c-N_1 b_1)/b_2]$.

For equity, it is easy to see that $n_j = c/(2b_j)$, provided N_1 and N_2 are sufficiently large. If enough type j countries do not satisfy this condition, then the equity refinement is ill-defined.

These results can be summarized in the following proposition:

Prop. 1.   The model of linear payoffs described above has many coalitions that satisfy the conditions for stability articulated in Eqn. (5). Two refinements result in unique coalition sizes. With an efficiency refinement (coalition maximizes coalition payoff), the size is

$$\text{Efficiency:}\quad (n_1, n_2) = (\min[c/b_1, N_1], \max[0, (c-N_1 b_1)/b_2]). \tag{8a}$$

With an equity refinement, provided N_1 and N_2 are sufficiently large (i.e., $c/(2b_j) \leq N_j$), then

$$\text{Equity:}\quad (n_1, n_2) = (c/(2b_1), c/(2b_2)). \tag{8b}$$

The interpretation of these two results is straightforward. For the efficiency refinement, the countries with the smaller b_j/S_j constitute the bulk of the coalition (in fact it includes all of the coalition, if N_j is large enough). As described earlier, b_j represents the marginal damage from emissions for the entire country; b_j/S_j is closer to the per capita marginal damage.

In some sense, this is counterintuitive. The countries that suffer the greatest damage from emissions have the largest incentive to do something about it, but this result says that they stay in the fringe. With plenty of type 1 countries, then the coalition will consist entirely of type 1 countries, while all type 2 countries remain in the fringe. On the other hand, type 2 countries have stronger incentives to free ride, which leads them to be in the fringe.

In fact, the result is not at all counterintuitive given the structure of the problem. When countries with smaller marginal damage form a coalition, the stable coalition is larger and thus more effective. It is not surprising that on efficiency grounds, we would want as many low damage countries as possible in the coalition.

For the equity refinement, both types of countries will be in the coalition, and participation is independent of size (S_j) of the country. Countries with smaller marginal damage (b_j) will make up a larger proportion of the coalition.

C. Comparative Statics

The next question is what are the implications of heterogeneity? How does the effectiveness of an IEA change as countries become more heterogeneous? To answer this question, we compare the case of $b_1 = b_2$ to $b_1 < b_2$, preserving the expected value of b. We will hold $S_1 = S_2 = 1$. In particular, define

$$\gamma_1 = \gamma - \delta/N_1 \tag{9a}$$
$$\gamma_2 = \gamma + \delta/N_2 \tag{9b}$$

For any $\delta \geq 0$, the weighted sum of γ_1 and γ_2 will remain equal to γ. With all countries the same $(\delta = 0)$, the equilibrium size of an IEA will be $1/\gamma$, and welfare will be W_0:

$$W_0 = - (N\gamma - 1)^2/\gamma \quad for \; \delta = 0 \tag{10a}$$

and for the case of $\delta > 0$, assuming N_1 is sufficiently large, the equilibrium IEA will consist of only type 1 countries, including $1/(\gamma - \delta/N_1)$ of them, which is a larger IEA. Although that means more abatement, welfare may or may not be larger. Welfare (for $\delta > 0$) will be (following from Eqn. 4):

$$W_+ \quad = [N-1/(\gamma - \delta/N_1)] \, (N\gamma - 1) \tag{10b}$$

To determine whether W_+ is larger or smaller than W_0, for small δ, differentiate Eqn. (10b) with respect to δ and evaluate at $\delta = 0$:

$$d \, W_+/ \, d\delta = - (N\gamma - 1) \, (\gamma - \delta/N_1)^{-2}/N_1 \tag{11}$$

Clearly the right hand side of Eqn. (11) is negative when evaluated at $\delta = 0$. This implies that welfare decreases with small increases in heterogeneity. Consequently, $W_+ < W_0$.

What this says is that heterogeneity decreases welfare, even though the equilibrium size of the IEA may increase. This result is consistent with prior literature on IEAs. As the problem is characterized here, the IEA is dominated by type 1 countries. So the fact that there are type 2 countries is virtually irrelevant. The prior literature tells us that as γ decreases, the size of an IEA increases but the aggregate welfare decreases. Our result is an extension of this.

The B.E. Journal of Economic Analysis & Policy, Vol. 10 [2010], Iss. 2 (Symposium), Art. 3

III. CONCLUSIONS

Although theoretical models of the formation of international environmental agreements are highly stylized and cannot be used to predict outcomes in real treaty negotiations, they provide useful guidance in how incentives play out in settings in which multiple players must agree to coordinate action to improve the provision of environmental goods. It is in this arena that this paper makes a contribution.

In this paper we examine the important question of how heterogeneity among countries affects the development of international environmental agreements. The question is whether heterogeneity (as opposed to homogeneity) leads to larger or smaller IEAs, to higher or lower levels of aggregate welfare and to more or less abatement. Is heterogeneity a hinderance or a help to forming IEAs to solve environmental problems?

We have developed a simple theory here, extending work by Barrett (2001) on a similar topic. We conclude that heterogeneity of countries may increase the size of IEAs, but it weakens them in terms of the aggregate welfare of all participants. Furthermore, the efficiency criterion for formation of an IEA suggests that the agreement would primarily consist of countries with low per capita marginal damage, due to the nature of the equilibrium (small marginal damage supports a larger coalition). On the other hand, the equity equilibrium refinement suggests a coalition made up of a combination of high and low damage countries.

REFERENCES

Barrett, Scott, "Self-Enforcing International Environmental Agreements," *Oxford Economic Papers*, **46**:878-94 (1994).

Barrett, Scott, "Heterogeneous International Environmental Agreements," in Carlo Carraro (Ed), *International Environmental Negotiations: Strategic Policy Issues* (Edward Elgar, Cheltenham, UK, 1997a).

Barrett, Scott, "The Strategy of Trade Sanctions in International Environmental Agreements," *Resource and Energy Economics*, 19:345-61 (1997b).

Barrett, Scott, "International Cooperation for Sale," *European Economic Review*, **45**:1835-50 (2001).

Kolstad: Equity, Heterogeneity and International Environmental Agreements

Barrett, Scott, "Consensus Treaties," *Journal of Institutional and Theoretical Economics*, **158**:529-47 (2002).

Barrett, Scott, *Environment and Statecraft* (Oxford University Press, Oxford, 2003).

Barrett, Scott, *Why Cooperate? The Incentive to Supply global Public Goods* (Oxford University Press, Oxford, 2007).

Botteon, M. and C. Carraro, "Environmental Coalitions with Heterogeneous Countries: Burden Sharing and Carbon Leakage," in A. Ulph (Ed), *Environmental Policy, International Agreements, and International Trade* (Oxford University Press, New York, 2001).

Burger, Nicholas and Charles Kolstad, "Voluntary Public Goods Provision, Coalition Formation, and Uncertainty," National Bureau of Economic Research Working Paper 15543 (Cambridge, Mass., 2009).

Carraro, Carlo and Domenico Siniscalco, "Strategies for the International Protection of the Environment," *Journal of Public Economics*, **52**:309-28 (1993).

Chander, Parkash and Henri Tulkens, "Theoretical Foundations of Negotiations and Cost-Sharing in Transfrontier Pollution Problems," *European Economic Review*, **36**:288-99 (1992).

Chander, Parkash and Henri Tulkens, "A Core-theoretic Solution for the Design of Cooperative Agreements on Transfrontier Pollution," *International Tax and Public Finance*, **2**:279-93 (1994).

Cooper, Richard N., "International Cooperation in Public Health as a Prologue to Macroeconomic Cooperation," pp 178-254 in R. N. Cooper et al (ed.), *Can Nations Agree?* (Brookings Institution, Washington, DC, 1989).

d'Aspremont, C., A. Jacquemin, J. Jaskold-Gabszewicz and J. Weymark, "On the Stability of Collusive Price Leadership," *Canadian Journal of Economics*, **16**:17-25 (1983).

Donsimoni, M.-P., N.S. Economides, and H.M. Polemarcharkis, "Stable Cartels," *International Economic Review*, **27**:317-27 (1986).

The B.E. Journal of Economic Analysis & Policy, Vol. 10 [2010], Iss. 2 (Symposium), Art. 3

Ecchia, Giulio and Marco Mariotti, "The Stability of International Environmental Coaltions with Farsighted Countries: Some Theoretical Observations," Ch. 10 in Carol Carraro (Ed), *International Environmental Negotiations* (Edward Elgar, Cheltenham, UK, 1997).

Eyckmans, Johan, "On the farsighted stability of international climate agreements," University of Leuven working paper (Jan 2003).

Finus, Michael, *Game Theory and International Environmental Cooperation* (Edward Elgar, Cheltenham, UK, 2001).

Finus, Michael, M. Elena Saiz and Eligius M.T. Hendrix, "An Empirical Test of New Developments in Coalition theory for the Design of International Environmental Agreements," *Environmental and Development Economics*, **14**:117-37 (2009).

Fuentes-Albero, Cristina and Santiago J. Rubio, "Can International Environmental Cooperation be Bought?," *European Journal of Operations Research*, **202**:255-64 (2010).

Guesnerie, Roger and Henry Tulkens (Eds), *The Design of Climate Policy* (MIT Press, Cambridge, 2009).

Helm, Carsten, "International Cooperation Behind the Veil of Uncertainty," *Environmental and Resource Economics*, **12**:185-201 (1998).

Hoel, Michael, "International Environmental Conventions: The Case of Uniform Reductions of Emissions," *Environmental and Resource Economics*, **2**:141-59 (1992).

Isaac, R.M. and J. M. Walker, "Group Size Effects in Public Good Provision: The Voluntary Contributions Mechanism," *The Quarterly Journal of Economics*, **103**:179-99 (1988).

Kolstad, Charles D., "Piercing the Veil of Uncertainty in Transboundary Pollution Agreements," *Environmental & Resource Economics*, **31**:21-34 (2005).

Kolstad, Charles D. and Alistair Ulph, "Learning and International Environmental Agreements," *Climatic Change*, **89**:125-41 (2008).

Kosfeld, Michael, Akira Okada and Orno Riedl, "Institution Formation in Public Goods Games," *American Economic Review*, **99**:1335-55 (2009).

McGinty, Matthew, "International Environmental Agreements Among Asymmetric Nations," *Oxford Economic Papers*, **59**:45-62 (2007).

Na, Seong-lin, and Hyun Song Shin, "International Environmental Agreements under Uncertainty," *Oxford Economic Papers*, **50**:173-85 (1998).

Rubio, Santiago and Alistair Ulph, "An Infinite Horizon Model of Dynamic Membership of International Environmental Agreements", *Journal of Environmental Economics & Management*, **54**:296–310 (2007).

Saha, Sarani, "Heterogeneity in International Environmental Agreements," Ch. 1 in *Three Essays in Environmental and Public Economics*, PhD Dissertation, University of California, Santa Barbara (2007).

Ulph, Alistair, "Stable International Environmental Agreements with a Stock Pollutant, Uncertainty and Learning," *Journal of Risk & Uncertainty*, **29**:53-73 (2004).

Wagner, Ulrich J., "The Design of Stable International Environmental Agreements," *Journal of Economic Surveys*, **15**:377-411 (2001).

Young, Oran, *International Governance: Protecting the Environment in a Stateless Society* (Cornell University Press, Ithaca, NY, 1994).

The B.E. Journal of Economic Analysis & Policy

Symposium

Volume 10, Issue 2	2010	Article 4

DISTRIBUTIONAL ASPECTS OF ENERGY AND CLIMATE POLICY

Comment on "Equity, Heterogeneity and International Environmental Agreements"

Scott Barrett[*]

*Columbia University, sb3116@columbia.edu

Recommended Citation
Scott Barrett (2010) "Comment on "Equity, Heterogeneity and International Environmental Agreements"," *The B.E. Journal of Economic Analysis & Policy*: Vol. 10: Iss. 2 (Symposium), Article 4.
Available at: http://www.bepress.com/bejeap/vol10/iss2/art4

Barrett: Comment on "Equity, Heterogeneity & Int'l Env. Agreements"

In the standard model of an international environmental agreement with symmetric countries, the equilibrium treaty participation level is unique, while the identities of the participants and non-participants can vary enormously. Non-uniqueness of identities is normally of no interest. The focus of this literature has been on the effect of a treaty on global welfare, and on the distribution of net benefits between participating and non-participating states (free riders). When countries are symmetric, these measures are blind to the identities of participating and non-participating states.

When countries are asymmetric, the identities of these states matter, because they can affect both global welfare and the distribution of net benefits between participating and non-participating countries. Professor Kolstad's main conclusion is that these distinctions may play a role in deciding which countries will and will not participate. This is what he means by participation "refinement."

How might the identities of participating and non-participating countries be determined? Here is one way to think about this.[1] Consider the symmetric game. Suppose there are five countries and that a self-enforcing treaty can sustain a participation level of just two countries. Label the countries 1, 2, ..., 5, and consider a model in which the participation stage of the game is played in five periods. Now assume that there is a rule, enforced externally, which says that country 1 moves first, country 2 second, and so on. Then it is reasonable to believe that the equilibrium treaty will consist of countries 4 and 5. This is because the other countries can look forward and reason backwards. They can be pretty sure that if the first three countries don't participate, the fourth will participate, knowing that the fifth will participate in the next period. The first three countries have an "early mover" advantage.

This assumes that the rule for deciding participation is given and enforced externally. But there is no World Government able to play such a role; and it may be difficult for countries to agree among themselves which ones among them ought to make a sacrifice and participate in an agreement and which ones ought to be permitted to free ride. International law is silent on this question.

However, norms may play a role. An illustration from life-or-death situations is the evacuation rule, "women and children first." A recent paper has shown that the survival/death outcomes of the people on board the *Titanic* are consistent with this rule being used to ration places on the limited number of lifeboats (Frey, Savage, and Torgler 2010). In an international context, a norm like "common but differentiated responsibilities" may play a similar role. In the Kyoto Protocol, all countries were encouraged to participate, but only the richest countries faced "binding" emission caps. That is, only the richest countries were expected to make sacrifices.

---

[1] This example is from Barrett (2005), pp. 205-206.

The B.E. Journal of Economic Analysis & Policy, Vol. 10 [2010], Iss. 2 (Symposium), Art. 4

The difficulty with applying a normative rule such as this to the model used by Kolstad is that the equilibrium participation level (Kolstad's equation 5) is determined independently of the rule. As I shall explain later, a different model may be more suitable for exploring the identities of the participating states.

Kolstad considers two "rules" (he calls them "refinements") for determining participation. The first is "efficiency." As defined by Kolstad, the countries "selected" to participate are the countries whose participation maximizes the payoff to all signatories, given the participation constraint (again, his equation 5). It is a common assumption in this literature that the parties to a treaty should choose their abatement so as to maximize their collective payoff. This is also a reasonable assumption, since in this stage of the game, participation is a given. However, the decision of which countries should and which should not participate is made in a different stage. It seems to me more reasonable to assume that an "efficiency rule" would determine participation so as to maximize global welfare and not only the welfare of the parties.

The other "rule" suggested by Kolstad is "equity," defined as equating the aggregate payoff for type 1 and type 2 participating countries. Like the other criterion, this one ignores the implications for non-parties. It also defines equity in terms of the aggregate payoffs for the two types of party. Other definitions of equity would seem at least as compelling—such as the rule that the countries with the most well off populations should make the greatest sacrifices (to implement this rule using Kolstad's model would require further elaboration of the term, "per capita damage").

As I said before, it might make more sense to consider the participation rule jointly with the participation level. To do that, however, requires a different model.

I previously developed a model that provides the needed flexibility (Barrett 2002; and Chapter 11 of Barrett 2005). It is a repeated game model in which the rule for enforcement is relaxed ever so slightly. In the usual repeated game model, which sustains the same outcome as the static model examined by Kolstad, participating countries optimize collectively both on and off the equilibrium path. Optimization off the equilibrium path limits punishments, and so sustains only limited participation. In this different model, countries have a more restricted choice. They can either impose the penalty required by the treaty or they can impose no penalty. This small change creates flexibility. It is now possible to vary the participation level virtually at will. In particular, participation can now be determined by considerations of equity or fairness and not only credibility.

When countries are symmetric, the most compelling "fair" agreement is one in which every country participates—a "consensus treaty." In Barrett (2002) I show that such a treaty can be sustained as long as the obligations by each country to limit emissions are weakened. (In Kolstad's model, abatement is binary; in this model, it is continuous.) I compare two alternative configurations. The first is a "narrow but deep" treaty, in which few countries participate, but each reduces emissions substantially. This configuration is compatible with the one examined by Kolstad. The second is a "broad but shallow" treaty. In this case, every country participates, but each reduces emissions only slightly. I derive conditions in which, from an *ex ante* perspective (that is, before a treaty is negotiated), all countries prefer to negotiate a consensus treaty. This treaty is both "fair," because all like countries receive the same payoff in equilibrium, and "efficient," because all countries prefer this outcome to the alternative.

It would be an easy matter to apply this model to a situation in which countries are asymmetric. In particular, this approach could be used to model the Kyoto Protocol. To implement the rule, "common but differentiated responsibilities," Kyoto enables every country to participate, but requires that only one "type" of country limit emissions (the so-called Annex I countries). This approach has different obligations for different countries but is essentially a consensus treaty (cost-effectiveness can be assured by application of the Clean Development Mechanism).

Of course, we now know that the Kyoto Protocol was unable to sustain a consensus (the United States is a non-participant). And we can be pretty sure that it will not be enforced (Canada has already said that it will not comply with the treaty). The problem overall is not that our models cannot accommodate the participation rule used in the Kyoto Protocol but that the Protocol was designed without taking into account the most important insight from our models—that if the treaty is to sustain cooperation, it must incorporate an effective, credible punishment mechanism.

References

Barrett, Scott (2002). "Consensus Treaties," *Journal of Institutional and Theoretical Economics*, 158 (4): 529-547.

Barrett, Scott (2005). *Environment and Statecraft*, Oxford: Oxford University Press (paperback edition).

Frey, Bruno S., David A. Savage, and Benno Torgler (2010). "Interaction of Natural Survival Instincts and Internalized Social Norms Exploring the Titanic and Lusitania Disasters." *Proceedings of the National Academy of Sciences* 107(11): 4862-4865.